ACS SYMPOSIUM SERIES 333

High-Energy Processes in Organometallic Chemistry

Kenneth S. Suslick, EDITOR
University of Illinois at Urbana–Champaign

Developed from a symposium sponsored by
the Division of Inorganic Chemistry
at the 192nd Meeting
of the American Chemical Society,
Anaheim, California,
September 7–12, 1986

American Chemical Society, Washington, DC 1987

Library of Congress Cataloging-in-Publication Data

High-energy processes in organometallic chemistry.
  (ACS symposium series, ISSN 0097-6156; 333)

  "Developed from a symposium sponsored by the
Division of Inorganic Chemistry at the 192nd Meeting
of the American Chemical Society, Anaheim,
California, September 7-12, 1986."

  Includes bibliographies and indexes.

  1. Organometallic chemistry—Congresses.
2. Radiation chemistry—Congresses.

  I. Suslick, Kenneth Sanders. II. American Chemical
Society. Division of Inorganic Chemistry. III. American
Chemical Society. Meeting (192nd: 1986: Anaheim,
Calif.) IV. Series.

QD410.H54   1987   547'.07   86-32245
ISBN 0-8412-1018-7

ACS Symposium Series

M. Joan Comstock, *Series Editor*

1987 Advisory Board

Foreword

The ACS SYMPOSIUM SERIES was founded in 1974 to provide a medium for publishing symposia quickly in book form. The format of the Series parallels that of the continuing ADVANCES IN CHEMISTRY SERIES except that, in order to save time, the papers are not typeset but are reproduced as they are submitted by the authors in camera-ready form. Papers are reviewed under the supervision of the Editors with the assistance of the Series Advisory Board and are selected to maintain the integrity of the symposia; however, verbatim reproductions of previously published papers are not accepted. Both reviews and reports of research are acceptable, because symposia may embrace both types of presentation.

Contents

Preface

IN THIS VOLUME, an attempt is made to draw together a wide range of disciplines and applications that involve high-energy processes in organometallic chemistry. The reader will find here state-of-the-art descriptions of methodologies that, at first glance, are dramatically disparate. There is, however, an underlying commonality that is sometimes conspicuous and other times subtle. The cross-fertilization of these approaches should benefit them all.

The opening chapter gives some historical background to the investigation of high-energy processes. The discussions in the subsequent chapters proceed from current studies in the gas phase, to examination of liquid-phase techniques, and finally to applications in the solid state. The final chapters provide a perspective on current and future industrial applications of the field.

The symposium upon which this book is based was truly international with respect to both speakers and audience. One measure of the multidisciplinary aspects of the topic was its cross-listing in three ACS divisions other than the sponsoring division. The symposium also allowed a thorough intermingling of academic and industrial researchers: More than a quarter of this volume's contributions are from the latter.

Funding for the conference came from the ACS Division of Inorganic Chemistry; the Petroleum Research Fund, administered by the ACS; and the Alfa Products Division of Morton Thiokol, Inc. Portions of travel expenses were borne by AT&T Bell Laboratories; BBC–Brown, Boveri & Co.; Exxon Research and Engineering; and Imperial Chemical Industries, Ltd. The conference would not have occurred without the generous support of these organizations. Finally, I wish to thank all of the contributing authors, whose cooperation and timeliness have been greatly appreciated.

KENNETH S. SUSLICK
University of Illinois at Urbana–Champaign
Urbana, Illinois 61801

November 21, 1986

Chapter 1

High-Energy Processes in Organometallic Chemistry in Perspective

Peter L. Timms

School of Chemistry, University of Bristol, Bristol, BS8 1TS, United Kingdom

The relationship between "conventional" and "high energy" processes in organometallic chemistry is defined. The growth of metal atom chemistry is surveyed as a model for the impact of a "high energy process" on the field of organometallic chemistry.

Chemistry can be viewed as a balance between thermodynamic and kinetic factors which dictate the course of chemical reactions and the stability of compounds. Chemists seeking to achieve particular goals, manipulate these factors using chemical or physical means. The papers in this symposium on "High Energy Processes in Organometallic Chemistry" describe recent attempts to apply mainly physical means to get around the thermodynamic and kinetic constraints of conventional organometallic chemistry.

Definitions

The term "high energy process" needs explanation. Its meaning is seen most easily if we first state that "conventional processes" in organometallic chemistry are reactions which occur at a useful rate at some temperature within the range of about $-100°$ to $+250\ °C$ and which do not involve the injection of other forms of energy or of very shortlived reagents such as atoms or radicals. Then "high energy processes" are broadly those in which the energy input is not thermal, e.g. the absorption of light, or is non-equilibrium thermal, e.g. the effect of ultrasonic cavitation, or in which unstable gaseous species, produced by an external, energetic process, are added as reactants.

   Our definition of a "conventional process" covers an enormous range of organometallic chemistry. Many organometallic chemists will want to include photochemical methods as part of their armoury of "conventional" techniques. This can be taken as a measure of the success of photochemical methods in the field. Indeed, it is the ambition of many of us who have worked on the development of "high energy processes" that our process should be accepted within the core of basic synthetic techniques of organometallic chemists.

One of the first uses of a high energy process in organometallic chemistry was the work by Paneth (1) and then Rice (2) on the reaction of metal mirrors with free alkyl radicals. The alkyl radicals were formed by thermal decomposition of a zinc or lead alkyl in a flowing gas stream and the radicals then reacted with a film or finely divided form of another metal downstream to give a new organometallic compound. Only a little earlier, the work of Kraus and Callis (3) had established the very important synthesis of tetraethyllead from NaPb and chloroethane. This is properly classified as a "conventional" organometallic reaction and put alongside the many other reactions where alkali metals or organic derivatives of alkali metals are used as reducing or coupling reagents. Nevertheless, the work of Mishima (4) showed that the NaPb/EtCl reaction probably occurs by a free radical mechanism. This means that there is a very fine distinction between the "conventional" synthesis of tetraethyllead via *in situ* generation of ethyl radicals and their reaction with lead and a "high energy" synthesis of tetraethyllead from ethyl radicals and lead in a flowing gas stream. The distinctions get more blurred if the NaPb alloy is regarded as an activated metal and compared with the activated metals made chemically by Rieke (5), the use of which is often grouped with "high energy processes". A related case of the narrow but definite distinction between "conventional" and "high energy" processes, come in the search for dimethylsilene some 35 years later than Paneth's work. The intermediacy of dimethylsilene in a number of reactions in the condensed phase at temperatures up 250°C had been proposed for years (6), but the first convincing proof came from the work of Atwell and Weyenberg in 1966, using $[CH_3O(CH_3)_2Si]_2$ to release dimethylsilene *in situ* at 130°-200°C (7). Shortly before this, Skell and Goldstein (8) had reported the generation of dimethylsilene in a "high energy process" by reacting dichlorodimethylsilane with potassium vapour in a helium stream at 260°C, and allowing the resulting dimethyl- silene to react with trapping agents downstream. There are often less mechanistic ambiguities in the use of free gaseous species as reagents than in the *in situ* generation of the same species in the condensed phase.

Energy Considerations

As already implied, "high energy processes" can be divided into two groups. First, those in which energy is applied *in situ*, to reactants to excite molecules to high translational, vibrational or electronic states so that the excited molecules can undergo immediate intra- or inter-molecular reactions. Such processes include photochemistry, radiolysis, shock wave studies, sonication and some discharge reactions. Second, those in which high energy species such as free atoms, radicals, carbene-like molecules or ions are formed in the gas phase by a chemical, thermal, photo- chemical or discharge process, and then reacted with other molecules in the gas or condensed phase. Before we consider any of these in detail, it is instructive to see what useful amounts of energy they involve compared with the energies of "conventional" processes.

Bonds between many neutral ligands and transition metal atoms in low oxidation states, have mean bond disruption enthalpies in the range 80-200 kJ mol^{-1} (9). Bonds to charged ligands, e.g.

cyclopentadiene, are generally stronger. These bond energies are strong enough to allow many "conventional" chemical synthetic approaches to be successful but they are often weak enough to permit reactions of organometallics to occur at moderate temperatures by dissociative or associative mechanisms. Hence the success of "conventional" organometallic chemistry.

The *in situ* "high energy processes" vary in the form of energy they provide. For example, the use of ultrasonics (sonochemistry) gives short-lived hot spots in liquids with transient, non-equilibrium temperatures of thousands of degrees (10), giving local translational and vibrational excitation to molecules, with the possibility of bond breaking followed by thermalization and secondary reactions. Photolysis with light of around 300 nm wavelength, provides electronic excitation energy of around 400 kJ mol^{-1} sufficient to bring about ligand dissociations, rearrangements, reductive elimination and many other processes (11). If γ-radiolysis is used, the absorbed energy per quanta is vastly greater than bond energies but one of the useful secondary processes is electron capture and the formation of ionic species (12).

Gaseous transition metal atoms can be taken as representative of the class of endothermic species which are formed in one place and then reacted to form organometallic compounds in another place. The atoms have heats of formation of about 300-800 kJ mol^{-1}. Nevertheless, their real value in chemical synthesis depends as much on kinetic as thermodynamic factors. A reaction (where M is

$$M(gas) \quad + \quad nL(gas) \quad \longrightarrow \quad ML_n(solid) \qquad (i)$$

a transition metal atom and L is a ligand molecule) will be exothomeric if there is a positive M-L bond dissociation energy; this result is independent of the heat of formation of M(gas). Comparing reaction (i) with another reaction (ii) starting with the solid metal, (i) will be more exothermic than (ii) by an amount

$$M(solid) \quad + \quad nL(gas) \quad \longrightarrow \quad ML_n(solid) \qquad (ii)$$

proportional to the heat of atomisation of the metal. However, attempts to achieve direct reactions between ligands and solid transition metals often failed because the kinetics and sometimes the thermodynamics are unfavourable. It is more meaningful to compare reaction (i) with the common reductive process (iii) (where M' is an electropositive metal and X an halogen). The greater

$$MX_m \quad + \quad nL \quad + \quad mM' \quad \longrightarrow \quad ML_n \quad + \quad mMX' \qquad (iii)$$

stability of mM'X compared to MX_m means that the reaction will be exothermic even if the product ML_n is quite endothermic. So, thermodynamically, reductive routes can achieve almost as much as can be achieved using the reaction of metal atoms and ligands. In practise, atom-ligand reactions have such low activation energies that they will occur at lower temperatures than most reductive reactions, enhancing the chances of isolating unstable products, and there are less opportunities for unwanted side reactions in metal atom than in reductive reactions. So for these reasons, atom reactions are more successful than reductive reactions in some cases.

The Development of High Energy Processes in Organometallic Chemistry

The Period 1930-60. Following the work of Paneth ($\underline{1}$), there was a long period of active work on free radicals, mostly of importance in organic chemistry but with some relevance to organometallic chemistry. The radicals were generated by thermal, photochemical or discharge processes on organic or organometallic compounds and detection was often by reaction with metal mirrors ($\underline{13\text{-}14}$). In the early 1950's, the matrix isolation method was developed to study radicals ($\underline{15}$), and this spectroscopic technique has proved of great value subsquently to metal atom chemistry. Also in this decade, there was a resurgence of interest in discharge reactions and organosilicon compounds were made this way ($\underline{16\text{-}17}$) (chemistry in arcs and discharges has a long history and in 1907 Svedberg ($\underline{18}$) was making metal colloids by striking arcs between metal electrodes under liquid methane--it is interesting to speculate on what organometallic compounds he formed at the same time!).

The rapid growth of the scope of organometallic chemistry during the 1950's meant that by the early 1960's the situation was ripe for the application of "high energy processes". Here we will trace only one of these processes in detail, the use of free metal atoms in synthesis, but with the expectation that there are lessons to be learned from this of relevance to the role of other "high energy processes".

Work on Free Transition Metal Atoms

The Key Developments. The work initiated by Skell and his co-workers in the early 1960's on carbene chemistry, had a profound influence on experimental techniques. He was the first to use high temperatures to generate unstable gaseous species under vacuum inside a liquid nitrogen cooled reaction vessel. The condensation of radicals and related species at liquid nitrogen temperatures was not new ($\underline{19\text{-}20}$). However, the containment of a very hot radical generator within a liquid nitrogen cooled reactor and the attainment of a line-of-sight path from the generator to the cooled walls was new. His results on reactions of carbon vapour, generated from a carbon arc, with organic compounds on the liquid nitrogen cooled walls of the surrounding vacuum vessel ($\underline{21}$), paved the way for the subsequent development of metal atom chemistry.

Between 1965 and 1969, there were rapid developments in the use of other high temperature species, particularly the silicon dihalides ($\underline{22\text{-}23}$), boron monofluoride ($\underline{24}$), boron atoms ($\underline{25}$), silicon atoms ($\underline{26}$), and alkali metal atoms ($\underline{27\text{-}28}$), in reactions at liquid nitrogen temperatures. Clearly this experimental method had to be applied to transition metal atoms and this happend in 1968-69 ($\underline{29}$). All that was required was to harness the simple and long established methods for vacuum evaporation of metals used in thin film studies ($\underline{30}$), to make metal atoms to condense with organic compounds at low temperatures. As is often apparent in retrospect, these experimental methods could have merged anytime in the previous five years.

The demonstration that stable and unstable organometallic compounds could be made in gram quantities from transition metal atoms sparked off enormous interest. The resulting chemistry has been extensively reviewed (31-35) and need not be repeated. More appropriate is to highlight some points in the development of the technique which have relevance to its acceptance by organometallic chemists.

The first reactions with transition metal atoms were carried out in the quite large and sophisticated stainless steel apparatus built originally for work on electron bombardment evaporation of boron (25). However, the immediate next generation of apparatuses were smaller glass reactors. This simplification was prompted in two ways; the adaption by Skell of his glass carbon-arc reactors for metal atom synthesis (36), and the desire here at Bristol to create more apparatuses quickly and cheaply to put the method into the hands of students (37). The appearance of simple apparatus which could be assembled for $1000 or less, greatly stimulated the field.

Spectroscopists also saw the potential of reacting ligands with transition metal under matrix isolation conditions. Photolysis of metal carbonyls in organic (38) or inert gas matrices (39) had already been done, but atoms offered the possibility of step-wise addition of ligands. DeKock (40), Turner (41), and Moskovits and Ozin (42) made early contributions, but the work of the last two became dominant (32). By 1972, there were the two distinct branches in transition metal atom chemistry, the preparative and matrix spectroscopic studies.

The Years of Rapid Growth. In the period, 1972-75, many people entered the field of atom synthesis, Klabunde (43), von Gustorf (44), Green (45), and Lagowski (46) to mention a few. The result was that a very large number of new organometallic compounds was synthesised sufficient to convince even the most conventional organometallic chemists that this approach had something to offer.

Most of the work initially was with the more volatile transition metals, i.e. the first row metals plus palladium, silver and gold, because these were easy to evaporate in reasonable quantities in simple apparatus. However, efforts to use the less volatile metals of the second and third rows gained momentum. Skell used sublimation of resistively heated wires of molybdenum and tungsten to make the remarkably stable $[Mo(\eta^4\text{-}C_4H_6)_3]$ and $[W(\eta^4\text{-}C_4H_6)_3]$ (47). Green attempted to use electron bombardment evaporation but initially there were difficulties because of electron reflection off the target metal at ground potential, the electrons then destroying the organometallic products. By 1973, he could evaporate titanium (45) but electron damage obstructed work with less volatile metals. The problem was solved by putting the target metal at a high positive potential (48-49) and from that time on electron bombardment evaporation was a growing force in the field.

Also in this period, the technique of reacting metal atoms with ligands in cold solutions was developed (50) and this gave the possibility of reacting metal atoms with involatile ligands which was later exploited (51-52). An instructive failure was the largely abortive attempts by von Gustorf (53) to use a 200 W continuous wave YAG laser for metal evaporation. Reflection of the laser light from the metal surface meant that heat transfer was poor and evaporation much less than was hoped. The much more recent spectroscopic scale work with laser evaporation (54-55) is successful because of the higher energy input per laser pulse.

By 1977-78, the initial growth period of metal atom synthesis in organometallic chemistry was coming to an end. Various factors contributed to this slow-down. One was that a lot of the more obvious reactions involving the readily vaporised metals had been done. The chemistry of the atoms of many of the less volatile metals was hardly touched as this chemistry required more sophistication in the equipment and there was a reluctance to take on the necessary technical and financial commitment. Another factor was the disappointing performance of commercial metal atoms equipment. The apparatuses sold by G.V. Planer in England and by Kontes Martin in the USA were not as easy to use or as effective in use as they might have been--they were certainly less good than the best apparatuses available in the dedicated metal atom laboratories at that time. Those who bought them were often seeking an easy entry to metal atom chemistry but the equipment required a lot of skill to use. The result was dampening to the field and organometallic chemists went back to their conventional methods which had certainly progressed in scope with the stimulus of competition from metal atom methods. Another consideration was a change in the emphasis in research on organometallic chemistry from mononuclear to polynuclear and cluster compounds. Although there are exceptions, transition metal atoms are not generally good routes to cluster compounds (35), so chemists looked away from atom methods. Those working on the spectroscopic scale in matrices were more successful with metal clusters but these results tended to seem of greater relevance to heterogeneous catalysis than to the main stream of organometallic synthesis. Finally, the growth of interest in materials science attracted atom chemists to apply their methods in this area rather than in synthetic organometallic chemistry and this trend has gained further momentum recently (56).

The Recent Period. The most striking feature of the last six years has been the dominance in organometallic synthesis *via* atoms of the Oxford group led by Green. Using very sophisticated equipment of their own design they have been able to evaporate routinely many grams of the most refractory metals to make quite large quantities of the products of types which are still inaccessible by the best conventional methods. As some of the products have been polynuclear and their formation has involved activation of C-H bonds of alkane, this atom chemistry has been of great interest again in the current "centre stage" of organometallic chemistry (57-59).

However, the cost of setting up for metal atom chemistry like the Oxford group has become quite high--at least $40,000 for a do-it-yourself apparatus and at least $100,000 for a commercial apparatus. This cost factor is quite inhibiting and there are few chemists who can afford to "have a go" at that price without a very clear idea of the value of what they can achieve.

Of course, there have been other achievements in this period too. We should note the "tandem high energy process" approach of Lagow (60-61), reacting thermal or discharge generated free radicals with free atoms to give new metal-alkyl derivatives (an elegant up-date of Paneth's work). Also, photolytic activation of matrix isolated atoms to react with methane (62). Metal atom chemistry in simpler apparatus is still going on. It has many uses and it remains the best way of making some organometallic compounds and of exploring the synthesis of new compounds as illustrated by some recent publications (63-65). At both the simple and sophisticated level of equipment, atom chemistry has undoubtedly earned a permanent place in the range of synthetic techniques available to the organometallic chemist. However, the pattern of the growth of the technique and its ultimate level of acceptance form a useful model for predicting how newer "high energy processes" in organometallic chemistry may develop.

The Future Role of High Energy Processes in Organometallic Chemistry

We have said that photochemistry and now atom chemistry have established positions in the field of organometallic chemistry. But what of the other "high energy processes" discussed in the Symposium? All of them are making a useful contribution to our total understanding of organometallic chemistry but two seem to have features which may allow them to make a greater contribution than the others. These are sonochemistry and the pulsed laser evaporation of metals to give gaseous clusters.

Sonochemistry (10) has the promise of being a popular technique, capable of being used alongside simple apparatus for photochemistry as a routine tool for enhancement of reaction rates. Just as organometallic photochemistry has been able to borrow much from the longer established field of organic photochemistry, so sonochemistry can employ inexpensive commercial equipment used in biological studies. As Suslick has stated (10), for an outlay of around $2000, a whole apparatus can be put together for homogeneous or heterogeneous sonochemistry.

Laser evaporation of clusters will not be a popular technique, it is costly in equipment, demanding in know-how, and probably limited to a spectroscopic scale. Nevertheless, it has already shown itself capable of making new species in the organic (54) and organometallic fields (55) which are of such tantalizing interest that they will inspire chemists to make them by less esoteric routes. Indeed, the work "inspiration" is surely the clue to the development of future "high energy processes" in organometallic chemistry. However exotic the technique in the "high energy process" may be, it is worthwhile if its results inspire conventional chemists to fresh thoughts and to see broader horizons. It is a special bonus if the method is simple and many chemists choose to try it!

Literature Cited

1. Paneth, F.A.; Hofeditz, W.H. Ber. 1929, 62, 1335.
2. Rice, F.O.; Johnston, W.R.; Evering, B.L. J.Am.Chem.Soc. 1932, 54, 3529.
3. Kraus, C.A.; Callis, C.C. U.S. Patent, 1 612 131, 1926.
4. Mishima, S. Bull. Chem. Soc. Japan 1967, 40, 608.
5. Rieke, R.D. Acc. Chem. Res. 1977, 10, 301.
6. Gasper, P.P.; Herold, B.J. In Carbene Chemistry, 2nd Ed.; Kirmse, W., Ed.; Academic Press, 1971; p 504.
7. Atwell, W.H.; Weyenberg, D.R. J.Organomet.Chem. 1966, 5, 594.
8. Skell, P.S.; Goldstein, E.J. J.Am.Chem.Soc. 1964, 86, 1442.
9. Pilcher, G.; Skinner, H.A. In Chemistry of the Metal-Carbon Bond; Hartley, F.R.; Patai, S., Eds.; Wiley-Interscience, 1982; Vol. 1, p 43.

10. Suslick, K.S. Advan. Organomet. Chem. 1986, 25, 73.
11. Geoffroy, G.L.; Wrighton, M.S. In Organometallic Photochemistry; Academic Press: New York, 1979.
12. Eastland, G.W.; Symons, M.C.R. J.Chem.Soc., Dalton Trans. 1984, 2193.
13. Steacie, E.W.R. In Atomic and Free Radical Reactions; Reinhold: New York, 1954, 2nd Ed.
14. Bass, A.M.; Broida, H.P. In Formation and Trapping of Free Radicals; Academic Press: New York, 1960.
15. See Chapter 4 by G.C. Pimentel in ref. 14 for a careful review of the discovery of matrix isolation techniques.
16. Akerlof, G.C. U.S. Patent 2 724 692, 1955.
17. Kanaan, A.S.; Margrave, J.L. Advan.Inorg.Chem.Radiochem. 1964, 6, 143.
18. Svedberg, T. Koll. Zeit. 1907, 1, 229, 257.
19. Rice, F.O.; Freamo, M. J.Am.Chem.Soc. 1951, 73, 5529.
20. Pease, D.C. U.S. Patent 2 840 588, 1958.
21. Skell, P.S.; Westcott, L.D. J.Am.Chem.Soc. 1963, 85, 1023, and Skell, P.S.; Westcott, L.D.; Goldstein, J.P.; Engel, R.R. J.Am.Chem.Soc. 1965, 87, 2829.
22. Timms, P.L.; Stump, D.D.; Kent, R.A.; Margrave, J.L. J.Am.Chem.Soc. 1966, 88, 940.
23. Timms, P.L. Inorg.Chem. 1968, 7, 387.
24. Timms, P.L. J.Am.Chem.Soc. 1968, 90, 4585.
25. Timms, P.L. Chem.Commun. 1968, 258.
26. Skell, P.S.; Owen, P.W. J.Am.Chem.Soc. 1967, 89, 3933.
27. Andrews, L.; Pimentel, G.C. J.Chem.Phys. 1966, 44, 2527.
28. Mile, B. Angew.Chem.Int.Edit. 1968, 7, 507.
29. Timms, P.L. Chem. Commun. 1969, 1033.
30. Olsen, L.O.; Smith, C.S.; Crittenden, E.C. J. Appl. Phys. 1945, 16, 425.
31. Timms, P.L. Advan. Inorg. Chem. Radiochem. 1972, 14, 121.
32. Moskovits, M.; Ozin, G.A. In Cryochemistry; Wiley-Interscience: New York, 1976.
33. Blackborow, J.R.; Young, D. In Metal Vapour Synthesis in Organometallic Chemistry; Springer-Verlag: Berlin, 1979.
34. Klabunde, K.J. In Chemistry of Free Atoms and Particles; Academic Press: New York, 1980.
35. Timms, P.L. Proc. R. Soc. Lond. 1984, A396, 1.
36. Skell, P.S.; Havel, J.J. J.Am.Chem.Soc. 1971, 93, 6687.
37. Timms, P.L. J.Chem.Educ. 1972, 49, 782.
38. Stolz, I.W.; Dobson, G.R.; Sheline, R.K. J.Am.Chem.Soc. 1962, 84, 3589.
39. Rest, A.J.; Turner, J.J. Chem. Commun. 1969, 375.
40. DeKock, R.L. Inorg.Chem. 1971, 10, 1205.
41. Graham, M.A.; Poliakoff, M.; Turner, J.J. J.Chem.Soc.A 1971, 2939.
42. Kündig, E.P.; Moskovits, M.; Ozin, G.A. Can.J.Chem. 1972, 50, 3587.
43. Klabunde, K.J.; Key, M.S.; Low, J.Y.F. J.Am.Chem.Soc. 1972, 94, 999.
44. von Gustorf, E.A.K.; Jaenicke, O.; Polansky, O.E. Angew. Chem. Int. Ed. 1972, 11, 533.

45. Benfield, F.W.S.; Green, M.L.H.; Ogden, J.S.; Young, D. <u>Chem. Commun.</u> 1973, 866.
46. Graves, V.; Lagowski, J.J. <u>Inorg.Chem.</u> 1976, <u>15</u>, 577.
47. Skell, P.S.; Van Dam E.M.; Silvon, M.P. <u>J.Am.Chem.Soc.</u> 1974, <u>96</u>, 627.
48. Timms, P.L. <u>Angew. Chem. Int. Ed.</u> 1975, <u>14</u>, 273.
49. Cloke, F.G.N.; Green, M.L.H.; Morris, G.E. <u>J.Chem.Soc., Chem. Commun.</u> 1978, 72.
50. MacKenzie, R.E.; Timms, P.L. <u>Chem. Commun.</u> 1974, 650.
51. Francis, C.G.; Timms, P.L. <u>J.Chem.Soc., Dalton Trans.</u> 1980, 1401.
52. Francis, C.G.; Ozin, G.A. <u>J. Macromol. Sci., A.</u> 1981, <u>16</u>, 167.
53. von Gustorf, E.A.K.; Jaenicke, O.; Wolfbeis, O.; Eady, C. <u>Angew. Chem. Int. Ed.</u> 1975, <u>14</u>, 278.
54. Kroto, H.W.; Heath, J.R.; Brien, S.C.O.; Curl, R.F.; Smalley, R.E. <u>Nature</u> 1985, <u>318</u>, 162.
55. Trevor, D.J.; Whetten, R.L.; Cox, D.M.; Kaldor, A. <u>J.Am.Chem.Soc.</u> 1985, <u>107</u>, 518.
56. Spalding, B.J. <u>Chemical Week</u> July 16th 1986, 34.
57. Bandy, J.A.; Cloke, F.G.N.; Green, M.L.H.; O'Hare, D.; Prout, K. <u>J.Chem.Soc., Chem. Commun.</u> 1984, 240.
58. Green, M.L.H.; O'Hare, D. <u>J.Chem.Soc., Chem. Commun.</u> 1985, 355.
59. Green, M.L.H.; O'Hare, D.; Parkin, G. <u>J.Chem.Soc., Chem. Commun.</u> 1985, 356.
60. Juhlke, T.J.; Braun, R.W.; Bierschenk, T.R.; Lagow, R.J. <u>J.Am.Chem.Soc.</u> 1979, <u>101</u>, 3229.
61. Guerra, M.A.; Bierschenk, T.R.; Lagow, R.J. <u>J.Chem.Soc., Chem. Commun.</u> 1985, 1550.
62. Billups, W.E.; Konarski, M.A.; Hauge, R.H.; Margrave, J.L. <u>J.Am.Chem.Soc.</u> 1980, <u>102</u>, 7393.
63. Gigugho, J.J.; Sneddon, L.G. <u>Organometallics</u> 1986, <u>5</u>, 327.
64. Vituli, G.; Uccello-Barretta, G.; Pannocchia, P.; Raffnelli, A. <u>Organometal Chem.</u> 1986, <u>302</u>, C21.
65. Elschenbroich, C.; Kroker, J.; Massa, W.; Mursch, M.; Ashe, A.J. III <u>Angew. Chem. Int. Ed.</u> 1986, <u>25</u>, 57

RECEIVED November 24, 1986

Chapter 2

Gas-Phase Organometallic Ion Chemistry

J. L. Beauchamp

Arthur Amos Noyes Laboratory of Chemical Physics, California Institute of Technology, Pasadena, CA 91125

Gas phase studies of the reactions of small molecules at transition metal centers have provided a rich harvest of highly speculative reaction mechanisms. Traditional mechanistic probes, such as isotopic labeling, have served mainly to indicate the complexity of "simple" processes such as the dehydrogenation of of alkanes. Ion beam techniques, ion cyclotron resonance spectroscopy, and laser activation methods have been employed to characterize the energetics and mechanisms of gas phase organometallic reactions in considerable detail. For example, product kinetic energy distributions reveal important features of the potential energy surfaces associated with the formation and rupture of H-H, C-H and C-C bonds at transition metal centers. In complex systems where several products are formed competitively, IR multiphoton activation of stable reaction intermediates allows the lowest energy pathway to be identified. Several applications of these techniques in studies of the reactions of alkenes and alkanes with atomic transition metal ions are presented to illustrate endeavors in the growing field of gas phase organometallic ion chemistry.

An impressive array of experimental methods are available which permit the use of extreme conditions to promote chemical reactions. In studying the reactions of ionic species with small molecules in the gas phase, it is possible to vary reagent translational energy over a wide range. In addition, the possibility of trapping ions in appropriate configurations of electromagnetic fields facilitates the use of single and multiphoton excitation processes to effect specific chemical reactions. These activation techniques can provide relative translational and internal energies which correspond to stellar temperatures. Such studies would be a mere curiosity if extreme conditions were always necessary to promote interesting reactions. Fortunately this is not the case, and the observation of processes

0097-6156/87/0333-0011$09.00/0

such as the conversion of cyclohexane to benzene in a single bimolecular encounter with an atomic transition-metal ion at room temperature (1) clearly demonstrate that it is not necessary to "nuke" the system to identify interesting candidates for detailed mechanistic and dynamic studies.

Considerable interest in the subject of C-H bond activation at transition-metal centers has developed in the past several years (2), stimulated by the observation that even saturated hydrocarbons can react with little or no activation energy under appropriate conditions. Interestingly, gas phase studies of the reactions of saturated hydrocarbons at transition-metal centers were reported as early as 1973 (3). More recently, ion cyclotron resonance and ion beam experiments have provided many examples of the activation of both C-H and C-C bonds of alkanes by transition-metal ions in the gas phase (4). These gas phase studies have provided a plethora of highly speculative reaction mechanisms. Conventional mechanistic probes, such as isotopic labeling, have served mainly to indicate the complexity of "simple" processes such as the dehydrogenation of alkanes (5). More sophisticated techniques, such as multiphoton infrared laser activation (6) and the determination of kinetic energy release distributions (7), have revealed important features of the potential energy surfaces associated with the reactions of small molecules at transition metal centers.

The purpose of this article is to review some of the current endeavors in this developing field. To maintain brevity, the focus is on recent studies carried out in our own laboratory and in conjunction with Professor M.T. Bowers at the University of California at Santa Barbara, with emphasis on the use of kinetic energy release distributions and infrared laser multiphoton excitation to probe potential energy surfaces for the reactions of atomic metal ions with alkenes and alkanes.

Experimental Methods

The majority of studies in our laboratory of the reactions of atomic transition metal ions with small molecules have been conducted using the ion beam apparatus shown in Figure 1 (8). The key to the success of this apparatus has been the use of a surface ionization source shown in the inset of Figure 1. Metal halide salts are evaporated from a heated tube furnace onto a resistively heated rhenium ribbon (1800 - 2500 °K) where they thermally decompose and the resulting metal atoms are surface ionized. Atomic metal ions are extracted, mass analyzed in a magnetic sector, and allowed to interact with neutral molecules in a collision cell with relative kinetic energies in the range 0.2-10.0 eV. Products scattered in the forward direction are mass analyzed using a quadrupole mass spectrometer and counted with a channeltron detector. Armentrout and co-workers have built an improved version of this apparatus with the important difference being the substitution of an octopole ion guide for the collision chamber to improve product ion collection efficiencies (9).

With the surface ionization source it is generally assumed that the reactant ion internal state distribution is characterized by the source temperature and that the majority of the reactant ions are in their ground electronic state. This contrasts with the uncertainty in reactant state distributions when transition metal ions are generated by electron impact fragmentation of volatile organometallic precursors (10) or by laser evaporation and ionization of solid metal targets (11). Many examples have now been recorded of modified excited state reactivities of atomic transition metal ions (12).

The apparatus and techniques of ion cyclotron resonance spectroscopy have been described in detail elsewhere. Ions are formed, either by electron impact from a volatile precursor, or by laser evaporation and ionization of a solid metal target (14), and allowed to interact with neutral reactants. Freiser and co-workers have refined this experimental methodology with the use of elegant collision induced dissociation experiments for reactant preparation and the selective introduction of neutral reactants using pulsed gas valves (15). Irradiation of the ions with either lasers or conventional light sources during selected portions of the trapped ion cycle makes it possible to study ion photochemical processes (16). In our laboratory we have utilized multiphoton infrared laser activation of metal ion-hydrocarbon adducts to probe the lowest energy pathways of complex reaction systems (6). Freiser and co-workers have utilized dispersed visible and uv radiation from conventional light sources to examine photochemical processes involving organometallic fragments (17).

A very powerful technique for obtaining information relating to potential energy surfaces for organometallic reactions involves the determination of kinetic energy release distributions for product ions (7). Experiments of this type have been conducted in collaboration with Professor M.T. Bowers and his students at the University of California at Santa Barbara. The instrument used for these studies, a VG ZAB-2F reversed geometry double focusing mass spectrometer, is shown schematically in Figure 2. Adducts of atomic metal ions with neutral reactants are extracted from a high pressure ion source and mass analyzed using the magnetic sector. Translational energies of products formed by adduct dissociation in the second field free region between the magnetic and electric sectors are determined from an analysis of the metastable peak shape obtained by scanning the electric sector (18).

In addition to the experimental methods pertinent to the work described herein, Farrar and co-workers have recently published a crossed beam investigation of the decarbonylation of acetaldehyde by atomic iron ions (19). Studies of this type, which provide product energy and angular distributions, will provide detailed insights into the mechanisms of organometallic reactions. Several investigators, including Depuy (20), McDonald (21), Squires (22), and Weissharr (23), have utilized flowing afterglow techniques to study organometallic reactions. These experimental methods all complement each other in terms of the type of information they provide and the accessible range of experimental parameters such as pressure and temperature.

Experimental Studies of the Reactions of Atomic Metal Ions with Hydrocarbons

Endothermic Reactions and the Determination of Bond Dissociation Energies for Organometallic Fragments. The reaction of atomic nickel ion with molecular hydrogen to yield NiH^+ is substantially endothermic. Reaction cross sections for this process, measured using the ion beam apparatus shown in Figure 1, are displayed in Figure 3 for reactions 1 and 2 with HD as the neutral.

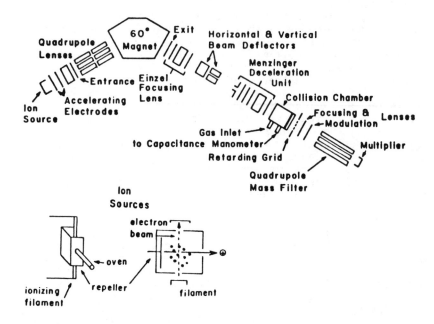

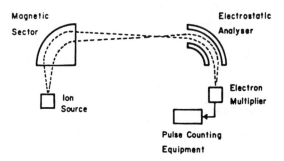

Figure 1. Schematic view of ion beam apparatus and surface ionization source.

Figure 2. Schematic view of reversed geometry double focusing mass spectrometer used for kinetic energy release measurements.

$$Ni^+ + HD \quad \begin{cases} \rightarrow NiH^+ + D & (1) \\ \rightarrow NiD^+ + H & (2) \end{cases}$$

An analysis of the threshold region for reaction 1 yields a Ni^+-H bond dissociation energy (Table I). Determination of a threshold involves assuming a functional form for the variation of the cross section with energy in excess of the threshold value and folding in the velocity distribution of the neutral target molecules (Doppler broadening) to fit the experimental data (25). Examination of endothermic reactions in more complex systems has yielded a wide range of bond dissociation energies for neutral and ionic organometallic fragments. For example, the threshold identified for the formation of $CoCH_2^+$ in reaction 3 yields a cobalt ion carbene bond dissociation energy (26).

$$Co^+ + C_2H_4 \rightarrow CoCH_2^+ + CH_2 \tag{3}$$

Representative bond dissociation energy data for organometallic fragments are summarized in Table I. Several trends in these data warrant comment. Not surprisingly, Cr^+ forms unusually weak bonds to ligands such as H and CH_3. This results from the disruption of the unusually stable half filled 3d shell in the atomic ion to form a sigma bond. Periodic trends of the metal hydride ion bond dissociation energies are shown in Figures 4 and 5 for the first and second row metals, respectively, where they are compared to the results of ab initio generalized valence bond dissociation consistent configuration interaction (GVB-DC-CI) calculations (27). The agreement is generally very good, with the experimental results tending to be slightly higher than the theoretical values. A careful analysis of the wave functions indicates that first row metals use a mixture of the 4s and 3d orbitals to form strong sigma bonds. The second row metals form strong sigma bonds with only minor participation of the 5s orbitals. It has been noted that the bond strengths can be correlated with the promotion energy from the ground state to the lowest state derived from a $4s^13d^n$ configuration (which is thus assumed to form a strong bond to hydrogen) (28). The correlation is somewhat better with a more appropriately defined promotion energy in which exchange terms are accounted for (29) and suggests an intrinsic metal hydrogen bond dissociation energy of around 60 kcal/mol.

It is of particular interest to note that the metal ion methyl bond dissociation energies in Table I are typically greater than the metal hydrogen bond energies. In part this can be attributed to the greater polarizability of the methyl group compared to hydrogen and the effect which this has on stabilizing the positive charge. Since $D°[H-H]$ and $D°[CH_3-H]$ are approximately equal, the relative nickel ion hydrogen and methyl bond strengths are confirmed by the observation that reaction 4 of nickel hydride ion with methane is a fast process (30).

$$NiH^+ + CH_4 \rightarrow NiCH_3^+ + H_2 \tag{4}$$

According to the data in Table I this process is exothermic by 9 kcal/mol. Reaction 4 is one of the few processes in which methane is known to undergo a facile reaction at a transition metal center. Interestingly, the analogous reaction with FeH^+ is not observed, even though it is exothermic by 10 kcal/mol (31).

Table I. Bond Dissociation Energies (eV) at 298 K for Organometallic
Fragments

M	$D°(M^+\text{-}H)$	$D°(M^+\text{-}CH_2)$	$D°(M^+\text{-}CH_3)$	$D°(M\text{-}H)$	$D°(M\text{-}CH_3)$
Ca	1.95 ± 0.2[a]				
Sc	2.44 ± 0.10[b]	≥4.05 ± 0.2[b]	2.56 ± 0.2[b]		1.41 ± 0.3[a]
Ti	2.43 ± 0.10[c]	3.69 ± 0.2[a]	2.45 ± 0.2[a]		2.0 ± 0.3[a]
V	2.09 ± 0.06[d]	3.30 ± 0.2[j]	2.17 ± 0.2[j]		2.05 ± 0.3[a]
Cr	1.40 ± 0.11[e]	2.21 ± 0.2[a]	1.28 ± 0.2[a]	1.8 ± 0.2[o]	1.76 ± 0.3[a]
Mn	2.10 ± 0.15[f]	4.16 ± 0.2[k]	2.2 ± 0.2[m]		0.5 ± 0.3[m]
Fe	2.12 ± 0.06[g]	4.2 ± 0.2[l]	2.56 ± 0.2[a]	1.9 ± 0.2[p]	2.76 ± 0.3[a]
Co	2.02 ± 0.06[h]	3.7 ± 0.3[l]	2.6 ± 0.2[l]	2.3 ± 0.4[p]	
Ni	1.72 ± 0.08[h]	3.7 ± 0.2[l]	2.13 ± 0.2[a]	2.8 ± 0.2[p]	
Cu	0.96 ± 0.13[h]			2.9 ± 0.3[q]	
Zn			3.03 ± 0.2[n]		0.86 ± 0.3[n]
Y	2.52 ± 0.15[c]				
Zr	2.34 ± 0.15[c]				
Nb	2.31 ± 0.15[c]				
Mo	1.80 ± 0.15[c]			2.3 ± 0.2[p]	
Ru	1.78 ± 0.15[i]		2.34 ± 0.2[i]	2.4 ± 0.2[p]	
Rh	1.56 ± 0.15[a]		2.04 ± 0.2[i]	2.6 ± 0.2[p]	
Pd	2.06 ± 0.15[a]		2.56 ± 0.2[i]	2.4 ± 0.2[p]	
Ag	0.65 ± 0.15[c]				

[a]Elkind, J. L.; Aristov, N.; Georgiadis, R.; Sunderlin, L.; Armentrout, P. B., to be published.
[b]Sunderlin, L.; Aristov, N.; Armentrout, P. B. J. Am. Chem. Soc., submitted.
[c]Elkind, J. L.; Armentrout, P. B. Inorg. Chem., 1986, 25, 1078.
[d]Elkind, J. L.; Armentrout, P. B. J. Phys. Chem., 1985, 89, 5626.
[e]Elkind, J. L.; Armentrout, P. B. J. Phys. Chem., submitted.
[f]Elkind, J. L.; Armentrout, P. B. J. Chem. Phys., 1986, 84, 4862.
[g]Elkind, J. L.; Armentrout, P. B. J. Phys. Chem., in press.
[h]Elkind, J. L.; Armentrout, P. B. J. Phys. Chem., submitted.
[i]Mandich, M. L.; Halle, L. F.; Beauchamp. J. L. J. Am. Chem. Soc., 1984, 106, 4403.
[j]Aristov, N.; Armentrout, P. B. J. Am. Chem. Soc., 1986, 108, 1806.
[k]Stevens, A. E.; Beauchamp, J. L. J. Am. Chem. Soc., 1979, 101, 6449.
[l]Halle, L. F.; Armentrout, P. B.; Beauchamp, J. L. Organometallics, 1982, 1, 963.
[m]Armentrout, P. B., in "Laser Applications in Chemistry and Biophysics", M. A. El-Sayed, Ed., Proc. SPIE, 1986, 38, 620.
[n]Georgiadis, R.; Armentrout, P. B. J. Am. Chem. Soc., 1986, 108, 2119.
[o]Sallans, L.; Lane, K. R.; Squires, R. R.; Freiser, B. S. J. Am. Chem. Soc., 1985, 107, 4379.
[p]Tolbert, M. A.; Beauchamp, J. L. J. Phys. Chem., 1986, 90, 5015.
[q]Burnier, R. C.; Byrd, G. D.; Freiser, B. S. Anal. Chem., 1981, 103, 784.

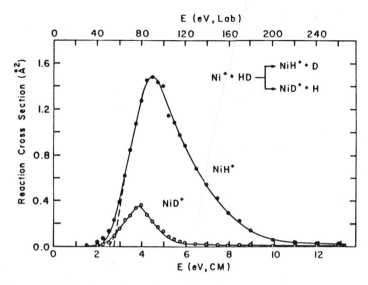

Figure 3. Variation with relative kinetic energy of cross sections for reaction of Ni$^+$ with HD. Data from reference 24.

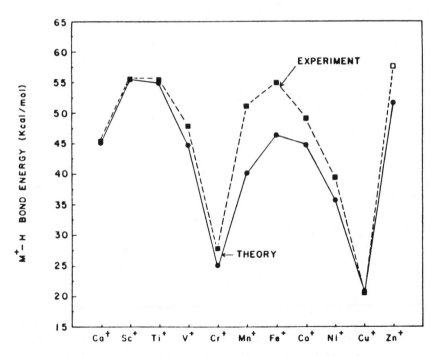

Figure 4. Comparison of theoretical and experimental bond dissociation energies for first row diatomic metal hydride ions. Data from reference 27.

The addition of several ligands to the metal system can reverse relative metal methyl and metal hydrogen bond dissociation energies. For example, reaction 5 is observed to be a fast exothermic process,

$$CpRh(CH_3)(CO)^+ + H_2 \rightarrow CpRhH(CO)^+ + CH_4 \qquad (5)$$

indicating that the rhodium hydride bond energy in the product exceeds the rhodium methyl bond energy in the reactant ion (32).

Exothermic Reactions of Transition Metal Ions with Hydrocarbons. Cross sections for the formation of product ions resulting from the interaction of Ni^+ with n-butane are shown in Figure 6 for a range of relative kinetic energies between 0.2 and 4 eV. In contrast to the results shown in Figure 3, several products (reactions 6-8) are formed with large cross section at low energies. These cross sections decrease with

$$
\begin{array}{ll}
\phantom{Ni^+ + n\text{-}C_4H_{10}} \rightarrow Ni(C_4H_8)^+ + H_2 & (6) \\
Ni^+ + n\text{-}C_4H_{10} \longrightarrow Ni(C_3H_6)^+ + CH_4 & (7) \\
\phantom{Ni^+ + n\text{-}C_4H_{10}} \rightarrow Ni(C_2H_4)^+ + C_2H_6 & (8)
\end{array}
$$

increasing energy and do not exhibit any apparent threshold at low energies. The products observed in the limit of low energy are identical to those observed in comparable abundance at thermal energies using ICR techniques. Processes 6 - 8 are clearly exothermic reactions and every encounter of Ni^+ with n-butane at low energy leads to the formation of one of these products. Unravelling the mechanisms of these reactions in which not only C-H but also C-C bonds are cleaved in what are apparently facile processes has proven to be an intriguing and challenging problem (33).

Although the list is certainly not exhaustive, a variety of questions can be asked in studying hydrocarbon activation processes. How does reactivity depend on easily varied factors such as the metal ion, the hydrocarbon, and reactant translational and internal energy? In considering various reactive intermediates, how many distinct species are formed? How fast do they decompose and what are the products arising from each? In examining products, what are the product structures? How is the available energy distributed among different degrees of freedom? These questions have all been answered in varying degrees and the highlights of these results will be summarized here.

Variation in Reactivity for Different Metals. Consider first how the reactivity varies with the metal ion. Table II presents low-energy product distributions for the reactions of first and second row metal ions with n-butane. Chromium, manganese and molybdenum ions are unreactive with n-butane (34). The first sigma bond energy is weak in the case of Cr^+ and Mo^+. The second bond is expected to be weak in the case of manganese ion, and facile activation of C-H bonds by these three ions is not anticipated or observed. The behavior of the first row group 8-10 metal ions Fe^+, Co^+, and Ni^+ is similar (4). Cleavage of C-C bonds is not observed for Ru^+ and Rh^+ and multiple dehydrogenation processes are prevalent for these ions (35). Only Pd^+ appears to react in the same fashion as the corresponding first row ion. The major reactions of Ti^+ and V^+ are not analogous to those of the remaining first row transition metal ions. Instead, there is a

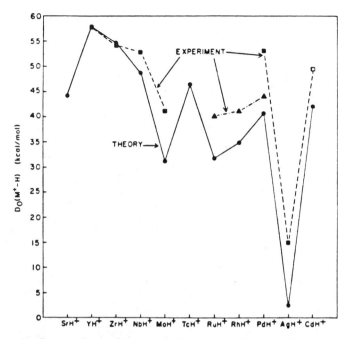

Figure 5. Comparison of theoretical and experimental bond dissociation energies for second row diatomic metal hydride ions. Data from reference 29.

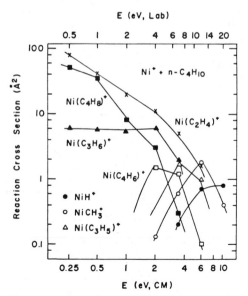

Figure 6. Variation with relative kinetic energy of cross sections for the formation of product ions resulting from the interaction of Ni$^+$ with n-butane. Data from reference 33.

similarity to the reactivity of the second row ions Ru^+ and Rh^+ ($\underline{36}$). Scandium ions are unique in that ethylene elimination to yield a $Sc(C_2H_6)^+$ product is an important reaction pathway ($\underline{37}$).

The only reaction in common to the metal ions included in Table II is the dehydrogenation of n-butane. This apparent similarity is seen to be superficial when the mechanism of dehydrogenation is probed using deuterium labelled substrates. Scandium, nickel and palladium ions dehydrogenate n-butane in highly specific and distinct processes (Scheme I). It was initially expected that the favorable dehydrogenation process would involve removal of H_2 from across the central C-C bond in a 1,2-process. For the three examples considered, this is observed only in the case of Pd^+ and has been attributed to the unusually high Lewis acidity of this species which results in hydride abstraction as a first step in the dehydrogenation process. Nickel ions very cleanly remove hydrogen from both ends of the molecule (1,4-mechanism) ($\underline{5}$), and Sc^+ undergoes a 1,3-dehydrogenation process ($\underline{37}$)! The proposed reaction mechanism for Ni^+ is considered in greater detail below. The unique reactivity of Sc^+ has been attributed to the availability of only two valence electrons on the metal center and the restrictions which this imposes on the stability of reaction intermediates and products ($\underline{37}$). Although the multiple dehydrogenation processes make an examination of reaction specificity difficult, the ions Ti^+, V^+, Ru^+, and Rh^+ all react predominately by a 1,2 process ($\underline{35},\underline{36}$). These dehydrogenation mechanisms are the preferred mode of reaction and may not be available for a particular substrate. Thus it is found that Co^+ ions, which preferentially remove hydrogen from longer chain hydrocarbons by a 1,4 process, dehydrogenate isobutane cleanly by a 1,2-mechanism to yield isobutylene ($\underline{4}$).

Reaction Intermediates and Their Lifetimes. Studies of product structures using ligand exchange reactions and collision induced dissociation processes reveal that the product of reaction 6 in which Ni^+ dehydrogenates n-butane is actually a bisethylene complex ($\underline{5}$). Other possibilities such as a metallocyclopentane or butene complex are ruled out by preparing these species and demonstrating that they exhibit distinct behavior in comparison to the product of reaction 6. The reaction coordinate diagram which was originally used to discuss this process is shown in Figure 7, with estimates of various bond energies being used to quantify the energetics of several proposed reaction intermediates. Competing with the dehydrogenation processes shown in Scheme I are the alkane elimination reactions illustrated in Scheme II. By invoking processes analogous to those shown in Schemes I and II it is possible to explain the majority of the reaction processes of atomic transition metal ions with more complex hydrocarbons. The major difference is that as the size of the hydrocarbon increases, further reactions occur and several small molecules may be eliminated to form stable adducts of the metal ion with a highly unsaturated hydrocarbon.

Returning to reaction 6, Schemes I and II and Figure 7 suggest that the loss of hydrogen and ethane result from a common intermediate and that a distinct intermediate is responsible for the elimination of methane. A comparison of the product distributions measured using the ion beam instrument with the relative metastable yields recorded with the reverse sector instrument supports this conjecture (Table III) in that the ratio of hydrogen to ethane loss is approximately the same and methane elimination is diminished in importance in the metastable data ($\underline{38}$). From

Table II. Comparison of the Reactions of Transition Metal Ions with n-Butane at a Relative Kinetic Energy of 0.5eV[a]

Neutral Product	Sc[+]	Ti[+]	V[+]	Fe[+]	Co[+]	Ni[+]	Ru[+]	Rh[+]	Pd[+]
H_2	0.37	0.17	0.39	0.20	0.29	0.48	0.20	0.27	0.38
$2H_2$	0.22	0.66	0.61				0.80	0.73	
CH_4	0.01	0.09		0.41	0.12	0.06			0.21
$CH_4 + H_2$	0.02	0.03							
C_2H_4	0.36	0.02							
C_2H_6	0.02	0.03		0.39	0.59	0.45			0.41
σ_{rxn}[b]	103	45	48	98	170	88	38	48	29

[a] Data from reference 36.
[b] Total reaction cross section, $Å^2$.

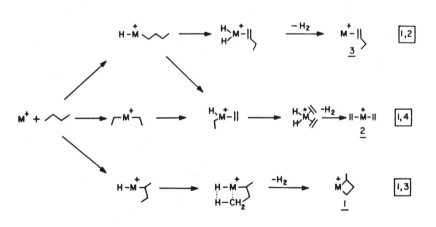

Scheme I

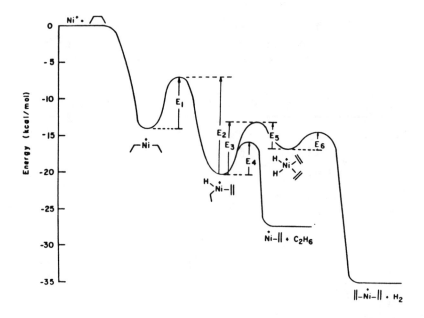

Figure 7. Reaction coordinate diagram for the dehydrogenation of n-butane by Ni^+.

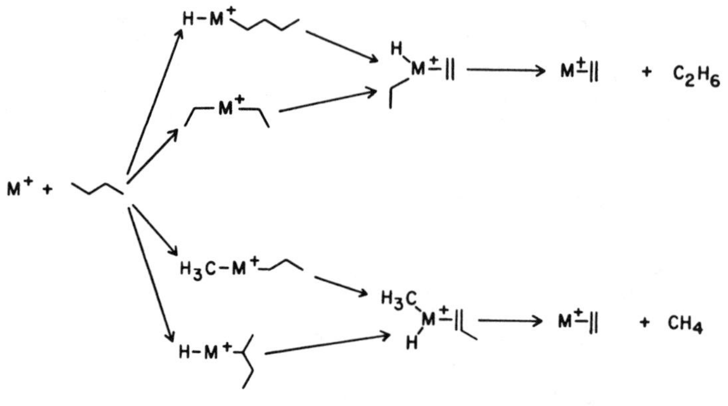

Scheme II

Table III. Relative Product Distributions for the
Reactions of Co^+ and Ni^+ with Alkanes [a]

Reaction		Product Distribution	
		Ion Beam	Metastable
Co^+ + propane	$\rightarrow Co(C_3H_6)^+ + H_2$	0.75	1.00
	$\rightarrow Co(C_2H_4)^+ + CH_4$	0.25	0.00
Co^+ + isobutane	$\rightarrow Co(C_4H_8)^+ + H_2$	0.27	0.48
	$\rightarrow Co(C_3H_6)^+ + CH_4$	0.73	0.50
	$\rightarrow Co(C_4H_8)^+ + H_2$	0.29	0.35
Co^+ + n-butane	$\rightarrow Co(C_3H_6)^+ + CH_4$	0.12	0.00
	$\rightarrow Co(C_2H_4)^+ + C_2H_6$	0.59	0.65
Ni^+ + isobutane	$\rightarrow Ni(C_4H_8)^+ + H_2$	0.07	0.08
	$\rightarrow Ni(C_3H_6)^+ + CH_4$	0.93	0.92
	$\rightarrow Ni(C_4H_8)^+ + H_2$	0.34	0.49
Ni^+ + n-butane	$\rightarrow Ni(C_3H_6)^+ + CH_4$	0.09	0.00
	$\rightarrow Ni(C_2H_4)^+ + C_2H_6$	0.58	0.50

[a] Data from reference 38.

this it can be inferred that the lifetime of the intermediate leading to methane loss is shorter and more extensive decomposition occurs in the source. Similar results are observed for the reaction of Co^+ with n-butane (Table III). The results for isobutane are different in that methane elimination remains prominent in the metastable yield data (Table III). This suggests that loss of hydrogen and methane may result from the same intermediate.

Quantitative studies of the lifetimes of reaction intermediates have been carried out by examining the collisional stabilization of adducts using a variety of buffer gases (36). Intermediates in the case of Ti^+ and V^+ could be collisionally stabilized, with measured upper limits to the unimolecular decomposition rates of 1.2×10^7 sec^{-1} and 1.5×10^5 sec^{-1}, respectively. Studies of isotope effects have led to the inference that the rate limiting step is the initial insertion of the metal ion into a C-H bond. Although further studies are warranted, this leads to the suggestion that the distinct intermediates formed in the interaction of Ni^+ with n-butane may involve the interaction of the metal ion with either the primary (slow insertion leading eventually to loss of H_2 and C_2H_6) or secondary (fast insertion leading eventually to loss of CH_4) C-H bonds of the molecule. For this explanation to remain consistent with all of the results, it is necessary to invoke beta-alkyl shifts in preference to beta-hydrogen shifts when the two processes can occur competitively in the case of Ni^+. Another possible explanation for the observed pattern of reactivity would involve insertion of Ni^+ into either of the distinct C-C bonds of n-butane as a first step followed exclusively by beta-hydrogen rather than beta-alkyl shifts. Extensive studies of the reactions of first row group 8-10 metal ions as well as the highly selective reactions of Sc^+ cannot be explained without invoking beta-alkyl shifts. In addition, there are well documented cases in which beta-alkyl shifts have been observed in condensed phase studies (39). In contrast, reactive metal centers have not been observed to oxidatively add unstrained C-C bonds. Even though endothermic processes at high translational energies may well proceed by insertion of the metal ion into C-C bonds, there is no clear evidence that this occurs at low energies.

Kinetic Energy Release Distributions and the Disposal of Energy to Products. Figure 7 suggests that the last step in the elimination of H_2 from the dihydrido bisethylene intermediate involves a potential energy surface with a large barrier for the reverse oxidative addition reaction, with the overall reaction being exothermic by 34 kcal/mol (1.4 eV). Where does this energy appear in product degrees of freedom? The portion that appears in product translation can be determined using the reverse geometry double focusing mass spectrometer shown in Figure 2.

To illustrate how the amount of energy released to product translation for a given reaction pathway may reflect specific details of the potential energy surface, consider the two hypothetical surfaces in Figure 8. The interaction of a metal ion M^+ with a neutral molecule A can result in the formation of an adduct, MA^+, which contains internal energy, E^*. In the absence of collisions, the internal excitation in this chemical activation process may be utilized for molecular rearrangement and subsequent fragmentation. In Figure 8, the adduct MA^+ is depicted fragmenting to MB^+ and C along two different potential energy surfaces designated Type I and Type II. For a reaction occurring on a Type I surface, it is assumed that there is no barrier, excluding a centrifugal barrier, to the reverse

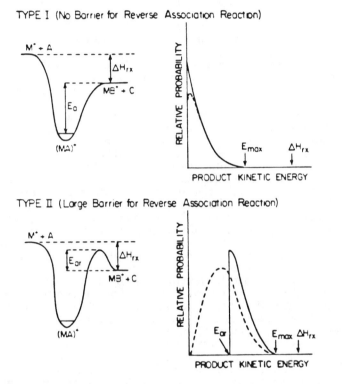

Figure 8. Characteristic shapes of kinetic energy release distributions for different model potential energy surfaces.

association reaction. The transition state resembles very loosely associated products, and very little interaction occurs between products after the transition state has been passed. Phase-space theory has been successfully used in modeling translational energy release distributions for reactions occurring on this type of potential energy surface ($\underline{40}$). A central assumption of these theories is that the statistical partitioning of energy between the reaction coordinate and all internal degrees of freedom at the transition state will be retained as the products separate. A consequence of this assumption is that the probability of a given energy being partitioned to relative product translation will decrease rapidly with increasing energy as shown in the upper right-hand portion of Figure 8. For rotating molecules ($J > 0$), angular momentum constraints may lead to a distribution such as the one indicated by the dashed line. Since the energy of the system in excess of that necessary for dissociation will be statistically divided between all the modes, the average kinetic energy release for a large molecule will be much less than the total reaction exothermicity.

As shown in Figure 8, a Type II surface involves a barrier with activation energy (E_{ar}) for the reverse association reaction. This type of surface is often associated with complex reactions which involve the simultaneous rupture and formation of several bonds in the transition state. In the absence of coupling between the reaction coordinate and other degrees of freedom after the molecule has passed through the transition state, all of the reverse activation energy would appear as translational energy of the separating fragments. Accordingly, the translational energy release would be shifted from zero by the amount E_{ar} and may again be peaked to higher kinetic energy due to angular momentum constraints (as shown by the solid line in the lower right half of Figure 8). The multicenter decomposition of ethyl vinyl ether to yield ethylene and acetaldehyde exhibits a distribution indicative of a Type II surface ($\underline{41}$). In practice, broad distributions such as the one indicated by the dashed line in the lower right half of Figure 8 are often observed and attributed to exit channel effects that distort the translational energy distribution of the products. In such cases it is not sufficient to know the energy distribution at the maximum of the potential energy barrier. The evolution of the system as it proceeds to products must be considered.

Studies of kinetic energy release distributions have implications for the reverse reactions. Notice that on a Type II surface, the association reaction of ground state MB^+ and C to form MA^+ cannot occur. In contrast, on a Type I potential energy surface the reverse reaction can occur to give the adduct MA^+. Unless another exothermic pathway is available to this species, the reaction will be nonproductive. However, it is possible in certain cases to determine that adduct formation did occur by observation of isotopic exchange processes or collisional stabilization at high pressures.

Kinetic energy release distributions for several dehydrogenation reactions are shown in Figure 9. The elimination of HD from Co(2-methylpropane-2d$_1$)$^+$, Figure 9a, and the elimination of D_2 from Ni(butane-1,1,1,4,4,4-d$_6$)$^+$, Figure 9b, are representative of 1,2- and 1,4-elimination processes, respectively. The kinetic energy release for the 1,4-elimination (Figure 9b) is very broad with a maximum kinetic energy release of 1.4 eV. The distinctive distribution for the 1,2-dehydrogenation of 2-methylpropane in Figure 9a is much narrower with a maximum

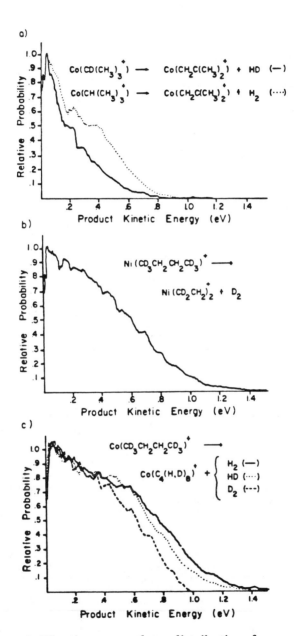

Figure 9. Kinetic energy release distributions for several dehydrogenation reactions. Data from reference 38.

release of 0.8 eV. In both cases the maximum release is close to the estimated reaction exothermicities for these processes.

Kinetic energy release distributions for loss of H_2, HD, and D_2 from Co(butane-1,1,1,4,4,4-d_6)$^+$ are presented in Figure 9c. The similarity between all three of the kinetic energy distributions in Figure 9c and the resemblance to the distribution for 1,4-elimination of D_2 from Ni(butane-1,1,1,4,4,4-d_6)$^+$ suggest that dehydrogenation of n-butane by Co$^+$ proceeds predominantly by a 1,4-mechanism and that scrambling processes are responsible for the elimination of H_2 and HD. If the 1,2-process were dominant, a much narrower distribution would be expected. CID and reactivity studies of the Co(C$_4$H$_8$)$^+$ product ion are consistent with a bis-olefin structure and a predominately 1,4-elimination process (approximately 10% of the reaction was attributed to a 1,2-elimination process) (42).

Applications of Phase Space Theory and the Determination of Ion Thermochemical Data from Studies of Exothermic Reactions. An attempt to fit the experimental distribution for dehydrogenation of n-butane by Co$^+$ by phase-space theory is shown in Figure 10a. The observed disagreement supports a Type II surface for this process. All of the studies to date of the dehydrogenation of alkanes by group 8, 9 and 10 first row metal ions exhibit kinetic energy release distributions which are characterized by large barriers for the reverse association reactions. This is supported by the failure to observe the reverse reaction as isotopic exchange processes when D_2 interacts with metal olefin complexes.

In contrast to the results obtained for dehydrogenation reactions, kinetic energy release distributions for alkane elimination processes can usually be fit with phase space theory. Results for the loss of methane from reaction 9 of Co$^+$ with isobutane are shown in Figure 10b. In fitting the

$$Co^+ + i\text{-}C_4H_{10} \rightarrow Co(C_3H_6)^+ + CH_4 \qquad (9)$$

distribution calculated using phase space theory to the experimental distribution the single important parameter in achieving a good fit is the reaction exothermicity, which in the case of reaction 9 depends on the binding energy of propylene to the cobalt ion in the product. As shown in Figure 10b, a best fit is achieved with a bond dissociation energy of 1.91 eV at 0 °K (2.08 eV or 48 kcal/mol at 298 °K). An analysis of the kinetic energy release distribution for reaction 10 of Co$^+$ with cyclopentane yields an identical value for the binding energy of propylene to an atomic cobalt ion.

$$Co^+ + \text{cyclo-}C_5H_{10} \rightarrow Co(C_3H_6)^+ + C_2H_4 \qquad (10)$$

As is the case for reaction 10, elimination of a pi-donor or n-donor base from a coordinately unsaturated metal center would generally be expected to proceed with a Type I potential surface (no barrier for the reverse association reaction). The validity of the phase space analysis for such processes in not surprising. The results for alkane elimination processes (e.g. reaction 9) are more remarkable in that the statistical analysis indicates that the excess energy in the activated complex is approximately equal to the reaction exothermicity, suggesting a loose transition state for the disruption of an adduct in which the intact alkane to be eliminated is interacting strongly with the metal center. This suggests an appropriate modification of the reaction coordinate diagram as

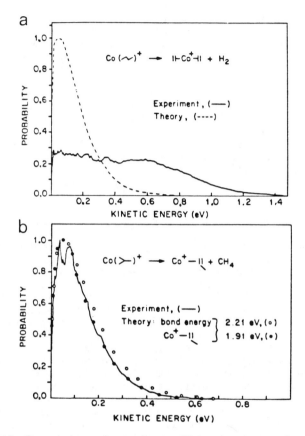

Figure 10. Comparison of experimetnal kinetic energy release distributions to phase-space calculations for (a) dehydrogenation of n-butane by Co^+ and (b) loss of methane in reaction of Co^+ with isobutane. Data from reference 38.

indicated in Figure 11 for the reaction of Co$^+$ with isobutane. The kinetic energy release distribution for methane elimination is determined entirely by the dissociation of the methane adduct, and is not useful in identifying the presence or determining the height of a reverse activation barrier. For this reason, the features in the exit channel for methane elimination are represented by a dashed line. The substantial kinetic energy release observed for hydrogen elimination and the failure of a statistic analysis to reproduce the observed distribution is a clear signature of the barrier shown in the exit channel for this process in Figure 11.

Theoretical investigations of similar reactions at neutral metal centers are generally consistent with the present results. Recent ab initio calculations by Low and Goddard (43) found the barriers for oxidative addition of H_2, CH_4, and C_2H_6 (C-C bond) to Pd to be 5.1, 30.5, and 38.0 kcal/mol, respectively. The increase in the barriers for alkane addition is due to the directionality of the bonding orbitals in CH_3 as compared to H. Two groups have independently performed calculations on the reductive elimination reactions of H_2 from $Pt(H_2)(PH_3)_2$ and of CH_4 from $Pt(H)(CH_3)(PH_3)_2$ (44,45). Obara, et al. calculated activation barriers of 8 and 28 kcal/mol, respectively, for H_2 and CH_4 elimination. Both studies found that, for hydrogen elimination, the barrier is "late" with essentially a completely formed H-H bond at the transition state. In contrast, an earlier transition state was suggested for CH_4 elimination in which the metal-carbon bond is significantly elongated, but the metal-hydrogen bond is still close to the equilibrium length. In a related study of methane elimination for nickel hydridomethyl, a similar transition state was determined. Again the barriers for reductive elimination were found to increase in the order $H_2 > CH_4 > C_2H_6$. If the transition states for methane elimination from the Ni$^+$ and Co$^+$ complexes discussed above are also early, then energy flow from the reaction coordinate into other modes after passage through the tight transition state is expected before passage through the final orbiting transition state. In contrast, little energy redistribution is anticipated for a late transition state which is essentially a hydrogen molecule associated with a metal-olefin complex.

The stability of metal ion-alkane adducts such as shown in Figure 11 remains an interesting question. The bonding in such systems can be regarded as intermolecular "agostic" interactions (46). Similar adducts between metal atoms and alkanes have been identified in low-temperature matrices (47). In addition, weakly associated complexes of methane and ethane with Pd and Pt atoms are calculated to be bound by approximately 4 kcal/mol (43). The interaction of an alkane with an ionic metal center may be characterized by a deeper well than in the case of a neutral species, in part due to the ion-polarization interaction.

For consistency, an initial well associated with the formation of an adduct of the metal ion with the alkane should be included in Figure 11. The chemical activation associated with the formation of such an adduct is likely to be essential in overcoming intrinsic barriers associated with insertion into C-H bonds. In comparison to larger hydrocarbons, the weaker interaction of ethane with first row group 8-10 metal ions may be insufficient to overcome intrinsic barriers for insertion. This would explain the failure to observe dehydrogenation of ethane by these metal ions, even though the process is known to be exothermic. The well depths could be determined from high pressure equilibria. Studies in our laboratory and elsewhere have indicated the ease with which many of these adducts can be generated, and "switching" reactions of the intact alkanes are observed to be facile processes (48).

The success of the phase space theory in fitting kinetic energy release distributions for exothermic reactions which involve no barrier for the reverse reaction have led to the use of this analysis as a tool for deriving invaluable thermochemical data from endothermic reactions. This is an important addition to the studies of endothermic reactions described above. As an example of these studies, consider the decarbonylation reaction 11 of Co^+ with acetone which leads to the formation of the

$$Co^+ + (CH_3)_2CO \rightarrow Co(CH_3)_2^+ + CO \qquad (11)$$

dimethyl cobalt ion as a product (49). The kinetic energy release distribution for this process (Figure 12) can be fit with phase space theory using a sum of the first and second metal methyl bond dissociation energies of 4.55 eV at 0 °K (4.77 eV or 110 kcal/mol at 298 °K) (38). From Table I the first bond dissociation energy is 61 kcal/mol, giving a somewhat weaker second bond energy of 49 kcal/mol. Significantly, the sum of the two bond energies is substantially greater than typical C-C bond strengths in saturated hydrocarbons, and insertion of Co^+ into a C-C bond is exothermic by 25 kcal/mol! The fact that the reaction is exothermic does not guarantee that the process will be observed.

Selective Probing of Potential Energy Surfaces for Organometallic Reactions using Multiphoton Infrared Laser Excitation. Many of the examples noted above involve the formation of chemically activated adducts as reaction intermediates. The internal energy of these adducts often exceeds the activation energies for several reaction pathways and multiple products result. It is the accessibility of multiple and sequential competitive pathways which makes it difficult to select out a single process for study. As a basis for discusssion, again consider the formation of a chemically activated species $(MA)^+$ which can decompose by two exothermic pathways generalized in equations 12 and 13.

$$M^+ + A \rightarrow (MA^+)^* \rightarrow C^+ + D \qquad (12)$$

$$M^+ + A \rightarrow (MA^+)^* \rightarrow F^+ + G \qquad (13)$$

Three possible potential energy surfaces for this system are presented in Figure 13. If C^+ and F^+ are both observed from the reaction of M^+ and A, all that can be inferred is that no point along the potential energy surface connecting MA^+ to products is higher in energy than the initial energy E^*. No information is obtained regarding the relative heights of the barriers to reactions, E_{a1} and E_{a2}, and hence it is not possible to distinguish between the three distinct cases depicted in Figure 13. A more complete description of the potential energy surfaces can often be deduced from the observed changes in product ratios as the internal energy, E^*, of the collision complex is continuously varied. One method for achieving this is to use translational energy as a variable. More often than not this procedure is not very informative. This is the case for the exothermic reactions of Ni^+ with n-butane shown in Figure 6. As another example, consider the cross sections for the exothermic reactions of Co^+ with isomeric pentenes in Figure 14 (50). The initially formed adduct in each case has an internal excitation of around 50 kcal/mol, and the same four

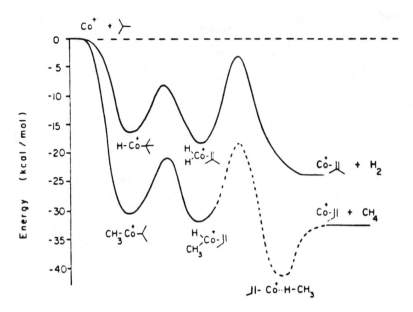

Figure 11. Reaction coordinate diagram for elimination of H_2 and CH_4 in reaction of Co^+ with isobutane.

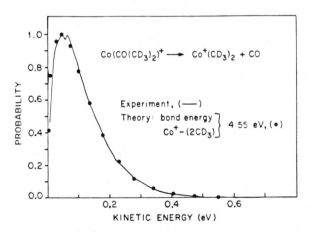

Figure 12. Comparison of experimental kinetic energy release distribution to phase-space calculations for decarbonylation of acetone by Co^+. Data from reference 38.

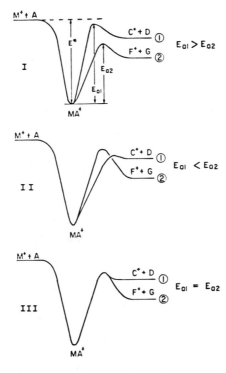

Figure 13. Model potential energy surfaces for competitive dissociation of a chemically activated reaction intermediate to yield two products.

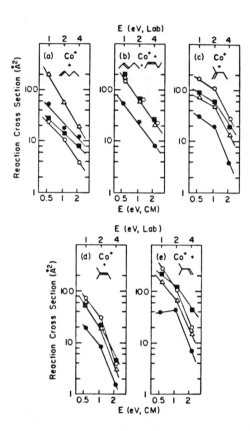

Figure 14. Variation with relative kinetic energy of cross sections for the formation of product ions resulting from the interaction of Co$^+$ with isomeric pentenes. Symbols represent products corresponding to elimination of H$_2$ (open circle), CH$_4$ (solid square), C$_2$H$_4$ (open triangle), and C$_3$H$_6$(solid circle). Data from reference 50.

products are observed (loss of hydrogen, methane, ethylene or propylene as indicated in Scheme III for the 2-pentene case). As is typical for exothermic reactions, the total cross sections decrease with increasing energy. The product ratios, however, do not vary greatly with energy. This result suggests that the $Co(C_5H_{10})^+$ adducts contain internal excitation which is considerably in excess of the activation energies for all four processes (which in our simple example shown in Figure 13 corresponds to the condition $E^* >> E_{a1}, E_{a2}$).

To reduce the high level of internal excitation which results from association of the metal ion with the olefin, adducts are prepared using the ligand exchange reaction 14 and the product is allowed to further

$$CoCO^+ + C_5H_{10} \rightarrow Co(C_5H_{10})^+ + CO \tag{14}$$

relax by infrared emission over the long time scale of trapped ion ICR experiments. Starting at the bottom of the well in Figure 13 it should in principle be possible to use collisional activation to identify the lowest energy pathway. Such studies have been used by Freiser and co-workers to identify isomeric ion structures, but the control of the relative kinetic energy and the number of collisions in such experiments is only semi-quantitative at best (51).

Absorption of infrared photons provides a relatively easy method for depositing small increments (2.5 kcal mol-1 photon-1) of energy into a molecule. Using low-power continuous wave infrared laser radiation, gas-phase ions have been shown to undergo decomposition at energies near threshold (52). This technique has been utilized to identify the lowest energy pathway, which may be different for different isomers, in complex systems. With reference to Figure 13, if case I best represents the reaction coordinate, only the thermodynamically most stable product would be observed. In case II the higher energy products which have the lower activation energy would be observed. Finally, if a common transition state were involved as in case III, both sets of produts would be formed. In addition, isomeric ions may exhibit dissociation yields which vary with wavelength in a manner which allows them to be distinguished (16). Thus, infrared mutltiphoton decomposition can yield details relating to activation parameters as well as structural information.

Figure 15 shows the results of infrared laser excitation of the Co(2-pentene)$^+$ adduct (6). The decrease in the abundance of the adduct ion is matched by an increase in the abundance of only a single product in which methane is eliminated. Compared to the bimolecular collision process, a high degree of selectivity is achieved in the IR activation experiment. This information can thus be used in constructing the reaction coordinate diagram shown in Figure 16, where only the elimination of hydrogen and methane are considered. Elimination of methane is the most exothermic process and in addition has the lowest activation energy. The lowest energy process available to the adduct involves cleavage of the allylic C-C bond in preference to the allylic C-H bond and subsequent rearrangement and elimination of methane.

The analogous experiment with 1-pentene also yields a single product as indicated in Table IV. In addition, the product, in which ethylene is eliminated, is different from that observed in the case of 2-pentene! In this case, the process with the lowest activation energy does not correspond to the most exothermic process (Table IV). Again the process of lowest energy involves cleavage of the allylic C-C bond in preference to the allylic C-H

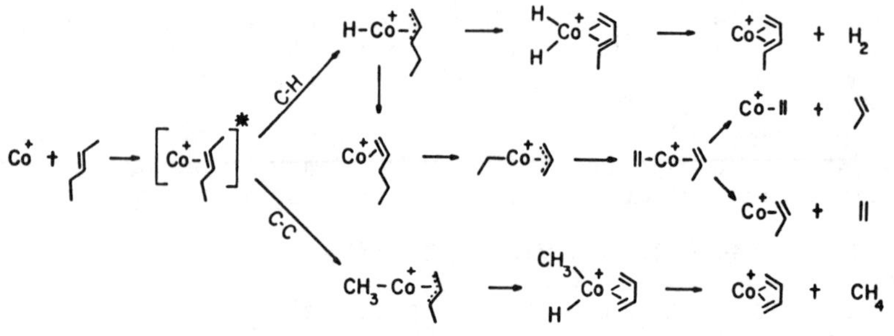

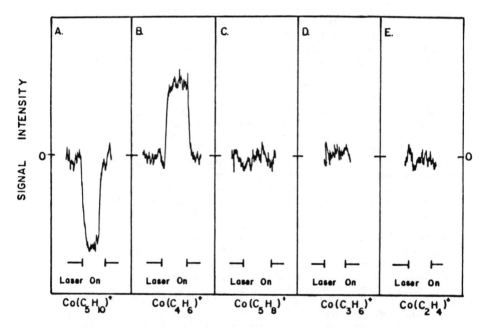

Scheme III

Figure 15. Changes in reactant and product ion yields resulting from cw infrared laser multiphoton activation of the $\dot{C}o$(2-pentene)$^+$ adduct. Data from reference 6.

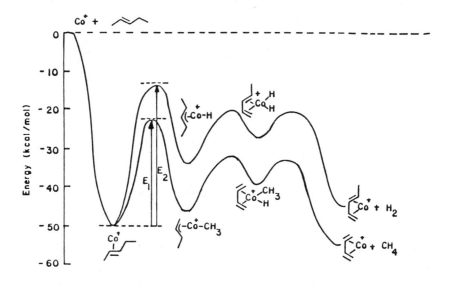

Figure 16. Reaction coordinate diagram for the elimination of hydrogen and methane in the reaction of Co^+ with 2-pentene.

Table IV. Product Distribution for Exothermic Reactions of Co^+ with C_5H_{10} Isomers

C_5H_{10} Isomer	Neutral Lost	ΔH^a (kcal/mol)	Product Distributions		
			ion beam	metastable	IR activation
1-pentene	H_2	-46	0.11	0.02	0
	CH_4	-55	0.13	0.01	0
	C_2H_4	-29	0.58	0.95	1.00
	C_3H_6	-25	0.18	0.01	0
2-pentene	H_2	-43	0.28	0.47	0
	CH_4	-52	0.33	0.43	1.00
	C_2H_4	-27	0.29	0.09	0
	C_3H_6	-23	0.10	0.01	0

^{a} Data from reference 6.

bond. The suggested mechanism for the elimination of ethylene starting from the 1-pentene adduct is included in Scheme III.

Also included in Table IV are the metastable product yields for comparison to the ion beam and IR activation results. From these data it appears that the processes involving elimination of hydrogen and methane involve a competitive dissociation from a common intermediate as shown in Figure 16. However, a common intermediate may not be involved in the elimination of ethylene and propylene (the latter product appears to be formed in a faster process), and Scheme III is overly simplistic.

A prerequisite for the success of the multiphoton IR laser activation experiment is that the adduct must absorb in the 10 micron region. Unfortunately, none of the adducts of atomic metal ions with saturated hydrocarbons which we have examined to date absorb in this region. As the tunability and intensity of infrared sources improve, the general applicability of this experimental methodology for probing the potential energy surfaces for organometallic reactions will be greatly increased.

Future Studies

Considering the extensive efforts which have been directed at understanding the reactions of atomic transition metal ions with hydrocarbons, it is somewhat surprising that our understanding of these processes is rudimentary at best. It is imperative that continued studies be coupled with high quality ab initio theory to provide a more fundamental understanding of not only the molecular and electronic structure of organometallic fragments but complete potential energy surfaces for some of the simpler processes.

Studies utilizing state selected reactants clearly represent an important frontier. Benchmark quantitative cross section data obtained by Armentrout and co-workers (53) demonstrate that with an energy difference of 0.25 eV, excited state Fe$^+$ (^4F derived from the $3d^7$ configuration) is much more reactive with H_2 than is the ground state (^6D derived from the $4s3d^6$ configuration). These studies utilized a surface ionization source with a calculated reactant state distribution that could be modified by quenching higher excited states in a drift cell. The use of single photon (54) and resonance enhanced multiphoton ionization processes (55) to prepare state selected reactants will allow for extensive studies of this type in the future. Collisional excitation can also be used to prepare state selected reactants. Beyond an examination of the effects of electronic excitation, a study of the effects of vibrational excitation in the hydrocarbon on reactivity would be of particular interest. Enhanced reactivity of a species such as ethane would verify the idea that intrinsic barriers for insertion of metal ions into C-H bonds are responsible for the low reactivity of this species.

Complementing studies of the effect of reagent excitation are studies which determine energy disposal to products. Measurements of kinetic energy release distributions are only a first step in this direction. The examples of kinetic energy release distributions presented above are selected from measurements for several hundred reactions (56). These data are still being digested to extract not only thermochemical data but mechanistic information as well. In numerous cases bimodal distributions have been obtained. The dehydrogenation of propane by Fe$^+$ and the elimination of propylene in the reaction of Co$^+$ with 1-pentene are examples of processes which yield bimodal distributions. These could result from distinct reactivity of different states of the metal ion, from the formation of two different internal energy states of the products, or from the formation of dissimilar transition states (which may then yield either the same or distinct products).

The application of newer methods to studies of gas phase organometallic reactions will lead to the development of routine techniques for determination of the thermochemistry of organometallic species. The examples discussed above demonstrate that an analysis of kinetic energy release distributions for exothermic reactions yields accurate metal ligand bond dissociation energies. This can be extended to include neutrals as well as ions. For example, reaction 15 has been used to determine accurate bond dissociation energies for Co-H and Co-CH$_3$ (57).

$$Co^+ + RSi(CH_3)_3 \rightarrow CoR + Si(CH_3)_3^+ \qquad (15)$$

Related reactions can be used to determine metal hydrogen and metal carbon bond dissociation energies in coordinately saturated neutral species. Experiments in progress in our laboratory using magnetic bottle techniques (58) will extend investigations of kinetic energy release distributions to include measurements for bimolecular in addition to unimolecular reactions. This will be particularly useful in the examination of direct processes which do not involve long lived intermediates.

From the examples given above it is clear that the development of a novel approach for real time measurements of the dissociation kinetics of reaction intermediates would greatly assist the unravelling of complex multistep processes associated with the transformation of even simple molecules at transition metal centers. While techniques are available for such measurements over a limited range of times, none of the methods are sufficiently general to be useful for extensive measurements.

An examination of the reactions of small molecules with organometallic fragments and species which possess only a single coordination vacancy will permit more direct comparisons with condensed phase studies. Studies of specific organometallic functional groups such as metal carbenes (59), metal alkyls (60) and metal oxides (61) have already revealed some exciting results. The addition of one or more ligands to a metal center greatly modifies the reactivity toward hydrocarbons. A species such as CpNi$^+$ will readily dehydrogenate hydrocarbons but does not exhibit any reactions in which C-C bonds are cleaved (62).

Extensions of studies of the reactivity of hydrocarbons at transition metal centers to include a variety of functional groups have been reported and remain to be explored in greater detail. Of particular interest are processes which lead to single and multiple bonding of elements other than carbon to the transition metal center. Little is known about the energetics of such bonds and their functional group chemistry. We were recently surprised to find that metal silylenes ($M = SiH_2^+$) are exceptionally stable, with bond dissociation energies as high as 70 kcal/mol (63). The observation and characterization of metal silylenes in the gas phase has now inspired attempts by several groups to prepare these species in more conventional condensed phase studies.

Anionic and cationic species are particularly easy to study in the gas phase using mass spectrometric techniques. Studies of organometallic reactions involving neutral species are becoming more prevalent. The elegant study of the catalytic hydrogenation of ethylene by photochemical activation of Fe(CO)$_5$ by Grant and co-workers is an example of the type of investigation that can be carried out in the gas phase (64).

Crossed beam studies involving both neutrals and ions will provided detailed information relating to the potential energy surfaces and reaction mechanisms for organometallic reactions (19). These studies are especially revealing when the reactions are direct and involve intermediates whose lifetime does not exceed a rotational period.

Finally, the general field of metal cluster chemistry deserves mention. Techniques for the preparation of metal clusters of virtually any size and composition are virtually routine (65). An ultimate objective in studying these species is to characterize the structure, properties, and reactions of clusters of a particular size. In addition to physisorption and chemisorption processes, more complex reactions such as the extensive dehydrogenation of alkanes by metal clusters have been reported for cluster of varying size (66). The use of ion beam (67) and ICR techniques (68) facilitates such studies since particular clusters can be mass selected and examined. Considering the complexity of processes occurring on a single metal center, it will be a real challenge to extract mechanistic details from cluster studies. To the extent that this can be accomplished the results may well revolutionize our understanding of reactions on catalyst surfaces. In the meantime it seems worthwhile to answer simpler questions. Can atomic metal ions directly insert into C-C bonds? What determines the ability of a distal hydrogen or alkyl group to transfer to a metal center? How important are multicenter reactions in the elimination of small molecules from organometallic reaction intermediates?

Acknowledgments

This work was supported in part by the National Science Foundation under Grant CHE-8020464 and by the Petroleum Research Foundation.

Citations in Literature

1. Byrd, G.D.; Burnier, R.C.; Freiser, B.S. J. Am. Chem. Soc. 1982, 104, 5944.
2. For a recent review, see Crabtree, R.H., Chem. Rev. 1985, 85, 245.
3. Muller, J.; Goll, W. Chem. Ber. 1973, 106, 1129.
4. Houriet, R.; Halle, L.F.; Beauchamp, J.L. Organometallics 1983, 2, 1818 and references contained therein.
5. Halle, L.F.; Houriet, R.; Kappes, M.M.; Staley, R.H.; Beauchamp, J.L. J. Am. Chem. Soc. 1982, 104, 6293.
6. Hanratty, M. A.; Paulson, C.M.; Beauchamp, J.L. J. Am. Chem. Soc. 1985, 107, 5074.
7. Hanratty, M.A.; Illies, A.J.; Bowers, M.T.; Beauchamp, J.L. J. Am. Chem. Soc. 1985, 107, 1788.
8. Armentrout, P.B.; Beauchamp, J.L. J. Amer. Chem. Soc. 1981, 103, 784.
9. Ervin, K.M.; Armentrout, P.B. J. Chem. Phys. 1985, 83, 166.
10. Halle, L.F.; Armentrout, P.B.; Beauchamp, J.L. J. Amer. Chem. Soc. 1981, 103, 962.
11. Kang, H.; Beauchamp, J.L. J. Phys. Chem. 1985, 89, 3364 and references contained therein.
12. For a recent example, see Mandich, M.L.; Halle, L.F.; Beauchamp, J.L. J. Am. Chem. Soc. 1984, 106, 4403.

13. Beauchamp, J.L. Ann. Rev. Phys. Chem. 1971, 22, 527.
14. Wise, M.B.; Jacobson, D.B.; Huang, Y.; Freiser, B.S. Organometallics, in press.
15. Cody, R.B.; Burnier, R.C.; Freiser, B.S. Anal. Chem. 1982, 54, 96.
16. Thorne, L.R. Beauchamp, J.L. in "Ions and Light" (Gas Phase Ion Chemistry, Vol. 3), Bowers, M.T., ed., Academic Press, New York, 1983.
17. Hettich, R.L.; Freiser, B.S. J. Amer. Chem. Soc., in press.
18. Jarrold, M.F.; Illies, A.J.; Kerchner, N.J.; Wagner-Redeker, W.; Bowers, M.T.; Mandich, M.L.; Beauchamp, J.L. J. Phys. Chem. 1983, 87, 2313, and references therein.
19. Sonnenfroh, D.M.; Farrar, J.M. J. Amer. Chem. Soc. 1986, 108, 3521.
20. Walba, D.M.; Depuy, C.H.; Grabowski, J.J.; Bierbaum, V.M. Organometallics 1984, 3, 498.
21. McDonald, R. N.; Schell, P.L.; McGhee, W.D. Organometallics 1984, 3, 182.
22. Lane, K.R.; Sallans, L.; Squires, R.R. J. Am. Chem. Soc. 1986, 108, 4368.
23. Tonkyn, R.; Weisshaar, J.C. J. Phys. Chem. 1986, 90, 2305.
24. Armentrout, P.B.; Beauchamp, J.L. J. Chem. Phys. 1980, 50, 37.
25. Elkind, J.L.; Armentrout, P.B. J. Phys. Chem. 1984, 88, 5454.
26. Armentrout, P.B.; Beauchamp, J.L. J. Chem. Phys. 1981, 74, 2819.
27. Schilling, J.B.; Goddard, W.A., III; Beauchamp, J.L. J. Am. Chem. Soc. 1986, 108, 582.
28. Armentrout, P.B.; Halle, L.F.; Beauchamp, J.L. J. Am. Chem. Soc. 1981, 103, 6501.
29. Schilling, J.B.; Goddard, W.A., III; Beauchamp, J.L., to be published.
30. Carlin, T.J.; Sallans, L.; Cassady, C.J.; Jacobson, D.B.; Freiser, B.S., J. Am. Chem. Soc. 1983, 105, 6320.
31. Halle, L.F.; Klein, F.S.; Beauchamp, J.L. J. Am. Chem. Soc. 1984, 106, 2543.
32. Beauchamp, J.L.; Stevens, A.E.; Corderman, R.R. Pure and Appl. Chem. 1979, 51, 967.
33. Halle, L.F.; Armentrout, P.B.; Beauchamp, J.L. Organometallics 1982, 1, 963.
34. Schilling, J.B.; Beauchamp, J.L., to be published.
35. Tolbert, M.A., Mandich, M.L., Halle, L.F.; Beauchamp, J.L. J. Am. Chem. Soc. 1986, 108, 5675.
36. Tolbert, M.A.; Beauchamp, J.L. J. Amer. Chem. Soc., in press.
37. Tolbert, M.A.; Beauchamp, J.L. J. Amer. Chem. Soc. 1984, 106, 8117.
38. Hanratty, M.A.; Beauchamp, J.L.; Illies, A.J.; van Koppen, P.; Bowers, M.T. J. Amer. Chem. Soc. submitted for publication.
39. Watson, P.L.; Roe, D.C. J. Am. Chem. Soc. 1982, 104, 6471.
40. Chesnovich, W.J.; Bass, L.; Su, T.; Bowers, M.T. J. Chem. Phys. 1981, 74, 2228.
41. Huisken, F.; Kranovich, D.; Zhang, Z.; Shen, Y.R.; Lee, Y.T. J. Chem. Phys. 1983, 78, 3806.
42. Jacobson, D.B.; Freiser, B.S. J. Am. Chem. Soc. 1983, 105, 5197.
43. Low. J.J.; Goddard, W.A. J. Am. Chem. Soc. 1984, 106, 8321.
44. Low, J.J.; Goddard, W.A. J. Am. Chem. Soc. 1984, 106, 6928.
45. Obara, S.; Kitaura, K.; Morokuma, K. J. Am. Chem. Soc. 1984, 106, 7482.
46. Brookhart, M.; Green, M.L.H. J. Organomet. Chem. 1983, 250, 395.

47. Turner, J.J.; Poliakoff, M. in "Inorganic Chemistry: Toward the 21st Century"; Chisholm, M.H., Ed. American Chemical Society, 1983, p. 35.
48. Tolbert, M.A.; Beauchamp, J.L., unpublished observations.
49. Halle, L.F.; Crowe, W.E.; Armentrout, P.B.; Beauchamp, J.L. Organometallics 1984, 3, 1694.
50. Armentrout, P.B.; Halle, L.F.; Beauchamp, J.L. J. Am. Chem. Soc. 1981, 103, 6624.
51. Jacobson, D.B.; Freiser, B.S. J. Am. Chem. Soc. 1983, 105, 736.
52. Bomse, D.S.; Beauchamp, J.L. J. Am. Chem. Soc. 1981, 103, 3292.
53. Elkind, J.L.; Armentrout, P.B. J. Phys. Chem., in press.
54. Dyke, J.M.; Gravenor, B.W.J.; Lewis, R.A.; Morris, A. J. Phys. B: At. Mol. Phys. 1982, 15, 4523.
55. Nagano, Y.; Achiba, Y.; Sato, K.; Komurer, K., Chem. Phys. Lett. 1982, 93, 510.
56. Jacobson, D.B.; Bowers, M.T.; Beauchamp, J.L., unpublished results.
57. van Koppen, P.; Jacobson, D.B.; Bowers, M.T.; Beauchamp, J.L., to be published.
58. Dearden, D.V.; Beauchamp, J.L., unpublished results.
59. Jacobson, D.B.; Freiser, B.S. J. Am. Chem. Soc. 1985, 107, 4373.
60. Jacobson, D.B.; Freiser, B.S. J. Am. Chem. Soc. 1985, 107, 3891.
61. Kang, H.; Beauchamp, J.L. J. Am. Chem. Soc. 1986, 108, 5663.
62. Corderman, R.R., Ph.D. Thesis, California Institute of Technology, 1979.
63. Kang, H.; Jacobson, D.B.; Shin, S.K.; Beauchamp, J.L.; Bowers, M.T. J. Am. Chem. Soc. 1986, 108, 5668.
64. Miller, M.E.; Grant, E.R. J. Amer. Chem. Soc. 1985, 107, 3386.
65. See, for example, Hopkins, J.B.; Langridge-Smith, P.R.R.; Morse, M.D.; Smalley, R.E. J. Chem. Phys. 1983, 78, 1627.
66. Trevor, D.J.; Whetten, R.L. Cos, D.M.; Kaldor, A. J. Amer. Chem. Soc. 1985, 107, 518.
67. Loh, S.K.; Hales, D.A.; Armentrout, P.B., Chem. Phys. Lett. 1986, in press.
68. Alford, J.M.; Weiss, F.D.; Laaksonen, R.T.; Smalley, R.E. J. Phys. Chem., in press.

RECEIVED November 12, 1986

Chapter 3

Gas-Phase Transition Metal Cluster Chemistry

D. J. Trevor[1] and A. Kaldor

Corporate Research Laboratory, Exxon Research and Engineering, Annandale, NJ 08801

This chapter reviews reactions of Fe, Nb, V, Co, Ni, and Pt clusters with H_2, Co and Nb clusters with N_2, a variety of transition metals clusters with CO, Fe and Pt with O_2 and Pt with a variety of C_6 hydrocarbons carried out with laser vaporization sources and fast flow chemical reactors. A trend is observed in which only the dissociative reactions appear to depend strongly upon cluster size. This suggests an explanation for certain size dependent behavior. The charge transfer model for bond activation is further developed. The issue of metal cluster structure and its influence on reactivity is discussed as well as the possibility of elucidating this structure through chemical reactivity studies.

Gas phase transition metal cluster chemistry lies along critical connecting paths between different fields of chemistry and physics. For example, from the physicist's point of view, studies of clusters as they grow into metals will present new tests of the theory of metals. Questions like: How itinerant are the bonding electrons in these systems? and Is there a metal to non-metal phase transition as a function of size? are frequently addressed. On the other hand from a chemist point of view very similar questions are asked but using different terminology: How localized is the surface chemical bond? and What is the difference between surface chemistry and small cluster chemistry? Cluster science is filling the void between these different perspectives with a new set of materials and measurements of physical and chemical properties.

[1]Current address: AT&T Bell Laboratories, Murray Hill, NJ 07974

0097-6156/87/0333-0043$07.75/0

Cluster chemistry has long been an active area of inorganic chemistry. This field has developed a large data base and a good understanding of the chemistry of the stable or partially unsaturated complexes. The newer area of gas phase transition metal cluster chemistry starts with the fully coordinatively unsaturated metal cluster. This species is highly reactive yet, demonstrates some interesting selectivity. The origin of this selectivity is kinetic and sets the perspective of this review.

Three laboratories have been the primary contributors in the area of gas phase transition metal cluster chemistry and their work forms the center piece of this mini-review. These are the Rice(1) group, the Argonne(2) group, and the Exxon(3) group. Other groups, however are rapidly entering this active field. The majority of the references are similarly grouped together. Photo- and thermal-chemistry of stable gas phase cluster complexes are not covered in this paper. Attempts are made to make connections and use of this data whenever possible. In addition, cluster ion chemistry is not covered. These reactions have just recently been extended to clusters of a variety of sizes and already shows some surprising similarity to neutral chemistry(4-7).

This overview is organized into several major sections. The first is a description of the cluster source, reactor, and the general mechanisms used to describe the reaction kinetics that will be studied. The next two sections describe the relatively simple reactions of hydrogen, nitrogen, methane, carbon monoxide, and oxygen reactions with a variety of metal clusters, followed by the more complicated dehydrogenation reactions of hydrocarbons with platinum clusters. The last section develops a model to rationalize the observed chemical behavior and describes several predictions that can be made from the model.

Cluster Source

Previously, intense beams of metal clusters could only be produced for the most volatile metals. The limitation arose from significant materials problems involved in the construction of high temperature ovens. The development of a source that utilizes laser vaporization and subsequent condensation in a rapidly flowing gas eliminated the materials problem and has enabled just about any material to be studied(1a,8).

Method. The laser vaporization source eliminates the material constraints inherent in conventional oven sources. This is accomplished by localizing the heating to a very small area at the surface of the sample and by entraining the vapor produced in a rapid flow of high pressure gas.

Figure 1 is a schematic of the laser vaporization source. This diagram depicts a pulsed valve on the left which supplies high pressure helium flow directly towards the right. Several workers have also chosen to use continuous helium flows(2,6,9). In general these sources are modifications of conventional supersonic beam sources.

A variety of lasers have been successfully employed is this type of source. The only important criterion is that the laser must have sufficient intensity to heat the surface for vaporization. A goal at one time in this area was to maximize the amount of material removed by the laser. Previous studies in laser welding and drilling clearly described how this can be achieved. Use of longer pulses avoids the creation of a plasma that is opaque to the laser, thus shielding the target surface. Currently, however, the limitation appears to be the amount of material that can be converted into clusters and not the amount vaporized. Doubled Q-switched YAG(1,3) excimer(2,9), and more recently copper vapor lasers(6) are the most popular.

Sample. This source places no restrictions on target material. Clusters of metals, produced. For example, polyethylene and alumina have been studied as well as refractory metals like tungsten and niobium. Molecular solids, liquids, and solutions could also be used. However the complexity of the vaporization process and plasma chemistry makes for even more complex mixtures in the gas phase. To date the transition metals(1-3) and early members of group 13 (IIIA) and 14 (IVA)(11-16) have been the most actively studied.

Clustering. After a small piece of the metal sample is vaporized by the laser, the next step is to control its condensation into clusters without significant loss to the walls. This is achieved by vaporizing the metal into a high pressure (10-3000 torr) gas that is flowing (10^5-10^3 cm/sec) down the extender tube shown in Fig. 1. The majority of the small (3 to ~6 atoms) and all of the large clusters are produced by condensation in this tube. The high pressure not only allows for rapid cooling but also limits the diffusion of the metal atoms keeping the high density necessary for rapid cluster formation. Helium is the best gas to use because of its high thermal conductivity, chemical inertness and its superior properties for producing supersonic beams.

Cluster growth appears to be dominated by atom or small cluster addition onto a few seed "clusters" as follows:

$$M_{n-1} + M \overset{k_n}{\underset{k_{-n}}{\Longleftrightarrow}} M_n^* \qquad (1)$$

$$M_n^* + He \overset{k_{sn}}{\dashrightarrow} M_n \qquad (2)$$

If the reverse of Reaction 1 is slow compared to 2 (the collisional stabilization step) then overall cluster growth will not depend strongly upon the total helium pressure. This is found to be the case using RRK estimates for k_{-n} and hard sphere collision cross sections for k_{sn} for all clusters larger than the tetramer. The absence of a dependence on the total pressure implies that the product of [M] and residence time should govern cluster growth. Therefore, a lower pressure can be compensated for by increasing the residence time (slower flow velocities).

Reactor

The reactor design is similar to conventional flow tubes(17). The main difference is use of higher pressures and faster flow velocities. Both these quantities shift the flow conditions from laminar towards turbulent. To date all workers have interpreted their data using a plug flow approximation which is likely to be equally valid for turbulent or laminar flows. Unfortunately turbulence can cause shock heating, streaming, and further instabilities. None of these effects are believed to be currently limiting these experiments(1d).

An estimate of the temperature of the clusters in the reactor tube can be obtained from the velocity of the atomic species in the beam. This velocity, assuming insignificant slippage, is related to the "stagnation" temperature T_i by(18)

$$T_i = (m/5k_B)(v_f^2 - v_i^2)$$

where m is the mass of the carrier gas, v_f is the final beam velocity and v_i is the initial flow velocity. This translational temperature characterizes the internal temperature of the clusters in the reactor only to the extent that the large number of collisions in the high pressure reactor retain this equilibrium. In the Exxon apparatus v_f can only be measured to within 20% which results in a 40% temperature uncertainty. These measurements gave an upper bound to T_i of 600K (19). The group at Rice has reported a T_i of 400±50 K from more accurate measurements on velocities of ions coming directly from the source.

Kinetics. To date only addition reactions have been reported. These reactions produce products or adducts that are the result of complete addition, or addition and subsequent elimination. An example of the later reaction is dehydrogenation of hydrocarbons on platinum clusters. These addition reactions are in many ways analogs to the chemisorption process on metal surfaces.

The extent of a reaction in these measurements is determined by bare metal cluster ion signal depletion. In most cases products are also observed. Some systems show multiple adducts indicating comparable or higher rates for each successive step up to a saturation level. For other systems the fully saturated product is observed almost as soon as the reaction starts. This later behavior is characteristic of an early rate-limiting step. Due to this complexity kinetics have only been reported on the formation of the first adduct, i.e. for the initial chemisorption step.

The following reactions describe the general behavior for addition of a reactant to form the first adduct:

$$M_n + R \overset{k_n}{\underset{k_{-n}}{\rlap{\Longleftarrow}\Longrightarrow}} M_nR^* \qquad (3)$$

$$M_nR^* + He \overset{k_{sn}}{-\!-\!-\!-\!\rightarrow} M_nR \qquad (4)$$

Assuming steady state for the energized $[M_nR^*]$ specie, the rate of production of product M_nR in terms of the consumption of unreacted cluster is written as follows:

$$-\frac{1}{[M_n]}\frac{d[M_n]}{dt} = k_n[R]\left(\frac{k_{sn}[He]}{k_{sn}[He] + k_{-n}}\right)$$

This equations can be simply integrated over the residence time in the reactor t because the reactant and helium are both in excess. This yields

$$R_n = - \ln\left(\frac{[M_n]_f}{[M_n]_i}\right) = t\, k_n[R]\left(\frac{k_{sn}[He]}{k_{sn}[He] + k_{-n}}\right) \quad (5)$$

where the reactivity R_n is calculated form the depletion of the bare cluster signal. The ratio in parentheses on the right hand side represents competition between collisional stabilization and unimolecular decay of $[M_nR^*]$. The collisional stabilization rate should increase with cluster size due to the increase in the number of low frequency modes as well at the growth in size of the clusters. This increase is mild compared with a simple RRK prediction for k_{-n},

$$k_{-n} \sim \omega\left(\frac{s\, k_BT}{E_0 + s\, k_BT}\right)^{s-1}$$

where ω is the frequency factor typically around $10^{13}\,\text{sec}^{-1}$, s is the number of modes, $3n-6$ ($3n-5$ for linear clusters), and E_0 is the critical energy for decomposition. For a given E_0 a minimum cluster size is required in order to form a collision complex that will live long enough to be collisionally stabilized by helium. For diatomic absorbates with bonds to the cluster of the order 0.5-1 eV the critical cluster size is in the 3-5 range. Therefore addition reactions should show a threshold type onset in reactivity with cluster size. Above this threshold cluster size k_{-n} becomes so small that the term in parentheses becomes unity and Eq. (5) is simplified to

$$R_n = - \ln\left(\frac{[M_n]_f}{[M_n]_i}\right) = t\, k_n[R] \quad (6)$$

This is the conventional form used to interpret the depletion measurements. For the reactions that involve elimination of part of the reactant such as dehydrogenation, collisional stabilization is not necessary, and Equ. (6) applies even for the atom.

The reverse of reactions (3) and (4) in the high pressure limit is the desorption process described by

$$M_nR_m \xrightarrow{k_{dnm}} M_nR_{m-1} + R$$

which will be endothermic by $E_d(n,m)$. Since M_nR_m must survive long enough to exit the reactor, which on the average takes $t/2$, an upper bound to the desorption rate constant k_{dnm} is set. Using the high pressure RRK form for k_{dnm} (Arrhenius)

$$2/t = k_{dnm} = \omega \exp(-E_d(n,m)/k_BT)$$

a minimum bond strength of 16 kcal/mole is set for detectable products. Any reaction products bond less strongly than 16 kcal/mole will likely desorb before exiting the reactor and are thus not detected.

Detection

Photoionization time-of-flight mass spectrometry is almost exclusively the method used in chemical reaction studies. The mass spectrometers, detectors and electronics are almost identical. A major distinction is the choice of ionizing frequency and intensity. For many stable molecules multiphoton ionization allowed for almost unit detection efficiency with controllable fragmentation[20]. For cluster systems this has been more difficult because high laser intensities generally cause extensive dissociation of neutrals and ions[21]. This has forced the use of single photon ionization. This works very well for low ionization potential metals ($\leq$ 7.87 eV) if the intensity is kept fairly low. In fact for most systems the ionizing laser must be attenuated. A few very small clusters and a lot more fully saturated cluster adducts are not detectable due to the available laser sources which limit the upper ionization potential (IP) to 7.87 eV. Future developments using higher energy photon sources will eliminate this problem bringing around new ones. Using too hard of an ionizing photon can also induce fragmentation by dissociative ionization. Therefore a series of photon energies will be necessary before we can be confident in parent ion identification.

Simple Addition Reactions

Hydrogen. The very surprising observation by Geusic, Morse and Smalley[1b] that cobalt and niobium clusters react in a very selective fashion with deuterium was rapidly followed by similar studies on a variety of other metal systems. Three groups simultaneously reported similar dramatic behavior for iron clusters[2b,3b]. Dihydrogen or dideuterium addition reactions have been reported for vanadium[3e], iron[1c,2a-d,3b], cobalt[1b], nickel[1c], niobium[1b,c,3f], platinum and aluminum[3g] clusters in some cases up to 40 atoms in size. Iron clusters have been studied containing greater than 100 atoms by the Argonne group[2b-d]. In all cases, under conventional reactor conditions only simple addition reactions have been observed. This is based upon the observation that the mass spectrum of the products contained only even numbers of hydrogen or deuterium atoms.

The transition metals iron, cobalt, niobium, and vanadium, as well as aluminum, show dramatic cluster size dependent behavior. Of the ones reported this leaves nickel and platinum clusters, which exhibit a more constant reactivity pattern. As described above (see reactor conditions) the depletion measurements can be related for at least the pentamer and larger clusters, to bimolecular rate constants. Figure 2 is a plot of these rate constants from the three laboratories work on the iron + hydrogen system. The data plotted from the Exxon group only included studies of clusters larger than the pentamer and indicates reactivity to begin at Fe_8. The work from the Argonne group is the most extensive and likely more precise.

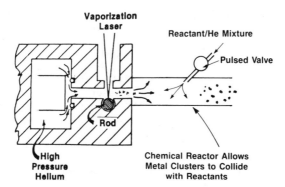

Figure. 1.   Schematic of essential components of the Exxon group cluster laser vaporization source and fast flow tube chemical reactor. On the far left is a 1 mm diameter pulsed nozzle that emits an ~200 µsec long pulse of helium which achieves an average pressure of approximately one atmosphere above the sample rod.   Immediately before the sample rod position the tube is expanded to 2 mm diameter. The length of this extender section can be varied form 6 mm to 50 mm depending upon the desired integration time for cluster growth. The reactor flow tube is 10 mm in diameter and typically 50 mm long. The reactants diluted in helium are added and mixed with the flow stream via the second pulsed valve.

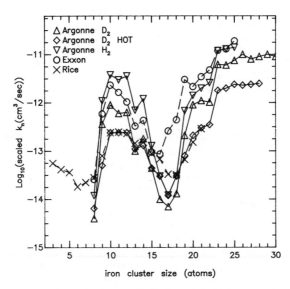

Figure 2. Logarithm of the scaled reactivities of iron clusters with hydrogen/deuterium from the Exxon (circles), Argonne (triangles and diamonds), and Rice (crosses) groups. The data are normalized to Fe_{10} of the Argonne groups.

From the mid-size to large clusters the size variation is in general agreement. The cluster sizes which exhibit the minimum and maximum reactivity are similar in each study. This is good agreement given the diversity in techniques. Rice and Exxon used 7.87 eV ionizing photons while Argonne used 6.42 eV. Also the Argonne flow reactor has a lower pressure and flow velocity, and the Rice instrument had approximately three times higher pressure in the reactor than the Exxon study. The agreement between the three laboratories indicates the generality of this technique, at least over the parameter space sampled.

The major disagreement in the data rests in the reactivity of the very small clusters. Originally only the Rice group found clusters smaller than Fe_8 to react. Subsequent studies at Exxon have found Fe_4 to react at least as rapidly as Fe_9. One possible explanation is that higher pressure reactor conditions are necessary to stabilize these intermediates. This might have been achieved in the later but not the initial Exxon work(22), and would rationalize why the Argonne group did not report reactions on these clusters. Other possibilities can rationalize this discrepancy such as higher cluster temperatures in the Exxon and Rice studies or diffusion of the very small clusters out of the reaction zone in the Argonne apparatus.

Clusters larger than 25 all appear to react without further dramatic variations. The magnitude of the large cluster rate constants are a few percent of the hard sphere collision cross section. The low coverage sticking coefficient of hydrogen on iron near room temperature is also a few percent(23). This agreement and the smooth reactivity suggests that iron clusters larger than twenty-five atoms in size are bulk-like with respect to their reactions with hydrogen and deuterium. The Argonne group data would place this distinction at Fe_{23}.

The Argonne group also reported a crude, but very insightful temperature dependence for this reaction(2b). They found the most unreactive cluster's reactivity to increase with an increase in "temperature" while that of the highly reactive clusters to decrease. The increase in reactivity with temperature is characteristic of an activated process, while the decrease might suggest that the measured reactivity for the very large clusters is non-activated and the increase in temperature is enhancing the reverse steps. The observed isotope effect on the reaction rate also suggests that breaking the H-H bond is a critical step.

Once a moderately reactive iron cluster activates the first hydrogen molecule, it rapidly reacts to adsorb more hydrogen, eventually reaching a saturation level. The saturation studies by the Argonne group suggest a surface stoichiometry of one hydrogen atom per surface iron atom(2c). This is the first direct use of chemistry to extract cluster structural information. This data ruled out planar and linear structures and was consistent with a globular geometry. In addition, medium-sized clusters rapidly saturate once they start to react. This behavior implies the first step is rate limiting and in a thermalized environment the rate constant for the first hydrogen adsorption is much smaller than for any subsequent ones. However, the exothermicity of the first step will heat the clusters and increase the rate of reaction, thus also rationalizing the rapid conversion.

The Argonne group observed that the ionization laser desorbed hydrogen from the cluster adducts (2d). Modeling this effect yielded a desorption enthalpy of 1.3 eV. This is slightly higher than the value for close-packed surfaces of bulk iron (1.2 eV)(23,24). An increase over the flat metal should be expected due to the higher degree of coordinative unsaturation and a lower steric hindrance due to the positive radius of curvature. Given this, the small difference is a bit surprising, and indicates the highly localized character of the metal hydrogen bond and the rapid approach to bulk properties.

On the bulk surface of iron, hydrogen chemisorbs dissociatively. The measured desorption energy(2d), observation of an isotope effect, and a significant increase in ionization potentials(3i) each individually support the proposition that dissociative chemisorption also takes place on metal clusters. Arguments bellow, based on cluster stabilization add further support. From the desorption energy mentioned above the bond enthalpy at ~400 K is 67 kcal/mole. This is more than half the heat of vaporization, 99.5 kcal/mole, for an iron atom from the solid. It is also approximately three times the dimer bond energy, 23 kcal/mole(25). Therefore the two metal hydrogen bonds formed in this chemical reaction represent a significant contribution to the overall stability of the clusters. In addition the exothermicity of the reaction is sufficient to break a single metal-metal bond in the cluster and induce most imaginable isomerizations needed to optimize the stronger metal-ligand bonds. This will be the small cluster analog of chemically induced surface reconstruction. Although a single metal-metal bond can be sacrificed, the exothermicity is insufficient to break all of the bonds an atom makes to its neighbors. Thus desorption of an atom from the cluster is not expected for this reaction.

Niobium cluster reactivities have been reported by both the Rice(1b,c) and Exxon(3f) groups. The maxima and minima agree and relative variations are in general agreement as shown in Fig. 3. Nb_8, Nb_{10}, and Nb_{16} are distinctly inert towards hydrogen. Niobium clusters even out to twenty-five atoms in size still exhibit strong size selective behavior. The reactivity of Nb_n as a function of cluster size is strikingly different from that for iron clusters. Very recently the Rice group has reported Nb_8^+ to also be significantly less reactive towards hydrogen than Nb_7^+ (1g). The similar reactivity pattern of the neutrals and positively charged clusters is quite surprising.

Hydrogen chemisorption on niobium films(26) has a temperature dependence that is consistent with an activation energy for chemisorption of ~2 kcal/mole H_2, and a sticking coefficient of 0.92. This larger sticking coefficient, even with a small activation barrier is consistent with both the higher reactivities of niobium clusters relative to Fe_x and the observation that even the large (>25 atoms) clusters are showing size selectivity.

The Exxon group has also reported reactivities of vanadium(3e). These are shown in Fig. 3. Again the pattern is size selective but not identical to niobium clusters. Vanadium has specific inert clusters like V_6, which is similar to niobium's 8 and 10 but the pattern is better described by an even/odd alternation. This suggests possible analogies with the one electron metals like the alkalis or copper. The reactivity of V_n levels off around n = 15, which is earlier in size than for Fe_n. The larger rate constants for vanadium and niobium suggest smaller activation energies than found for iron.

Both vanadium and niobium metals form dihydrides only at high pressures(27), and numerous phases with hydrogen compositions less than one(28). Experiments were performed to saturate vanadium clusters with deuterium. Figure 4 is a plot of the number of deuterium molecules found in the products. The solid straight lines are for D:V ratios of 1 and 2. The corresponding curved dashed lines include corrections for some bulk atoms(2c). The best fit to the data including only surface atoms indicate a stoichiometry of 1.5. It is likely that this high surface stoichiometry is an indication of bulk incorporation of deuterium.

Cobalt shows a dramatic size dependence(1b,c) that resembles the behavior of iron more so than that of vanadium or niobium. The smallest cluster to react is the trimer and the 5-9 atom clusters are significantly more inert than any of the larger clusters. Cobalt also has a significant dip in reactivity between 19 and 22. A theoretical calculation rationalized the onset in reactivity at the trimer to be associated with energetic stability of the products(29).

Nickel reactivities are relatively flat with cluster size(1c). For clusters smaller than the decamer a weak even odd pattern exists down to the trimer, the smallest cluster reported to react. This pattern might be suggestive of a one-electron bonding scheme as used in very early calculations on Ni_n clusters(30).

The Exxon group has attempted studies of deuterium chemisorption on platinum clusters (3k). The high mass, large number of naturally occurring isotopes and high ionization potentials make reactivity experiments impossible for platinum. However an estimate of the extent of the reaction near saturation can be made by looking at peak broadening with least-squares-fitting procedures. The average number of deuterium atoms chemisorbed per platinum atom is found to be less than one for the clusters in this study. No reaction is observed on the atom through the tetramer. Hydrogen chemisorption on platinum surfaces is generally weak and even on a stepped surface(31) H_2 is desorbed by 450 K. The expected stoichiometry for platinum is one hydrogen per surface atom(32). This is a commonly used number in saturation chemisorption measurements of the dispersion of supported metal catalysts. It is likely that the temperature of the clusters in this study is sufficient to cause significant desorption thus explaining their low hydrogen affinity.

Copper clusters, as reported by the Rice group(1c), do not react with hydrogen. Hydrogen chemisorption on copper surfaces is also an activated process. Surface beam scattering experiments place this barrier between 4-7 kcal/mole(33). This large value is consistent with the activated nature of hydrogen chemisorption on metal clusters, and the trend toward bulk behavior for relatively small clusters (>25 atoms in size).

Nitrogen. The Rice group reported that the reactions of dinitrogen with niobium and cobalt clusters(1c) exhibit reactivity patterns very similar to dihydrogen. Iron has not yet been observed to react in the gas phase with dinitrogen. The Rice group did report some

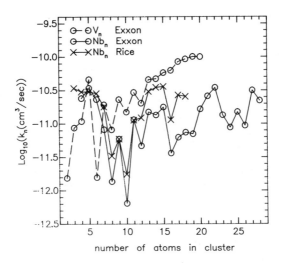

Figure 3. Logarithm of the reactivities of niobium and vanadium clusters. The Exxon data (circles) are scaled relative to the Argonne Fe_{10} reactivity, and the Rice data (crosses) are normalized to Nb_9.

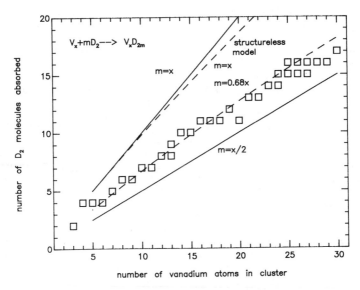

Figure 4. The number of deuterium molecules found in products on V_x clusters produced while attempting to saturate the reaction. The solid lines are plots of $D_2:V$ ratios of 1 and 0.5, including all vanadium atoms. The dashed lines are corrected, assuming globular shaped clusters. The best least-squares fit to the data, $D_2:V = 0.68$, is also plotted as a dashed line.

depletion(1c) but at only very high reactant flows and products were not observed. Even iron hydride clusters did not react with dinitrogen in an attempt to study low pressure ammonia synthesis on gas phase iron clusters(34).

The similarity of the reactivity patterns for niobium and cobalt and the non-reactivity of iron with nitrogen suggests that dissociative chemisorption is taking place. Dissociation of molecularly chemisorbed nitrogen is an activated process on all metals(35) and is most exothermic for the early metals in the periodic table(36). The limited observations on clusters seems to be consistent with these trends.

Methane. Both iron(3b) and aluminum(3g) clusters are inert towards methane. Molecular absorption is likely too weak for the products to be detected, making this technique only sensitive to dissociative absorption. Although the C-H bond is comparable in strength to H-H, activation of the former should be further constrained by steric effects(37). In addition, for the reaction to be possible the sum of the bond strengths of M_n-CH_3 and M_n-H must exceed that of CH_3-H. For iron M_n-H is 67 kcal/mole requiring the M_n-CH_3 bond to at least be 33 kcal/mole. Cryogenic matrix studies have found electronic excitation of metal atoms essential for their insertion into methane(38). These observations imply at least a highly activated process, if not one that is endothermic. Based on the limited literature values of M-CH_3 bond energies Co and Ni are the only 3d transition metal atoms that form sufficiently strong bonds for this reaction to be exothermic(39). Recent molecular beam studies also find large barriers for methane activation on tungsten(40) and nickel(41).

Carbon monoxide. This chemisorption was first studied on niobium and cobalt clusters by the Rice group(1c). Although nitrogen and CO are isoelectronic they behave quite differently on these metals. The distinct size selective reactivity patterns displayed by niobium and cobalt for hydrogen and nitrogen chemisorption were completely absent for CO where all clusters studied reacted with comparable rates. CO reactions were recently extended to a total of twelve transition metals by the Exxon group(3j). No wide variation of reaction rate across the periodic table was observed. Figure 5 is a bar graph of these results. The height of each rectangle represents the measured reactivity of a specific size cluster with CO. Almost all the metal clusters with more than three atoms react in a facile fashion.

Unlike hydrogen these reactions do not appear to be activated. In addition the products distributions observed indicate comparable rates for multiple adduct formation. The mass complexity, relatively high ionization potentials, and the known prevalent dissociative ionization of the fully saturated carbonyls(42) has possibly caused the failure of some initial saturation experiments(43). The ability to synthesize the stable carbonyl complexes will help this field significantly due to the vast amount of information available, especially their structures.

The small cluster threshold behavior, suggested by the kinetics scheme presented earlier, is apparent in this data set. Assuming (1) the RRK form for k_{-n}, (2) the addition reactions are not activated, and (3) the number of participating modes are independent of the element type, the M-CO bond strengths can be grouped by energy. For

M_2-CO the first adduct bond energies are greater for Co, Ru, Pd, W, Ir, and Pt than V, Fe, Ni, Nb, and Mo. For the trimers, iron has a weaker bond than the other metals studied. For the tetramer and larger clusters the reactivity is controlled by the value of k_n and no longer by the competition between unimolecular decomposition and collisional stabilization. The large cluster regime is not covered by this model used to make the correlation between kinetics and energetics.

This ordering of M_2-CO bond strengths is consistent with the trend of increasing M-CO average bond strengths in metal carbonyl complexes as one goes down a column in the periodic table(44). The non-reactivity of Ni_2 is inconsistent with the trend of the carbonyl complexes going across a row. However, for polycrystalline films the heat of adsorption of CO generally decrease going across a row(45). In fact, groups 3 (IIIB) and 4 (IVB) have the largest values. These large values are associated with the trend of early transition elements to dissociate CO(46). By groups 8-10 (VIIIB) the surface and organometallic complexes have comparable values. The average M-CO bond dissociation energy in $Ni(CO)_4$ is 35 kcal/mole(44) which lies near the top of the range of heats of adsorption of CO on nickel surfaces(47). Therefore, these measurements indicate that with the exception of nickel the bond energies for the first CO bonded to the metal dimers studied follow a similar global trend as the average bond dissociation energies of the metal carbonyls.

Carbon monoxide eventually dissociates at room temperature on all but some of the group 8-10 (VIIIB) metals(44). This dissociation occurs only for metal surfaces which form sufficiently strong metal-carbon plus metal-oxygen bonds to break the 257 kcal/mole CO bond. The known values for gas phase metal atoms predict the same trend(48). The similarity in the behavior of surfaces and atoms implies for the most part that the clusters should behave likewise. This also implies highly localized bonding. The non-reactivity observed for V_2, Nb_2 and Mo_2, which based on these prior assumptions should dissociate CO, could mean that the dimers are too small to form a strong bond to both carbon and oxygen atoms.

Dioxygen and Carbon Dioxide. Remarkable in a way, are the very few reaction studies reported with dioxygen(3b). For most transition metals the monooxides produced from surface oxidation or purity problems are ubiquitous and difficult to eliminate. In early experiments the Argonne group reported reactions of oxygen with iron clusters(2a) by having the reactant present during the vaporization and clustering process. This allowed for very high energy plasma processes to possibly dominate, and no reactivity measurements were reported. In addition the ionization laser intensity was sufficiently high to cause significant fragmentation. They found the composition of the products to be oxygen poor. More recently the Exxon group reported iron reactions with dioxygen. The data was presented assuming equilibrium in the flow reactor. This data even with this questionable assumption showed no distinct size selective behavior.

Platinum clusters, n = 2-11 react with dioxygen at a rate that is within an order of magnitude of gas kinetic. There is no distinct size selective behavior. Products of these gas phase reactions observed with 7.87 eV ionization laser, are $Pt_nO_{2m}^+$ where for m=1, n=5,7-11 and for m=2, n=8-11. There was a very weak product for $Pt_6O_2^+$ although the depletion of the metal was equivalent to the other metal clusters. An increase in ionization potentials upon oxidation may be responsible for this pattern of detected products.

Preliminary studies of carbon dioxide reactions with niobium and cobalt clusters by the Rice group(1e) have found another size selective reaction. The reaction proceeds on small clusters (Nb_{3-7}) via dissociation, producing $Nb_{3-7}O$ + CO. For larger clusters Nb_nCO_2 is found. Dissociation implies a metal-oxygen bond strength > 127 kcal/mole, which is consistent with the heat of absorption on polycrystalline films(49) and the dissociation energy of Nb-O diatomic(50). Studies such as these extended to other reactants such as N_2O and NO will act to survey metal cluster-oxyen bond strengths across the transition metal series.

Lewis Bases. A variety of other ligands have been studied, but with only a few of the transition metals. There is still a lot of room for scoping work in this direction. Other reactant systems reported are ammonia(2e), methanol(3h), and hydrogen sulfide(3b) with iron, and benzene with tungsten(1f) and platinum(3a). In a qualitative sense all of these reactions appear to occur at, or near gas kinetic rates without distinct size selectivity. The ammonia chemisorbs on each collision with no size selective behavior. These complexes have lower ionization potential indicative of the donor type ligands. Saturation studies have indicated a variety of absorption sites on a single size cluster(51).

The Exxon data for iron with methanol also does not show size selective behavior(52). The Exxon group has been able to show by infrared multiple photon dissociation that the O-H bond breaks forming methoxy on these small iron clusters(3h). This is consistent with the behavior of methanol at room temperature on iron surfaces(53). This is the first example in which the chemisorption is confirmed to be dissociative and the reaction is not size selective. However the Lewis acidity of the oxygen atom lone pair in methanol is likely sufficient to cause the initial product to be molecularly chemisorbed.

Benzene reacts at gas kinetic rates with platinum clusters(3a). The products produced have significantly lower IP's than the reactants. This change can be very significant. For example both the platinum atom and benzene have (-IP's-) greater than 8.9 eV while the $Pt(C_6D_6)_2$ species has an IP less than 7.87 eV, indicating at least a drop of one electron volt in ionization potential. Arene complexes all have similarly low IP's(54). With 7.87 eV ionization laser no dehydrogenation is observed for benzene reactions with platinum clusters out to Pt_7 (the largest cluster for which the adduct mass could be studied reliably) and the low ionization potential of the products, suggest π bonded structures versus oxidative addition to C-H bonds. Figure 6 shows the mass spectrum of platinum at the early stages of its reaction with benzene. The initial report (3a) of platinum cluster chemistry with C_6 hydrocarbons showed minor dehydrogenation of benzene starting beyond the trimer. These new studies suggest that the dehydrogenation was caused by fragmentation induced by either too high of an ionization laser intensity (work was carried out using only 100-400 μJ of 6.42 eV photons) or high vaporization laser

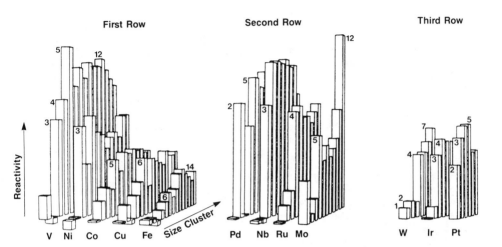

Figure 5. A bar graph of metal cluster reactivities with CO on a linear scale. Cluster size increases going into the page and metal types across. Once beyond a few atoms in size most all clusters react at rates within an order of magnitude of each other.

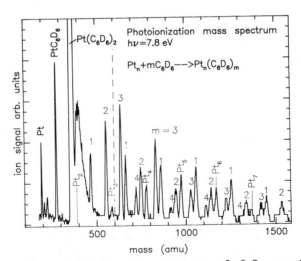

Figure 6. Time-of-flight mass spectrum of C_6D_6 reacting with platinum clusters. The peak labels represent m in $Pt_n(C_6D_6)_m{}^+$. There is a scale change between n=2, m=2 and m=3.

intensity. The latter may yield clusters of high enough temperature
to cause thermally induced dehydrogenation.

In summary, reactions with ligands that require a bond to be
broken before chemisorption is completed show size selective
behavior. This applies to hydrogen and nitrogen and likely
methane. Ligands that can form bonds with these highly
coordinatively unsaturated metal clusters do so with fast rates and
with little size discrimination. This applies even to systems that
eventually do dissociate like methanol and most likely the CO adducts
on the early transition metals. Once there is a significant
attractive interaction any barrier in the entrance channel is abated,
thus eliminating the size selective behavior. The cluster
reactivities are neither identical to surface chemistry nor cluster
inorganic chemistry, but smoothly interpolate between these fields
with the notable exception of the dramatic size selective chemistry
associated with bond activation on the small metal clusters. These
very simple but very informative measurements are sensitive to just
the initial chemisorption step and not subsequent rearrangements and
isomerizations. Techniques to measure of the following steps in
reactions on gas phase clusters are just now being developed(3h).

Platinum dehydrogenation reactions

Platinum films, clusters and surfaces under the conditions in the
flow reactor are expected to rapidly dehydrogenate most
alkanes(55). The rate limiting step is the initial chemisorption of
the alkane to the metal(56). These studies are quite sensitive to
the first step and thus are expected to be informative. Unlike most
of the reactions already reported these dehydrogenation reactions
involve desorption of hydrogen from the metal cluster adduct. This
gives the collision complex a way to remove its excess energy and
become stabilized. However if this desorption process becomes
competitive with chemisorption then the reactivity observed will be
harder to interpret. For platinum the difficulty in making hydrogen
stick strongly suggest that this will not be a problem.

Cyclohexane. In qualitative terms small platinum clusters rapidly
and non-selectively chemisorbs cyclohexane. The products are highly
dehydrogenated with C:H ratios rapidly approaching one. Figure 7
shows the gain in mass upon reaction for the first adduct and the
second minus the first. The platinum atom within the errors
dehydrogenates C_6D_{12} to C_6D_6. The second adduct on the platinum atom
does not dehydrogenate significantly. However by Pt_6 the platinum
clusters appear to be able to convert two cyclohexane molecules into
chemisorbed species with a C:D ratio near one.

This facile dehydrogenation is consistent with the kinetic
models derived from catalytic conversion of cyclohexane to
benzene(57). These models predict an ensemble size for the active
site of only one atom. On the other hand facile dehydrogenation of
cyclohexane on metal surface under UHV conditions are described using
a model that extends over several metal atoms and suggests a specific
type of interaction is necessary for an efficient reaction(58). This
is obviously not the case for the very small platinum metal
clusters.

<u>Other C_6 hydrocarbons</u>. The dehydrogenation of normal hexane and 2,3-dimethyl butane also proceeds but not as voraciously on small platinum clusters. Figure 8 is a plot of the hydrogen content in the first adduct as a function of the size of the platinum metal cluster. The metal atom reacts via dihydrogen elimination to produce PtC_6H_{12} products. The platinum trimer is now the smallest cluster that will produce a C:H near one. The similarity of size dependent dehydrogenation of the normal hexane and the branched molecule suggest that these systems may not readily aromatize these alkanes. Further structural studies are needed to identify the reaction products.

Chemistry studies of alkanes on platinum surfaces under UHV conditions are limited by the very weakly bound molecularly chemisorbed state(59). The low surface temperatures required to adsorb the molecule are insufficient to activate C-H bonds. However at higher pressures at room temperature and above, extensive dehydrogenation is expected. This process is thermally activated on surfaces but in the gas phase on small platinum clusters occurs at approximately 1-10% gas kinetic. This rapid reaction suggests for very small platinum clusters that the activation barrier for alkanes in the gas phase is at most just a few kilocalories per mole, and does not involve a high degree of steric hindrance.

<u>Models of size selective reactivity</u>

The initial goal of a successful model is to rationalize the significant and surprising observations. The first objective will be to explain the size dependence found in this simple chemistry i.e. why in going from Fe_{17} to Fe_{22} does the rate constant increase by a factor of a thousand? The second goal is to rationalize the striking correlation, first made by the Exxon group, between the variation in ionization potentials and the logarithm of the rate constant for chemisorption of dihydrogen(3c). Finally since many of the reactions are facile the minimum criteria in ligand electronic structure will be sought that will assure non-activated chemisorption.

The above observation suggests that the electronic structure must play a significant role in determining size-selective chemistry. An ionization potential in a simple one-electon approximation measures only the orbital energy of the highest occupied molecular orbital (HOMO). This energy, especially as the system gets large, is only one of the numerous orbitals that contribute to the cluster's total stability. Both the success of frontier fragment-orbital analysis in organometallic reactivity studies(60) and the relatively small changes in activation energies necessary to explain the observed variations in rate constants, suggest electronic structure plays the pivotal role. In addition the recent successes in explaining alkali metal cluster(61) distributions and IPs based on a jellium model suggest electronic effects are directly apparent and do not necessarily depict themselves in terms of specific geometric structures.

An alternative approach is to relate variations in reactivity with metal clusters' structures. Unfortunately even less is known about their geometric arrangement than their electronic structure. Suggestions made in the literature will be summarized and simple ideas of the degree of coordinative unsaturation will be presented.

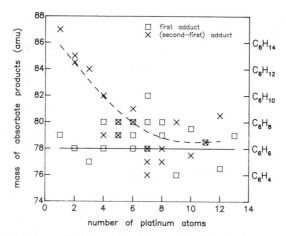

Figure 7.   A plot of the adduct masses (product - bare metal) produced in reacting platinum clusters with c-C_6D_{12}. The lines for the first adduct (solid) and second-first (dashed) adduct masses are drawn to help guide the eye through the scatter in the data.

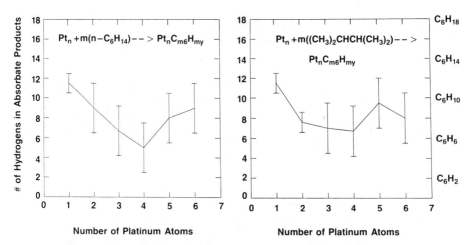

Figure 8.   Plots of the number of hydrogens retained in the products formed in reacting platinum clusters with n-C_6H_{14} and $(CH_3)_2CHCH(CH_3)_2$.

Empirical fit and Charge Transfer. Figure 9 is a plot of the logarithm of the bimolecular rate constants for deuterium chemisorption as a function of size for iron, niobium and vanadium clusters. The dashed line is a plot of the ionization potentials of these same clusters scaled by an empirical charge transfer model. The surprising observation of a significant anticorrelation between a metal clusters' ionization potential and their rate of reaction with deuterium/hydrogen strongly suggested charge transfer from the metal cluster to D_2 was essential in activating the bond. The IPs are scaled according to

$$k_n = A \exp(-\varepsilon(IP(n)-E_0)/k_BT)$$

where E_0 and ε are determined from a least-squares fit to the rate constants, assuming A is independent of cluster size. The A factor used is obtained by assuming that the large cluster plateau in the iron and vanadium cluster reactivities corresponds to zero activation energy. The same value is then used for niobium as for vanadium clusters. With this assumption any cluster having an ionization potential less than E_0 will have zero activation energy. The ε scale factor can be interpreted in terms of the amount of charge transfer or as will be shown later a cluster electronegativty difference. The inclusion of the k_BT factor in the exponent was chosen to draw a close analogy with the Arrhenius expression since the Argonne group has observed thermal activation for the iron system.

The correlation in Fig. 9 is best for iron, especially for cluster with more than eight atoms. The opposite behavior is observed for the smaller clusters of niobium and iron. This departure along with (1) observations by the Rice group that positively charged ions of niobium have similar reactivity patterns as the neutral clusters(1e,g) and (2) the report by the Argonne group of size selective behavior on ammoniated clusters that have significantly lower ionization potentials than the bare clusters(62), necessitates the interpretation of ε in terms more general than the fractional charge transfer.

Electronic structural model. The size selective reactivity of these metal clusters is surprising. Certainly the metal cluster are coordinatively unsaturated. It appears that coordinative unsaturation is essential to satisfy the energetic criteria but more specific aspect of the electronic structure must play a role in controlling the activation energy of the process.

A series of papers by Shustorovich(63) and/or Baetzold(64) summarized in a recent article(65) have addressed the problem of chemisorption on metal surfaces in terms of electron accepting and donating interactions. Saillard and Hoffmann(66) developed qualitatively identical pictures of these interactions but starting from fragment orbital type analysis. These papers are only a few of the theoretical discussions that consider hydrogen activation, however we will use their approach because it address the problem in a fashion that can interpolate between the organometallic cluster and the bulk.

Starting from the point of view of discrete molecules there are four frontier orbital and two primary interactions. The first interaction is donation from the metal cluster to the ligand through the cluster's HOMO and the reactant's lowest unoccupied molecular orbital (LUMO). The other interaction is for the cluster to act as the acceptor using its LUMO and the reactant's HOMO. The energies of these orbitals obtained by EHT for a variety of reactants and a canonical cluster are shown in Fig. 10. From this diagram it is immediately obvious why the transition metals are effective in bond rearragements. These metals supply electron density that bridges the HOMO-LUMO gap which is associated with the high stability of the closed shell reactants. Interactions involving these electrons supply paths that avoid high activation barriers. The electron rich character of metals imply that they should be very effective donors, in spite of the large energy separation between the metal-HOMO and the LUMO of the reactant.

Consider the hydrogen molecule approaching a metal cluster. The long-range interaction will be repulsive because the majority of the electrons in the cluster are spin-paired and holding the cluster together. This long-range repulsion creates a barrier to the reaction. As the hydrogen molecule gets closer, the cluster acts as an e^- donor, interacting with σ^* of H_2 and as an acceptor with the bonding molecular orbital of the absorbate. Both of these interactions weaken the absorbate bond and strengthen the bonding to the cluster. Thus if the attractive interaction overwhelms the repulsion, dissociative chemisorption will occur with a small activation barrier in the entrance channel. The activation energy will depend upon a compromise between the repulsion and the longest-range attraction. The donor interaction is longer-range and will increase as orbital separation decreases. Therefore the lower IP clusters are better long-range donors than acceptors and thus have smaller activation energies. This is exactly what is needed to rationalize the correlation between a cluster's IP and its ability to activate hydrogen.

This simple model predicts for a sufficiently low IP cluster that bonds can be broken without a barrier and that metals on the left end of the periodic table should be the most facile, both in agreement with our observations. If the antibonding σ type orbital in the reactant is significantly higher in energy, the barrier could be sufficient to prevent reactions from being detected. This is apparently the case for methane.

These arguments change when extended to π_* bonded reactants like N_2, CO, and C_6H_6 where now the LUMO has π^* character. Small populations in the LUMO will no longer guarantee dissociation. Instead a stable molecular chemisorption bond can form. Since these experiments have a much higher sensitivity for activation energies (a few kilocalories/mole) than for stable bonds (16 kcal/mole) molecular bonding dominates the reactivity. The LUMO of nitrogen is high enough that molecular states do not form (even on the bulk surface they need to be stabilized by alkali promotion) and follow the same pattern as in hydrogen. Iron clusters appear to be on the verge of reacting with nitrogen in these experiments-likely higher temperatures or alkali promotion will test this simple model. Finally molecular oxygen will look like a diradical to even the smallest metal cluster and should react without size selectivity even if the final products are dissociative.

This model for bond activation has already failed several simple

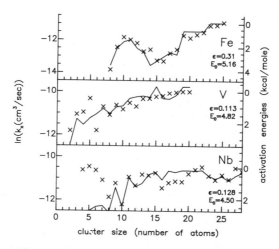

Figure 9. Plots of the rate constants (X) of iron, vanadium and niobium clusters reacting with dihydrogen/dideuterium, and their respective bare cluster ionization potentials (solid lines) scaled as described in the text.

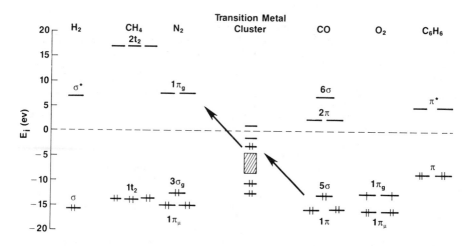

Figure 10. Indicated are the HOMO and LUMO orbital energies obtained form EHT calculations for a variety of reactants. In the center are estimated orbital energies for a canonical metal cluster.

tests. First, reactions of the charged clusters show similar patterns, which indicate simple charge transfer is inadequate. In addition the repulsive interaction that caused the barrier in the first place is likely insignificant compared to the attractive charge-induced dipole interaction. In the other direction the model fails as well. The Argonne group has found size-selective reactivity of hydrogen on iron clusters that are saturated with ammonias. These ammine-Fe_n clusters have significantly lower ionization potentials and still show size selectivity; again conflicting with a simple charge transfer model. The simplest extension of this model is to include both donor and acceptor interactions. The relevant parameter is the Mulliken electronegativity which averages the the IP and electron affinity(EA). Unfortunately at this time insufficient EA data is available for meaningfull tests.

As seen in Fig. 9 the correlation between IP and reactivity changes significantly for small clusters. In fact for many of the clusters under eight atoms in size the correlation is in the opposite sense. The considerations given so far have been based on energetics and have not included any detail valency or symmetry. The assumption is that the density of states is sufficiently high that at any energy there is an orbital that could satisfy any of these further restrictions. As the cluster gets smaller this becomes more difficult and thus departures from the "bulk" model should occur. In addition the cluster IPs also increase significantly making acceptor interactions more important.

Local coordinative saturation. For the very small cluster the idea of drawing on the field of stable cluster complexes is very attractive. Lauher(69) and Teo(70) have presented rules that describe the total bonding capabilities of the metal clusters. These work well for saturated and partially unsaturated metal clusters. However the high degree of coordinative unsaturation is significant leaving these rules with little predictive power. However L. Brewer has taken a slightly different approach and explained the structure of alloys and crystal habit of the transition elements. This work is able to predict the structure of most bulk metal systems. This points out that even for metals that are very malleable and have little preference for directional bonding simple rules can still be developed and might be very applicable to clusters.

For niobium and cobalt clusters structures have been proposed based upon the elements behavior(71). Niobium's specific inertness has been associated with structures that are analogous to close-packed surface of W(110) which also has an activation barrier for hydrogen chemisorption. Since the IPs are also expected to be higher for closed packed structures these two sets of observations are in agreement. This model at its current stage of development requires different structures for each system and as yet has not been useful in making predictions.

It is important to note that none of these arguments have addressed the first concern of why. But, have shifted the question to what in the electronic structure of these clusters is causing the IP, reactivity, and likely variations in structure with size.

Physical Characterization

Through out this mini-review correlations have been made between chemical and physical properties of the metal clusters. The simultaneous study of their chemistry and the application of standard molecular-beam chemical physics probes have cooperatively enabled the rapid growth of this new area. Physical properties currently measurable are ionization potentials(3d-f,3i,72) photoionization efficiencies(73), magnetic moments(19), photofragmentation of ions(74,75) and in some cases neutrals, polarizabilites(76), and some strong "infrared absorptions"(3h). The number of systems studied and the measurement methods are rapidly growing. This data base is essential in the understanding of the compromises that are involved in metal bonding, as the metallic state evolves from small clusters.

Theory has participated in all aspects of this area's development. The almost hand-waving arguments used to rationalize their chemical behavior need testing and will likely be replaced by more elegant quantitative discussions. The theoretical aspects(77) and most physical property measurements(78) of small metal clusters have been recently reviewed.

Summary

The field of gas-phase transition metal cluster chemistry has expanded rapidly due to the development of the laser vaporization source and the fast flow chemical reactor. The work from the three major laboratories have been reviewed. Many additional laboratories are developing cluster chemistry programs and will soon certainly make significant contributions.

In summary a few "generalizations" have been found. First, size selective chemistry is strongly associated with chemisorption that requires bond-breaking. Second, metal clusters react rapidly with ligands that molecularly chemisorb even when the eventual products involve dissociation of the ligand. Dehydrogenation of C_6-alkanes on small platinum clusters take exception to this.

The charge transfer model suggested to rationalize the correlation between ionization potential and reactivities of iron, vanadium, and niobium with dihydrogen fails for other systems. However a model that takes into account the frontier orbital interactions, although highly simplistic, does account for a variety of observations. This model suggests extensions that include electron affinities as well as IPs and the possibility of developing an electronegativity scale for clusters as a function of their size.

Geometric structure of the bare metal clusters and the complexes formed by reaction are unknown and present a significant experimental challenge. Chemical studies are starting to imply something about the structure of the products and will be invaluable until more direct chemical physics probes are available.

Acknowledgments

We wish to thank the current and past members of the cluster chemistry groups at Rice University, Argonne National Laboratory and Exxon Research and Engineering Co. Rick Smalley, Steve Riley, Eric

Parks, Ken Reichmann, Eric Rohlfing, Rob Whetten, Mitch Zakin, and Don Cox deserve thanks for communication of their data before publication and/or allowing its use for the first time in this mini-review. The open communication among these collaborators and competitors have made this an extremely enjoyable, exciting, and rapidly progressing area of research.

<u>Literature Cited</u>

1. Rice: a) J. Dietz, M.A. Duncan, D.E. Powers, R.E. Smalley J. Chem. Phys. <u>74</u>, 6511 (1981). b) M.E. Geusic, M.D. Morse, R.E. Smalley J. Chem. Phys. <u>82</u>, 590 (1985). c) M.E. Geusic, M.D. Morse, R.E. Smalley Rev. Sci. Instrum. <u>56</u>, 2123 (1985). d) M.D. Morse, M.E. Geusic, J.R. Heath, R.E. Smalley J. Chem. Phys. <u>83</u>, 2293 (1985). e) P.J. Brucat, C.L. Pettiette, S. Yang, L.-S. Zheng, M.J. Craycraft, R.E. Smalley J. Chem. Phys. <u>85</u>, 4747 (1986). f) M.D. Mores, M.E. Geusic, S.C. O'Brien, R.E. Smalley Chem. Phys. Lett. <u>122</u>, 289, (1985). g) J.M. Alford, F.D. Weiss, R.T. Laaksonen, R.E. Smalley, J. Phys Chem in press.

2. Argonne: a) S.J. Riley E.K. Parks, G.C. Nieman, L.C. Pobo, S. Wexler J. Chem.Phys. <u>80</u>, 1360 (1984); S.J. Riley, E.K. Parks, L.G. Pobo, S. Wexler Ber. Bunsenges. Phys. Chem. <u>88</u>, 287 (1984). b) S.C. Richtsmeier, E.K. Parks, K. Liu, L.G. Pobo, S.J. Riley J. Chem. Phys. <u>82</u>, 3659 (1985). c) E.K. Parks, K. Liu, S.C. Richtsmeier, L.G. Pobo, S.J. Riley J. Chem. Phys. <u>82</u>, 5421 (1985). d) K. Liu, E.K. Parks, S.C. Richtsmeier, L.G. Pobo, S.J. Riley J. Chem. Phys. <u>83</u>, 2882,5353 (1985). e). S.J. Riley, E.K. Parks, K. Liu "Int. Symp. Optical and Optoelectronic Appl. Sci. Eng.", Quebec, 1986.

3. Exxon: a) D.J. Trevor, R.L. Whetten, D.M. Cox, A. Kaldor J. Am. Chem. Soc. <u>107</u>, 518 (1985). b) R.L. Whetten, D.M. Cox, D.J. Trevor, A. Kaldor J. Phys. Chem. <u>89</u>, 566 (1985). c) R.L. Whetten, D.M. Cox, D.J. Trevor, A. Kaldor Phys. Rev. Lett. <u>54</u>, 1494 (1985). d) R.L. Whetten, M.R. Zakin, D.M. Cox, D.J. Trevor, A. Kaldor J. Chem. Phys. <u>85</u>, 1697 (1986). e) R.L. Whetten, M.R. Zakin, D.M. Cox, D.J. Trevor, A. Kaldor, in preparation f) A. Kaldor, D.M. Cox, M.R. Zakin, D.J. Trevor Z. Phys. Chem. D in press g) D.M. Cox, D.J. Trevor, R.L. Whetten, A. Kaldor in preparation. h) M.R. Zakin, R.O. Brickman, D.M. Cox, K.C. Reichman, D.J. Trevor, A. Kaldor, J. Chem. Phys. <u>85</u>, 1198 (1986). i) M.R. Zakin, D.M. Cox, R.L. Whetten D.J. Trevor, A. Kaldor J. Chem. Phys. submitted. j) D.M. Cox, D.J. Trevor, K. Reichmann, A. Kaldor J. Phys. Chem. submitted. k) D.J. Trevor, A. Kaldor unpublished results.

4. P.J. Brucat, C.L. Pettiette, S. Yang, L.-S. Zheng, R.E. Smalley J. Chem. Phys. <u>85</u>, 4747 (1986).

5. M. Mandich, W.D. Reents, Jr., V.E. Bondybey, J. Phys. Chem. in press; V.E. Bondybey, W.D. Reents, Jr., M. Mandich, J. Chem. Phys. in press.

6. S.K. Loh, D.A. Hales, P.B. Armentrout, Chem. Phys. Lett. in press.

7. L. Sallans, K.R. Lane, R.R. Squires, B.S. Freiser, J. Am. Chem. Soc. <u>107</u>, 4379 (1985) and references therein.

8. V.E. Bondybey, J.H. English Chem. Phys. Lett. <u>94</u>, 443 (1983).

9.   M. Jarold, J.E. Bower J. Chem. Phys. in press.
10. E.A. Rohlfing, D.M. Cox, R. Petkovic-Luton, A. Kaldor J. Phys. Chem. 88, 6277 (1984).
11. E.A. Rohlfing, D.M. Cox, A. Kaldor, J. Chem. Phys. 81, 3322 (1984).
12. H.W. Kroto, J.R. Heath, S.C. O'Brien, R.F. Curl, R.E. Smalley, Nature (London) 318, 162 (1985).
13. L.A. Bloomfield, R.R. Freeman, W.L. Brown Phys. Rev. Lett. 54, 2246 (1985); L.A. Bloomfield, M.E. Geusic, R.R. Freeman, W.L. Brown Chem. Phys. Lett. 121, 33 (1985).
14. J.R. Heath, Y. Liu, S.C. O'Brien, Q.-L. Zhang, R.F. Curl, F.K. Kittle, R.E. Smalley J. Chem. Phys. 83, 5520 (1985); S.C. O'Brien, Y. LIn, Q. Zhang, J.R. Neath, F.K. Tittel, R.F. Curl, R.E. Smalley J. Chem. Phys. 84, 4074 (1986).
15. D.M. Cox unpublished results.
16. R.G. Wheeler, K. LaiHing, W.L. Wilson, J.D. Allen, R.B. King, M.A. Duncan J. Am. Chem. Soc. in press.
17. L.F. Keyser J. Phys. Chem. 88, 4750 (1984).
18. J.B. Anderson, J.B. Fenn Phys. Fluids 8, 780 (1965).
19. D.M. Cox, D.J. Trevor, R.L. Whetten, E.A. Rohlfing, A. Kaldor Phys. Rev. B32, 7290 (1985).
20. D.A. Gobeli, J.J. Yang, M.A. El-Sayed Chem. Rev. 85, 529 (1985).
21. D.J. Trevor in preparation.
22. D.M. Cox unpublished results.
23. J. Benziger, R.J. Madix Surf. Sci. 94, 201 (1980).
24. G. Wedler, H.-P. Geuss, K.G. Colb, and G. McElhiney Appl. Surf. Sci. 1, 471 (1978); F. Bozso, G. Ertl, M. Grunze, M. Weiss Appl. Surf. Sci. 1, 103, (1977).
25. P.J. Brucat, L.-S. Zeng, C.L. Pettiette, S. Yang, R.E. Smalley J. Chem. Phys. 84, 3078 (1986)
26. D.I. Hagen, E.E. Donaldson Surf. Sci. 45, 61 (1974).
27. J.J. Reilly, R.H. Wiswall Inorg. Chem. 9, 1678 (1970).
28. M. Hirabayashi, H. Asano in "Metal Hydrides"; ed. G. Bambakidis; Plenum: New York, 1981, p 53. and M.A. Pick p 329.
29. D.P. Onwood, A.L. Companion J. Phys. Chem. 89 3777 (1985).
30. T.H. Upton, W.A. Goddard III, C.F. Melius J. Vac. Sci. Technol. 16, 531 (1979).
31. K. Christmann, G. Ertl, T. Pignet surf. Sci. 54, 365 (1976);B. Poelsema, L.K. Verheij, G. Comsa Surf. Sci. 152/153, 496 (1985).
32. J.E. Benson, M. Boudart J. Catal. 4, 704 (1965).
33. M. Balooch, M.J. Cardillo, D.R. Miller, R.E. Stickney Surf. Sci. 46, 358 (1074).
34. S.J. Riley private communication.
35. L.J. Whitman, C.E. Bartosch, W. Ho, G. Strasser, M. Grunze Phys. Rev. Lett. 56, 1984 (1986).
36. G. A. Somorjai, "Chemistry in Two Dimensions: Surfaces"; Cornell University:Ithaca, 1981.
37. C.B. Lebrilla, W.F. Maier J. Am. Chem. Soc. 108, 1606 (1986).
38. R.N. Perutz Chem. Rev. 85, 77 (1985).
39. R.G. Pearson Chem. Rev. 85, 41 (1985)
40. C.T. Rettner, H.E. Pfnur, D.J. Auerbach J. Chem. Phys. 84, 4163 (1986).
41. M.B. Lee, Q.Y. Yang, S.L. Tang, S.T. Ceyer J. Chem. Phys. 85, 1693 (1986).

42. M.A. Duncan, T.G. Dietz, R.E. Smalley Chem. Phys. 44, 415-419 (1979).
43. S.J. Riley and the Exxon group have made inital attempts, private communication.
44. G. Ertl, "The Nature of the Surface Chemical Bond"; Rhodin, T.N., Ertl, G. ed., North-Holland, New York, 1979, p 313; ref. 36.
45. G. Pilcher, H.A. Skinner "The Chemistry of the Metal-Carbon Bond"; F.R. Hartley and S. Patai ed., John Wiley & Sons, New York, 1982, p 43-90.
46. T.N. Rhodin, J.W. Gadzuk, "The Nature of the Surface Chemical Bond"; Rhodin, T.N., Ertl, G. ed., North-Holland, New York, 1979, p 113; ref. 36
47. G. Wedler, H. Poppa, G. Schroll Surf. Sci. 44, 463 (1974); C.R. Helms, R.J. Madix Surf. Sci. 52, 677 (1975).
48. J.A. Martinho Simoes, J.L. Beauchamp Chem. Rev. in press.
49. D.Brennan, D.O. Hayward Phil. Trans. Roy. Soc. (London) A258, 375 (1965).
50. K.P. Huber, G. Herzberg "Molecular Spectra and Molecular Structure IV. Constants of Diatomic Molecules"; Van Nostrand Reinhold: New York, 1979.
51. S.J. Riley private communication.
52. M.R. Zakin, D.M. Cox unpublished results.
53. P.H. McBreen, W. Erley, H. Ibac Surf. Sci. 133, 1469 (1983).
54. H.M. Rosenstock, K. Draxl, B.W. Steiner, J.T. Herron J. Phys. Chem. Ref. Data 6, sup 1 (1977).
55. G.A. Somorjai Chem. Soc. Rev. 13, 312 (1984);C.B. Lebrilla, W.F. Maier J. Am. Chem. Soc. 108, 1606 (1986).
56. L.E. Firment, G.A. Somorjai J. Chem. Phys. 66, 2901 (1977).
57. M.E. Ruiz-Vizcaya, O. Novaro, J.M. Ferreira, R. Gomez   J. Catal. 51, 108 (1978).
58. M.-C. Tsai, C.M. Friend, E.L. Muetterties J. Am. Chem. Soc. 104,2539,(1982);M.-C. Tsai, E.L. Muetterties J. Phys. Chem. 86, 5067 (1982).
59. S.M. Davis, F. Zaera, G.A. Somorjai J. Cat. 85, 206 (1984).
60. T.A. Albright Tetrahedron 38, 1339 (1982) and ref. therein.
61. W.D. Knight, K. Clemenger, W.A. deHeer, W.A. Saunders, M.Y. Chou, M.L. Cohen Phys. Rev. Lett. 52, 2141 (1984).
62. S.J. Riley and E.K. Parks private communication.
63. E. Shustorovich, R. Baetzold, E.L. Muetterties J. Phys. Chem. 87, 1100 (1983).
64. R.C. Baetzold J. Chem. Phys. 82, 5724 (1985).
65. E. Shustorovich, R.C. Baetzold Science 227, 879 (1985).
66. J.-Y. Saillard, R. Hoffmann J. Am. Chem. Soc. 106, 2006 (1984).
67. E. Shustorovich Surf. Sci. 150, L115 (1985).
68. H.S. Johnston, "Gas Phase Reaction Rate Theory"; Ronald: New York, 1966.
69. J.W. Lauher J. Am. Chem. Soc. 101, 2604 (1979).
70. B.K. Teo Inorg. Chem. 24, 1627 (1985)., ibid 24, 4209 (1985).
71. J.C. Phillips Chem. Rev. 86, 619 (1986).
72. D.E. Powers, S.G. Hansen, M.E. Geusic, D.L. Michalopoulos, R.E. Smalley J. Chem. Phys. 78, 2866 (1983).
73. E.A. Rohlfing, D.M. Cox, A. Kaldor J. Chem. Phys. 81, 3846 (1984).

74. P.J. Brucat, L.-S. Zheng, C.L.Pettiette, S. Yang, R.E. Smalley J. Chem. Phys. 84, 3078 (1986).
75. L.A. Bloomfield, R.R. Freeman, W.L. Brown Phys. Rev. Lett. 54, 2246 (1985).
76. K. Clemenger, W.D. Knight, W.A. deHeer, W.A. Saunders Phys. Rev. B31, 2539 (1985).
77. J. Kouteck'y, P. Fantucci Chem. Rev. 86, 539 (1986) .
78. M.D. Morse Chem. Rev. in press.; W. Weltner, Jr., R.J. Van Zee Ann. Rev. Phys. Chem. 35, 291 (1984).

RECEIVED November 12, 1986

Chapter 4

Photofragmentation of Transition Metal Cluster Complexes in the Gas Phase

V. Vaida

Department of Chemistry and Biochemistry, University of Colorado, Boulder, CO 80309-0215

The gas phase photofragmentation of transition metal cluster complexes is discussed. The information available for the gas phase dissociation of $Mn_2(CO)_{10}$, $Co_3(CO)_9CCH_3$ and $Mo_2(O_2CCH_3)_4$ is compared to the well established data base on cluster compounds in condensed phase. The gas phase results are discussed in light of the predictions of electronic structure theory concerning the photofragmentation of these cluster compounds.

The study of metal containing compounds proved to be a difficult conceptual and technical task. To the experimentalist the difficulty comes from the fact that standard high resolution spectroscopic methods are often not applicable to the study of organometallic complexes. As experimental chemical physics added to its repertoire sensitive time and energy resolved spectroscopic probes, it opened the way for the exploration of metal containing molecules. At the same time, theoretical chemistry is developing methods able to handle many electron systems necessary for the study of transition metal complexes. The rich photochemistry of transition metal complexes has been used extensively to address questions concerning reactions of molecules in excited electronic states (1). These studies are motivated by expectations that such information will find practical application in the systematic design of organometallic reactions and catalytic processes (2,3,4).

To date, most of the photochemical data available for transition metal complexes comes from condensed phase studies (1). Recently, the primary photochemistry of a few model transition metal carbonyl complexes has been investigated in gas phase (5). Studies to date indicate that there are many differences between the reactivity of organometallic species in gas phase (5,6) as compared with matrix (7-10) or solution (11-17) environments. In most cases studied, photoexcitation of isolated transition metal

0097-6156/87/0333-0070$06.00/0
© 1987 American Chemical Society

complexes leads to extensive fragmentation, in contrast to the
outcome of a photochemical experiment in condensed phase (1,7-16).
    Presently, the gas phase photofragmentation of several
transition metal cluster complexes is reviewed. The techniques
employed for these gas phase studies rely on sensitive ionization
detection and the use of a broad range of excitation energies.
The information available is discussed in light of the effects of
excitation energy and the environment on the photofragmentation
process of several transition metal cluster complexes. The
photochemical information provides a data base directly relevant
to electronic structure theories currently used to understand and
predict properties of transition metal complexes (1,18,19).

Experimental Methods

A battery of sensitive techniques is being developed to probe the
photofragments resulting from photolysis of metal complexes in
collision free conditions. The aim is to characterize the energy
content, structure and chemistry of the photoproducts. These
methods rely on ultraviolet (UV) laser photolysis followed by
detection methods based on UV absorption (20), chemical trapping
(21,22), IR absorption (23,24,25) and ionization (5,6,26,27).
    In this paper, the photofragmentation of transition metal
cluster complexes is discussed. The experimental information
presented concerning the gas phase photodissociation of transition
metal cluster complexes comes from laser photolysis followed by
detection of fragments by ionization (5). Ion counting techniques
are used for detection because they are extremely sensitive and
therefore suitable for the study of molecules with very low vapor
pressures (6,26,27). In addition, ionization techniques allow the
use of mass spectrometry for unambiguous identification of signal
carriers.
    The technique of multiphoton ionization (MPI) detection of
photofragments generated on photolysis of transition metal cluster
carbonyls has been reviwed recently (5). The mechanism operative
in an MPI experiment is described as electronic excitation of a
molecule followed by further absorption of photons leading to
ionization. The ions produced are detected and mass analyzed.
There are many documented experiments on organic molecules (28) as
well as metal complexes such as chromyl chloride (29) and metal
dimethyls (30), that illustrate this process. If however, the
excited electronic state initially accessed is highly reactive,
dissociation competes favorably with the up-pumping necessary for
ionization and the processes involved in an MPI experiment are
those outlined in Figure 1. All transition metal carbonyls and
carbonyl clusters fall into this category as their electronic
states are highly dissociative. The processes involved are
initiated by excitation to a well defined electronic state of the
parent molecule. During the lifetime of the excited state, the
parent dissociates efficiently into neutral fragments. These
fragments can absorb more photons and either dissociate further or
ionize. As excitation to neutral fragments is the rate limiting
step in the overall process, this experiment yields spectroscopic

information about the neutral fragment when the laser is tuned
over a state of this fragment. Figure 2 illustrates this process
by showing the atomic spectrum of the Mn fragment (31) obtained on
dissociation of $Mn_2(CO)_{10}$. An invaluable feature of such an
experiment is that the ions formed can be mass selected and,
therefore, the signal carriers can be unambiguously identified.
Figure 3 shows the mass spectrum obtained by MPI of the
photofragments (32) generated on excitation of $Mn_2(CO)_{10}$. This
spectrum shows ions corresponding to the metal and bare metal
dimer, with no partially decarbonylated fragments observed.
Similar experiments on a large number of transition metal
carbonyls have shown that this process favors dissociation to and
detection of metal clusters or atoms. Since most metal-$(CO)_n$
photofragments are themselves subject to efficient dissociation,
MPI experiments do not identify the primary photoproducts. This
situation contrasts sharply with electron impact ionization where
the parent ion is usually formed and daughter ions are seen as a
result of parent ion fragmentation. Figure 4 shows the electron
impact mass spectrum of $Mn_2(CO)_{10}$ (33), for comparison with the
MPI mass spectrum of Figure 3.

In an alternate experiment, primary photoproducts generated
by photolysis of a transition metal cluster complex are ionized by
electron impact. A comparison of the electron impact ion signal
with the laser tuned to an electronic state of the molecule and
the signal with the laser off, yields information about the
primary photofragments (33,34). The fragments detected on
photolysis of $Mn_2(CO)_{10}$ at 337 nm are the partially decarbonylated
manganese complexes $Mn(CO)_5^+$, $Mn_2(CO)_4^+$ and $Mn_2(CO)_5^+$. While this
experiment yields more quantitative and direct information about
the nature of photoproducts than that obtained using the MPI
technique, it is less sensitive and cannot be used to obtain
electronic spectra of photofragments.

The details of both instruments as well as their relative
advantages and problems are discussed elsewhere (5,33).

Examples

$Mn_2(CO)_{10}$. The first metal-metal bond to be characterized (35) is
the formally single Mn-Mn bond in $Mn_2(CO)_{10}$. This compound has
often been used as the model for developing electronic structure
theories (1,18,36,37). Extremely efficient photofragmentation is
responsible for the structureless electronic spectrum and the lack
of emission following excitation of this molecule. This
spectroscopic deficiency necessitates photofragmentation studies
to obtain data to verify theoretical models. Most of the
photochemical experiments in the past explored the reactions of
the lowest excited singlet state in the near ultraviolet.
According to the simple one-electron electronic structure
description, excitation into this strong $\sigma \to \sigma^*$ band should lead
to a reduction of the metal-metal bond order to zero resulting in
homolytic cleavage into $M(CO)_5$ units. While condensed phase
experiments did at first appear to agree with this prediction
(1,14,35-39), more recent time-resolved studies have painted a

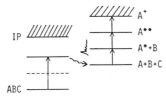

ⓐ MPE with efficient photochemistry
   taking place during the lifetime
   of the excited state

ⓑ Photodissociation to the neutral
   fragments

ⓒ MPE of neutral fragments

ⓓ Ionization of the excited neutral
   fragments

Figure 1. Schematic of a multiphoton ionization experiment for molecules with reactive excited electronic states.

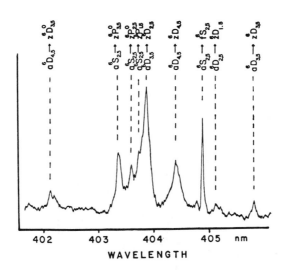

Figure 2. Electronic spectrum (32) of the Mn photofragment obtained on photolysis of $Mn_2(CO)_{10}$ in gas phase.

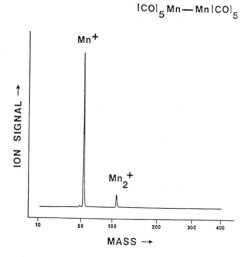

Figure 3. Photofragment MPI mass spectrum (<u>32</u>) obtained after gas phase photolysis of $Mn_2(CO)_{10}$.

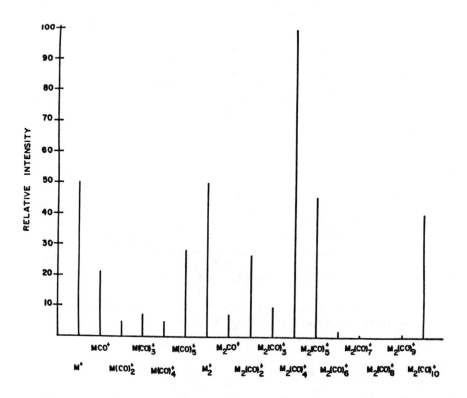

Figure 4. Electron impact mass spectrum (<u>33</u>) of $Mn_2(CO)_{10}$.

more complex picture (12-14,40). It has now been shown that in addition to metal-metal bond cleavage, CO loss and isomerization are important photochemical pathways in condensed phase.

The primary photofragmentation pathways of this molecule are more directly explored in the gas phase in the absence of any solvent or matrix effects. While early gas phase studies indicated exclusive metal-metal bond cleavage (30), recent experiments in our laboratory have shown evidence for both metal-metal and metal-CO cleavage on excitation into the $\sigma \to \sigma^*$ band (32,41). Excitation by a low flux N_2 laser (337nm) of the lowest singlet state of $Mn_2(CO)_{10}$ leads to dissociation which is probed by MPI of the ensuing fragments. The resulting mass spectrum is shown in Figure 3 and consists of Mn^+ and Mn_2^+ with no carbonylated species detectable. The lack of carbonylated fragments in this study agrees with results on all other metal carbonyl complexes and the expectation of high dissociation yields for most initial photofragments. These results indicate that excitation of the $\sigma \to \sigma^*$ band of this compound leads to both homolytic cleavage of the metal bond and ligand loss with retention of the metal bond.

Alternatively, the primary photofragments of $Mn_2(CO)_{10}$ are probed by electron impact ionization of fragments formed on photolysis (33). Excitation into the same near UV band leads to generation of $Mn(CO)_5^+$, $Mn_2(CO)_4^+$ and $Mn_2(CO)_5^+$. This experiment was also performed by excitation at 250 nm and 193 nm, of two higher electronic states of the molecule.

Table I gives the mass fragments observed by electron impact ionization at each one of these three excitation energies. The results confirm the fact that both metal-metal and metal-CO cleavage occurs in the low energy band, while excitation into higher excited states enhances the metal-CO channel at the expense of the Mn-Mn bond homolysis.

Table I. Mass Fragments Observed by Electron Impact Ionization as a Function of Photolysis Excitation Energy

Wavelength	ε	Assignment	Fragments
350 nm	33,700	$\sigma \to \sigma^*$	$Mn(CO)_5^+$ $Mn_2(CO)_4^+$ $Mn_2(CO)_5^+$
249 nm	8,200	$d\pi \to \pi^*$	$Mn_2(CO)_4^+$ $Mn_2(CO)_5^+$
193 nm	84,400	$M \to \pi^*$	$Mn_2(CO)_5^+$

$\underline{Co_3(CO)_9CCH_3}$. The Methinyltricobaltenneacarbonyls, $Co_3(CO)_9CX$, are built on the tetrahedral Co_3C unit which gives these compounds unusual stability and unique properties. Interest in these molecules is enhanced by their probable catalytic properties (<u>42,43</u>).

The solution photochemistry of these complexes has been studied extensively (<u>44,45</u>). The photoproducts observed in solution are species like $Co_4(CO)_{12}$ and $Co_2(CO)_8$. These photoproducts are consistent with a mechanism involving declustrification of the Co_3C core.

The photochemistry of a representative molecule of this class, $Co_3(CO)_9CCH_3$, was investigated in gas phase in our laboratory using laser photolysis followed by MPI detection of the photofragments (<u>41</u>). Figure 5 shows the photofragment mass spectrum of this compound obtained by MPI with photolysis at 450 nm and 337 nm.

The fragments observed on excitation of $Co_3(CO)_9CCH_3$ at 337 nm are Co_3CCH^+, Co_2CCH^+, $CoCCH^+$, Co_3C^+, Co_2C^+, CoC^+, Co_3^{+3}, Co_2^+ and Co^+ and at 450 nm are CoC^+, Co_2^+ and Co^+. In keeping with all other metal carbonyl molecules studied by MPI, it is assumed that these clusters are created as neutral species in the photofragmentation process and are ionized in the final MPI step. Unsaturated carbonyl fragments cannot be seen directly in this experiment since the remaining carbonyl ligands photodissociate efficiently.

For the photolytic wavelength of 450 nm, we observe extensive destruction of the Co_3 core with only the smallest cobalt fragments Co^+, CoC^+ and Co_2^+ being observed. This is consistent with the anti-bonding description for the electronic state involved and the interpretation of the solution-photchemistry experiment. However, when the photolysis wavelength is changed to 337 nm, fragments such as Co_2^+ and Co_2CCH^+ predominate in the mass spectrum, with fragments containing the Co_3C unit also being present. This higher degree of retention of the Co_3C central core suggests a photolytic event destabilizing the ligand-core structure. This result contrasts with the solution experiment in which no marked wavelength dependence was observed.

In conclusion, the gas phase photofragmentation of $Co_3(CO)_9CCH_3$ shows a wavelength dependence that, with the proper selection of wavelength, gives the experimentalist access to a range of interesting unsaturated metal and metal-carbon clusters.

$\underline{Mo_2(O_2CCH_3)_4}$. Metal compounds with multiple metal-metal bonds such as $Mo_2(O_2CCH_3)_4$ of D_{4h} symmetry, have attracted much experimental and theoretical attention focussed on the description of bonding and bond strength (<u>46-48</u>). Their electronic structure has been investigated experimentally by various methods such as resonance Raman, photoelectron spectroscopy, ultraviolet absorption and polarization studies of the matrix isolated sample (<u>49-56</u>).

We have undertaken a gas phase photodissociation study of $Mo_2(O_2CCH_3)_4$, whose results are viewed in light of the one electron electronic structure models available for this compound

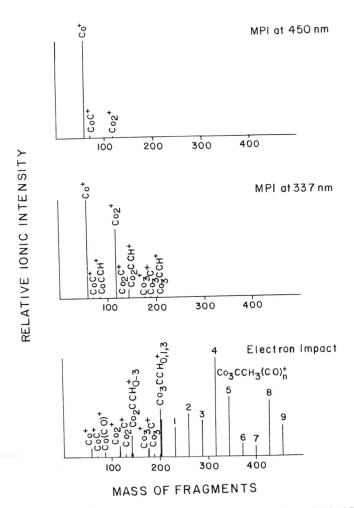

Figure 5. (a) Photofragment MPI mass spectrum of $Co_3(CO)_9CCH_3$ at 450 nm (**41**); (b) Photofragment MPI mass spectrum of $Co_3(CO)_9CCH_3$ at 337 nm (**41**).

(<u>41</u>). The spectrum of this compound displays a structured band at
437 nm. The vibrational structure was extensively studied and
assigned to vibronic levels of a $\delta \to \pi^*$ and a $\delta \to \delta^*$ state with
activity in the Mo-Mo and Mo-O stretching modes (<u>49-56</u>). The MPI
photofragment spectrum following excitation at 450 nm is dominated
by Mo^+ and MoO^+. In contrast, excitation at 337 nm accesses a
structureless band assigned to the $\delta \to \pi^*$ (carbon) transition.
Photofragmentation is dominated by $Mo_2O_{1-4}^+$ with a weaker MoO_{1-3}^+
series present. The data suggests a higher degree of retention of
the metal-metal bond at 337 nm than at 450 nm. This result is
consistent with the description of the electronic states as the
lower energy visible band leads to intense activity in the Mo-Mo
stretching mode.

Acknowledgments

Contributions to this work by D.G. Leopold, W.E. Hollingsworth,
and D.A. Prinslow are gratefully acknowledged. Thanks for help
with parts of these experiments go to S.P. Sapers and G.A.
Gaines.

This work is supported by grants from the NSF (CHE8318605)
and NIH (GM33335) and the University of Colorado Biomedical Fund.

Literature Cited

1. Geoffroy, G. L.; Wrighton, M. S. "Organometallic
    Photochemistry"; Academic: New York, 1979
2. Masters, C. "Homogeneous Transition Metal Catalysis"; Chapman
    and Hall: New York, 1981
3. Moggi, L.; Juris, A.; Sandrini, D.; Manfrin, M. F. <u>Rev. Chem.
    Intermed.</u> 1981, 4, 171
4. Wrighton, M. S.; Ginley, D. S.; Schroeder, M. A.; Moise,
    D. L. <u>Pure. Appl. Chem.</u> 1975, 41, 671
5. Hollingsworth, W. E.; Vaida, V. <u>J. Phys. Chem.</u> 1986, 90, 1235
6. Gedanken, A.; Robin, M. B.; Kuebler, N. A. <u>J. Phys. Chem.</u>
    1982, 86, 4096
7. Perutz, R. N.; Turner, J. J. <u>J. Am. Chem. Soc.</u> 1975, 97, 4791
8. Burdett, J. K.; Graham, M. A.; Perutz, R. N.; Poliakoff, M.;
    Rest, A. J.; Turner, J. J.; Turner, R. F. <u>J. Am. Chem. Soc.</u>
    1975, 97, 4805
9. Perutz, R. N.; Turner, J. J. <u>J. Am. Chem. Soc.</u> 1975, 97,4800
10. Perutz, R. N.; Turner, J. J. <u>Inorg. Chem.</u> 1975, 14, 262
11. Kelly, J. M.; Bent, D. V.; Hermann, H.; Schulte-Frohlinde,
    D.; von Gustorf, E. K. <u>J. Organomet. Chem.</u> 1974, 69, 259
12. Rothberg, L. J.; Cooper, N. J.; Peters, K. S.; Vaida, V.
    <u>J. Am. Chem. Soc.</u> 1982, 104, 3536
13. Hepp, A. F.; Wrighton, M. S. <u>J. Am. Chem. Soc.</u> 1983, 105,
    5934
14. Fox, A.; Poe, A. <u>J. Am. Chem. Soc.</u> 1979, 102, 2498
15. Hughey IV, J. L.; Anderson, C. P.; Meyer, T. J. <u>J. Organomet.
    Chem.</u> 1977, C49, 125
16. Sweany, R. L.; Brown, T. L. <u>Inorg. Chem.</u> 1977, 16, 421
17. Meyer, T. J.; Caspar, J. V. <u>Chem. Rev.</u> 1985, 85, 187

18. Hoffmann, R. Science 1981, 211, 995
19. Hirst, D. M. Adv. Chem. Phys. 1982, 50, 517
20. Breckenridge, W. H.; Sinai, M. J. J. Phys. Chem. 1981, 85, 3557
21. Nathanson, G.; Gitlin, B.; Rosan, A. M.; Yardley, J. T. J. Chem. Phys. 1981, 74, 361
22. Yardley, J. T.; Gitlin, B.; Nathanson, G.; Rosan, A. M. J. Chem. Phys. 1981, 74, 370
23. Ouderick, A. J.; Werner, P.; Schultz, N. L.; Weitz, E. J. Am. Chem. Soc. 1983, 105, 3354
24. Seder, T. A.; Church, S. P.; Ouderick, A. J.; Weitz, E. J. Am. Chem. Soc. 1985, 107, 1432
25. Fletcher, T. R.; Rosenfeld, R. N. J. Am. Chem. Soc. 1983, 105, 6358
26. Johnson, P. M. Acc. Chem. Res. 1980, 13, 20
27. Bernstein, R. B. J. Phys. Chem. 1982, 86, 1178
28. Gobeli, D. A.; Yang, J. J.; El-Sayed, M. A. Chem. Rev. 1985, 85, 529
29. Wheeler, R. G.; Duncan, M. A. J. Phys. Chem. 1986, 90, 1610
30. Yu, C. F.; Youngs, F.; Tsukiyama, K.; Bersohn, R.; Press, J. J. Chem. Phys. in press
31. Rothberg, L. J.; Gerrity, D. P.; Vaida, V. J. Chem. Phys. 1981, 74, 2218
32. Leopold, D. G.; Vaida, V. J. Am. Chem. Soc. 1984, 106, 3720
33. Prinslow, D. A.; Vaida, V. to be published
34. Freedman, A.; Bersohn, R. J. Am. Chem. Soc. 1978, 100, 4116
35. Wrighton, M. S.; Graff, J. L.; Luong, J. C.; Reichel, C. L.; Robins, J. L. "Reactivity of Metal-Metal Bonds" Chisholm, M. H.; Ed.; ACS: Washington, D.C., 1981
36. Levenson, R. A.; Gray, H. B.; Ceasar, G. P. J. Am. Chem. Soc. 1970, 92, 3653
37. Levenson, R. A.; Gray, H. B. J. Am. Chem. Soc. 1975, 97, 6042
38. Wegman, R. W.; Olsen, R. J.; Gard, D. R.; Faulkner, L. R.; Brown, T. L. J. Am. Chem. Soc. 1981, 103, 6089
39. Waltz, W. L.; Hackelberg, O.; Dorfman, L. M.; Wojcicki, A. J. Am. Chem. Soc. 1978, 100, 7259
40. Yesaka, H.; Kobayashi, T.; Yasufuku, K.; Nagakura, S. J. Am. Chem. Soc. 1983, 105 6249
41. Hollingsworth, W. E.; Vaida, V. to be published
42. Seyferth, D. Adv. Organomet. Chem. 1976, 14, 97
43. Penfold, B. R.; Robinson, B. H. Accts. Chem. Res. 1973, 6, 73
44. Geoffroy, G. L.; Epstein, R. E. Inorg. Chem. 1977, 16, 2795
45. Geoffroy, G. L.; Epstein, R. E. Adv. Chem. Ser. 1978, 168, 132
46. Cotton, F. A. Accts.Chem.Res. 1978, 11, 225
47. Chisholm, M. H.; Cotton, F. A. Accts. Chem. Res. 1978, 11, 356
48. Johnson, B. F. G. "Transition Metal Clusters"; John Wiley: New York, 1979
49. Manning, M. C.; Holland, G. F.; Ellis, D. E.; Trogler, W. C. J. Phys. Chem. 1983, 87, 3083
50. Martin, R. S.; Newman, R. A.; Fanwick, P. E. Inorg. Chem. 1979, 18, 2511

51. Fichtenberger, D. L.; Blevins, C. H. J. Am. Chem. Soc. 1984,
     106, 1636

52. Trogler, W. C.; Solomon, E. I.; Trajberg, I. B.; Ballhausen,
     C. J.; Gray, H. B. Inorg. Chem. 1977, 16, 828

53. Trogler, W. C.; Gray, H. B. Acc. Chem. Res. 1978, 11, 232

54. Manning, M. C.; Gray, H. B. Inorg. Chem. 1982, 21, 2797

55. Norman Jr., J. G.; Kolari, H. J.; Gray, H. B.; Trogler, W. C.
     Inorg. Chem. 1977, 16, 987

56. Dubicki, L.; Martin, R. L. Aust. J. Chem. 1969, 22, 1571

RECEIVED November 12, 1986

Chapter 5

Coordinatively Unsaturated Metal Carbonyls in the Gas Phase via Time-Resolved Infrared Spectroscopy

Tom Seder[1], Andrew Ouderkirk[2], Stephen Church[3], and Eric Weitz

Department of Chemistry, Northwestern University, Evanston, IL 60201

The spectroscopy, reaction kinetics, and photophysics of coordinatively unsaturated metal carbonyls generated in the gas phase via UV photolysis are probed via transient infrared spectroscopy. The parent compounds that have been used to generate coordinatively unsaturated species are $Fe(CO)_5$, $Cr(CO)_6$ and $Mn_2(CO)_{10}$. In contrast to what is observed in solution phase, photolysis of these compounds produces a variety of coordinatively unsaturated photoproducts. The rate constants for addition of CO to $Fe(CO)_x$ (x=2,3,4), $Cr(CO)_x$ (x=2,3,4,5) and $Mn_2(CO)_9$ are reported as is the rate constant for the reaction of two $Mn(CO)_5$ radicals to form $Mn_2(CO)_{10}$. The reasons for differences in magnitudes of the measured rate constants are discussed in terms of spin conservation and the nature of the reaction: whether it is an addition reaction or a displacement reaction. Spectra of all of the above species have been recorded and absorption peaks are assigned to specific vibrational modes. The spectra are generally compatible with structures for these species deduced from matrix isolation studies of the compounds. Coordinatively unsaturated fragments are observed to be formed with significant internal excitation. This observation and the trend toward an increase in the degree of coordinative unsaturation of the photofragments with increasing photolysis energy allows for the formulation of a proposed mechanism for photodissociation in these compounds.

[1]Current address: Physical Chemistry Department, General Motors Research Laboratories, Warren, MI 48090
[2]Current address: 3M, 3M Center Road, Central Research, Building 208-01-01, Process Technology, St. Paul, MN 55144
[3]Current address: Max Planck Institut für Strahlenchemie, D-4330, Mülheim a.d. Rühr, Federal Republic of Germany

0097-6156/87/0333-0081$06.00/0
© 1987 American Chemical Society

Over the last decade the spectroscopy, photochemistry and reactivity of metal carbonyls has been a subject of intense interest. As a result of this research it has been found that metal carbonyls undergo a wide range of facile photochemical reactions [1,2]. However, the pathways for these reactions, particularly in the gas phase, have been only partially characterized. In a wide variety of these reactions, coordinatively unsaturated, highly reactive metal carbonyls are produced [1-18]. The products of many of these photochemical reactions act as efficient catalysts. For example, $Fe(CO)_5$ can be used to generate an efficient photocatalyst for alkene isomerization, hydrogenation, and hydrosilation reactions [19-23]. Turnover numbers as high as 3000 have been observed for $Fe(CO)_5$ induced photocatalysis [22]. However, in many catalytically active systems, the active intermediate has not been definitively determined. Indeed, it is only recently that significant progress has been made in this area [20-23].

Much of the difficulty in characterizing either the metal carbonyl photoproducts or reaction intermediates stems from their exceedingly high reactivity. For example, it has been shown that $Cr(CO)_5$ coordinates a hydrocarbon solvent within a few picoseconds after it has been produced [24]. However, coordinatively unsaturated metal carbonyls have been spectroscopically observed in a variety of elegant studies involving photolysis in inert gas matrices, liquids, and hydrocarbon glasses [1,2,25-30]. Via prolonged photolysis in matrices, initial photoproducts can be induced to lose additional ligands leading to the production of a variety of coordinatively unsaturated species. Spectroscopic studies of these species have been very valuable in determining information on structure and bonding in this class of compounds. However, matrix isolation studies have their limitations. Because of the nature of the technique it is difficult to obtain kinetic information. Performing studies in "liquid matrices", solutions of compounds in noble gas liquids, alleviates this problem to some extent but this technique is also limited in terms of the solvents and temperature ranges that are accessible [26]. In addition, matrix studies have always raised the spectre of "matrix effects" altering the geometry of matrix isolated molecules versus molecules in the gas phase or even in solution.

Despite the considerable amount of information that has been garnered from more traditional methods of study it is clearly desirable to be able to generate, spectroscopically characterize and follow the reaction kinetics of coordinatively unsaturated species in real time. Since desired timescales for reaction will typically be in the microsecond to sub-microsecond range, a system with a rapid time response will be required. Transient absorption systems employing a visible or UV probe which meet this criterion have been developed and have provided valuable information for metal carbonyl systems [14,15,27]. However, since metal carbonyls are extremely photolabile and their UV-visible absorption spectra are not very structure sensitive, the preferred choice for a spectroscopic probe is time resolved infrared spectroscopy. Unfortunately, infrared detectors are enormously less sensitive and significantly slower than phototubes, thus time resolved infrared techniques have historically been plagued by a lack of speed and/or sensitivity. These problems can be somewhat overcome by a study of reactions in solution where much greater densities are possible than in the gas phase and fast bimolecular reaction are diffusion limited [1,28,29]. However, since coordinatively unsaturated metal carbonyls have shown a great affinity for coordinating solvent we felt that the appropriate place to begin a study of the spectroscopy and kinetics of these species would be in a phase where there is no solvent; the gas phase. In the gas phase, the observed spectrum is expected to be that of the "naked" coordinatively unsaturated species and reactions of these species with added ligands are addition reactions rather than displacement reactions. However, since many of the saturated metal carbonyls have limited vapor pressures, the gas phase places additional constraints on the sensitivity of the transient spectroscopy apparatus.

Nevertheless, we were able to develop a transient absorption apparatus involving IR probe radiation that is suitable for gas phase studies, as have a number of other groups either coincident with or subsequent to our work [1]. In the remainder of this article we will discuss the apparatus and the results of our studies on three prototypical metal carbonyl species; $Fe(CO)_5$, $Cr(CO)_6$ and $Mn_2(CO)_{10}$. The discussion in this article will center on the nature of the photolytically generated coordinatively unsaturated species, their kinetic behavior and photophysical information regarding these species. This latter information has enabled us to comment on the mechanism for photodissociation in these systems. Since most of the results that will be discussed have been presented elsewhere [3-10], we will concentrate on a presentation of data that illustrates the most important features that have come out of our research and directly related research regarding the kinetics, photophysics and photochemistry of coordinatively unsaturated metal carbonyls.

Experimental

The apparatus used for our transient absorption measurements has been described in detail elsewhere [3-10]. Briefly, the output of an excimer laser operating on either XeF, KrF or ArF makes a single pass through the photolysis cell after being directed through a cylindrical BaF_2 lens which produces a more homogeneous beam. The low (less than $5mj/cm^2$) energy pulses are made to fill the entire cell volume. This procedure is necessary to avoid spurious signals resulting from both temperature and photoproduct inhomogeneities. The glass photolysis cell has a radius of .75 cm and an active length of 10 cm. The CaF_2 windows of the cell are protected from photoproducts by a curtain of rare gas which flows over the windows and out the exhaust ports without mixing with the sample gases. After passing through computer controllable flow controllers and a water jacketed column, which can be used for temperature control, the sample gases enter the cell through a centrally located port and are pumped out of two symmetrically located exhaust ports. Photoproducts are observed to be deposited only in the region between the two exhaust ports. The sample gases are flowed at a rate such that they are replaced between the 1Hz excimer laser

pulses. The sample gas mixture contains a small quantity of metal
carbonyl parent (always less than 200 mtorr and typically less than
30 mtorr) to which is added variable quantities of inert gas and/or
reactant gases. The inert gas increases the overall heat capacity
of the cell which results in an attenuation of potential shock waves
induced via the UV pulse. It also acts to diminish the rate of
diffusion of material out of the region of the probe beam and acts
as a third body in recombination reactions. For the latter purpose,
the rare gas or reactant gas pressure is always kept high enough so
that third order recombination reactions are in a pseudo second
order regime. Thus the rate constants we measure do not depend on
the pressure of added gas and we do not see curvature in our plots
of rate of reaction versus added CO [3-10].

The transient species produced via UV photolysis are monitored
via the output of a home-built, liquid nitrogen cooled, line
tunable, carbon monoxide laser. This laser is capable of operating
on low quantum number vibrational transitions of CO including 1-0.
The c.w. infrared beam, after making a double pass through the flow
cell fills the entire area of the element of an indium antimonide
detector. For wavelength determination, the IR beam is split and a
portion is passed through a 0.5m monochromator equipped with a 10 μm
grating and calibrated for use in second order via a HeNe laser.

To obtain maximum linearity and detectivity, the photovoltaic
indium antimonide detector was equipped with a variable back biasing
circuit which allows operation of the detector at the origin of the
i-v curve of the diode. The output of the detector is amplified, fed
through a unity gain buffer amplifier and ultimately digitized with
a Biomation 8100 transient digitizer. Typically 64 waveforms are
averaged via simple addition using a Nicolet 1170 signal averager.
The resulting signals are stored on a Nova/4 minicomputer which is
in communication with a Harris super-minicomputer. The Harris
computer is used for signal analysis via a non-linear least squares
routine. The measured response time of the detector and associated
electronics is 35 nsec.

Transient spectra were constructed by recording transients at
desired wavelengths, all of which were normalized by the probe
energy, and having the computer assemble spectra by connected points
on each transient at a common delay time following the photolysis
pulse. Kinetic information was obtained by monitoring transients at
the desired wavelengths as a function of reactant gas pressure
and/or cell temperature. Unless otherwise stated all experiments
were carried out at 21 $\pm$ 1° C.

The parent metal carbonyls were obtained from Alpha Chemicals
at stated purities of >98%, 98%, and 95% for Fe, Cr and Mn,
respectively. In all cases volatile impurities were removed before
use. CO and rare gases were obtained from Matheson at stated
purities of >99.99+% and were used without further purification.

Results

Figure 1 depicts the time resolved spectrum generated by photolysis
of 30mtorr of Fe(CO)$_5$ with a KrF excimer laser. As with all
transient absorption experiments a major potential problem involves
assigning the observed absorptions to specific species. For the

$Fe(CO)_5$ system we were significantly aided by previous matrix work
on this system which assigned the infrared absorptions of $Fe(CO)_4$
and $Fe(CO)_3$ and by chemical trapping studies, which provided us with
information on which photofragments are produced at various
photolysis wavelengths [17,30,31]. However, it is undesirable to
have to rely on the prior existence of matrix and/or chemical
trapping data for assignments of absorptions. Thus we developed a
rather straightforward procedure which we refer to as a "kinetic
bootstrap" procedure which allows us to determine the nature of
specific absorption bands. This procedure is illustrated in figure 2
which shows the same spectral region as figure 1 but for KrF laser
photolysis of $Fe(CO)_5$ in the presence of a large excess of CO. On
this time scale the added CO has already reacted with lower
fragments to regenerate $Fe(CO)_4$, which can be observed to further
react with CO on the timescale that is depicted, to generate $Fe(CO)_5$
[8]. This "kinetic bootstrap" procedure will be further illustrated
with its application to the $Cr(CO)_6$ and $Mn_2(CO)_{10}$ systems (vide
infra). Using existing matrix data, chemical trapping data and our
"kinetic bootstrap" procedure, we have been able to assign the
absorption designated a,b,c in figure 1 to $Fe(CO)_x$ where x = 4,3,2
respectively. Feature d is due to depletion of parent and feature e,
as will be discussed in more detail , is due to vibrationally
excited CO formed in the photodissociation process [3,8].

Once specific absorption features are assigned, kinetic studies
can be performed via tuning the probe laser to a frequency absorbed
by the fragment whose reaction kinetics are of interest. Ideally,
it is also desirable to measure the rate of formation of the
reaction product and to verify that these two rates correlate with
each other. This has been done for the $Fe(CO)_x$ system with added CO
where the reaction can be schematically depicted as

$$Fe(CO)_x + CO \longrightarrow Fe(CO)_{x+1} \tag{1}$$

A typical example of the data is shown in figure 3 where the rate of
reaction of $Fe(CO)_3$ with added CO is depicted. Data is presented
for both the loss of $Fe(CO)_3$ and the regeneration of $Fe(CO)_4$. Data
for reaction of $Fe(CO)_3$ and the other $Fe(CO)_x$ species is presented
in table I [3,4,8].

Similar data have been obtained for the $Cr(CO)_6$ system and are
also presented in table I [5,9,12,13]. Time resolved spectra for
this system are shown in figure 4. Spectra are presented for XeF,
KrF and ArF photolysis. This system affords an excellent example of
the application of our kinetic procedure for assigning absorptions
to specific coordinatively unsaturated photofragments. Observation
of figure 4 clearly indicates that as photolysis energy is increased
additional absorption bands appear, primarily at lower frequency
than those observed on lower energy photolysis. Both matrix data
and chemical trapping data strongly imply that the primary species
produced on XeF laser photolysis is $Cr(CO)_5$ [16,32]. Figure 5
illustrates how other absorptions can be assigned via our kinetic
procedure. This figure depicts the change in the absorption
spectrum on addition of CO to a sample of $Cr(CO)_6$ that has been
photolyzed with a KrF excimer laser. From this figure it can be

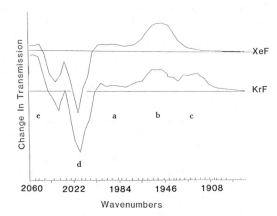

Figure 1. Transient IR spectra following photolysis of $Fe(CO)_5$
with XeF and KrF laser radiation. Traces are taken ~1 μsec after
photolysis. In addition to $Fe(CO)_5$ (30 mtorr for KrF, 200 mtorr
for XeF) the photolysis cell contained 5 torr Ar. The symbols
are defined in the text.

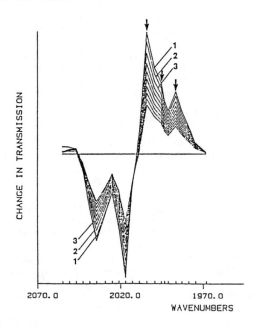

Figure 2. Transient IR spectra following KrF photolysis of 30
mtorr of $Fe(CO)_5$ + 100 torr CO. The spectrum (~2060-1920 cm^{-1})
is depicted over a 5 μs time range which has been segmented into
10 equal time intervals, the first three of which are designated.
The arrows indicate the now partially resolved A_1 and B_1 modes of
$Fe(CO)_4$. (Reproduced with permission from reference 8.
Copyright 1986 American Institute of Physics.)

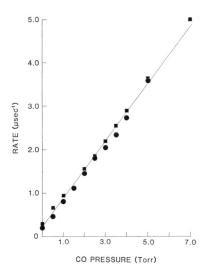

Figure 3. Plot of the pseudo first order rate for reaction of Fe(CO)$_3$ with CO. The rate of disappearance of Fe(CO)$_3$ at 1954 cm^{-1} (■) and the rate of appearance of Fe(CO)$_4$ at 1984 cm^{-1} (●) are plotted against pressure of added CO. (Reproduced with permission from reference 8. Copyright 1986 American Institute of Physics.)

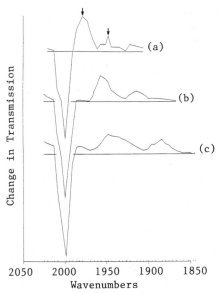

Figure 4. The transient absorption spectrum of Cr(CO)$_6$ 0.5 msec following a) XeF, b) KrF and c) ArF photolysis of gas phase Cr(CO)$_6$. The arrows in a) indicate the Cr(CO)$_5$ absorptions. (Reproduced with permission from reference 9. Copyright 1986 American Chemical Society.)

Table I

Summary of the Bimolecular Rate Constants for $M(CO)_x$-CO
Recombination Reactions

	Spin Allowed	Gas Phase Rate Constant $(10^{-13}$ cm^{3}mol^{-1}s$^{-1})$
$Cr(CO)_5 + CO \rightarrow Cr(CO)_6$	Y	1.5,[a,e] 2.2,[b]
$Cr(CO)_4 + CO \rightarrow Cr(CO)_5$	Y	2.4,[a,e] 2.6,[b] 14.0[c]
$Cr(CO)_3 + CO \rightarrow Cr(CO)_4$	Y	1.8[e]
$Fe(CO)_4 + CO \rightarrow Fe(CO)_5$	N	0.003[d]
$Fe(CO)_3 + CO \rightarrow Fe(CO)_4$	Y	1.2[d]
$Fe(CO)_2 + CO \rightarrow Fe(CO)_3$	Y	1.7[d]

[a]Ref. 5.
[b]Ref. 12.
[c]Ref. 13.
[d]Refs. 3, 8.
[e]Ref. 9.

seen that the initial absorption spectrum on KrF laser photolysis evolves into a spectrum very similar to that on XeF laser photolysis. This spectrum continues to evolve with the $Cr(CO)_5$ absorptions disappearing and $Cr(CO)_6$ being regenerated [9]. Measurements of the rate of change of the absorption of different peaks in the KrF and XeF spectra further indicates that the major peaks in each spectra all evolve at the same rate. This indicates that there is <u>primarily</u> one coordinatively unsaturated species produced at each of these wavelengths. Since the rate of loss of the species produced on KrF laser photolysis is the same as the rate of production of $Cr(CO)_5$ and $Cr(CO)_5$ is produced without an induction time, this strongly implies that by far the major product of KrF laser photolysis is $Cr(CO)_4$. This is consistent with chemical trapping data [16]. In addition, the fact that the loss of $Cr(CO)_5$ on reaction with CO leads to regeneration of $Cr(CO)_6$ at the same rate as $Cr(CO)_5$ is lost is further confirmation of our assignments. A similar procedure allows us to assign the additional peaks produced on ArF photolysis as being due to $Cr(CO)_3$ and $Cr(CO)_2$ [9]. The frequencies of the gas phase absorptions for these coordinatively unsaturated fragments is presented in table II. Data for the rates of reaction of each of these species with CO is summarized in table I.

The time resolved spectra produced on excimer laser photolysis of $Mn_2(CO)_{10}$ are shown in figure 6. Note that as in the case of iron pentacarbonyl and chromium hexacarbonyl photolysis, there is a distinct increase in the amplitude of the lower frequency absorption bands as the photolysis energy increases. By comparison with the frequency of matrix isolated and solution phase $Mn(CO)_5$, the band at ~1996 cm^{-1} is assigned to the gas phase $Mn(CO)_5$ radical [33]. This

Table II
Infrared Absorptions of Matrix Isolated and Gas Phase
$Fe(CO)_x + Cr(CO)_x$

| | Frequency (cm^{-1}) | | | | |
	Ar Matrix	Gas Phase[a]	Assignment	Symmetry	Ref.
Fe(CO)$_4$	1995	2000	A_1	C_{2v}	31
	1988		B_1		
	1973	1985	B_2		
Fe(CO)$_3$	~2042	(b)	A_1	C_{3v}	31
	1935.6	1950	E		
Fe(CO)$_3$[c]		1957	B_1	C_{2v}	
		1945	A_1		
Fe(CO)$_2$	1905[d]	1920			
Cr(CO)$_5$	2093.4		A_1	C_{4v}	39
	1965.4	1980	E		
	1936.1	1948	A_1		
Cr(CO)$_4$	1938		B_1 }	C_{2v}	40
	1932	1957	A_1 }		
	1891	1920	B_2		
Cr(CO)$_3$	1867	1880	E	C_{3v}	32c
Cr(CO)$_2$	1903[e]	1914	?	?	32c

a) Approximate values from references 8 and 9.
b) Not observed due to overlap with "hot" CO absorptions -- see text.
c) Tentative assignment of the bands observed upon ArF laser photolysis to the excited singlet state Fe(CO)$_3$. See reference 8.
d) Tentatively assigned as Fe(CO)$_2$ in reference 32c.
e) CH$_4$ matrix.

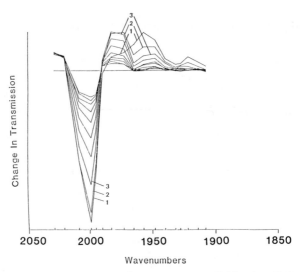

Figure 5. Transient time resolved spectrum following KrF photolysis of Cr(CO)$_6$ with 5.0 torr Ar and 0.5 torr CO. The spectrum is displayed over a 10 μs range which is segmented 10 equal time intervals. The first 3 intervals are labelled (Reproduced with permission from reference 9. Copyright 198 American Chemical Society.)

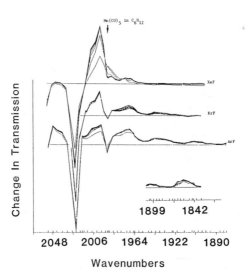

Figure 6. Transient time resolved spectra following excimer laser photolysis of Mn$_2$(CO)$_{10}$. The position of the Mn(CO)$_5$ absorption in C$_6$H$_{12}$ solution is indicated by an arrow. The at the bottom is an extension of the ArF spectrum.

assignment is further confirmed by kinetic studies. Observation of the rate of decay of this band indicates that its decay is second order (see figure 7). From this plot, I/I_0 measurements, and an estimated value for ε, the $Mn(CO)_5$ absorption coefficient, based on the strength of the CO absorption bands in $Mn(CO)_5Cl$ [33], a value for the rate constant for equation (2)

$$Mn(CO)_5 + Mn(CO)_5 \longrightarrow Mn_2(CO)_{10} \qquad (2)$$

has been determined to be $(2.7 \pm 0.6) \times 10^{10} \ \ell \ mole^{-1}s^{-1}$ [6,10]. Error limits refer to experimental uncertainties and do not include uncertainties in the choice of ε.

Figure 8 displays the transient absorption spectrum for $Fe(CO)_5$ photolyzed by a KrF laser on a shorter timescale than that displayed in figure 1. Note that on this timescale the absorptions ascribed to the $Fe(CO)_x$ features do not all develop at the same rate [3,8]. The absorption band assigned to $Fe(CO)_2$ clearly appears more rapidly than the absorption band for $Fe(CO)_3$ which in turn appears more rapidly than the absorption feature for $Fe(CO)_4$. The more rapid appearance of absorption bands belonging to the more highly coordinatively unsaturated photofragments is a general feature observed in all the photolysis experiments we have conducted to date. Its ramifications in terms of the photophysics of the processes we have studied will be discussed in more detail in the next section. It is interesting to note that the fact that more highly coordinatively unsaturated species appear more promptly than less coordinatively unsaturated species and with less shifting of the absorption line to higher energy can be used as a further aid in the assignment of absorption bands to specific species [9].

Discussion

Kinetics. Inspecting table I, it can be seen that the rate of reaction for CO addition to all $Fe(CO)_x$ and $Cr(CO)_x$ species is of the same order of magnitude except for the rate of reaction of $Fe(CO)_4$ with CO. Why is this reaction different than all the other reactions? The answer to this question can be found in studies of the electronic structure of $Fe(CO)_4$. The ground state of $Fe(CO)_4$ is a triplet whereas the ground state of $Fe(CO)_5$ is a singlet [34,35]. Thus the addition reaction of CO to $Fe(CO)_4$ is spin forbidden. This has further implications for the $Fe(CO)_x$ system. If a spin forbidden reaction is expected to be significantly slower than a spin conserving reaction, then the ground states of $Fe(CO)_3$ and $Fe(CO)_2$ are also triplets. This prediction has been previously made for $Fe(CO)_3$ and we have postulated, based on our kinetic data, that $Fe(CO)_2$ has a triplet ground state [8]. Note that in the $Cr(CO)_x$ system, the rates of reaction of all the coordinatively unsaturated fragments with CO are very similar. This is in accord with what is known about the electronic structure of the $Cr(CO)_x$ system: the ground state of each of the coordinatively unsaturated species are expected to be singlet states [9].

One could ask further whether the relatively slow rate of reaction of $Fe(CO)_4$ with CO manifests itself in the activation energy or preexponential. To answer this question, we have

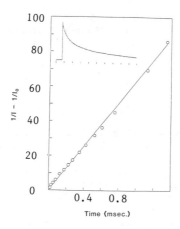

Figure 7.   The rate of loss of $Mn(CO)_5$ is plotted as a second
order decay. The data were obtained at 2004.3 cm^{-1}.   The inset is
a transient waveform at this frequency which covers a 2 ms time
range.

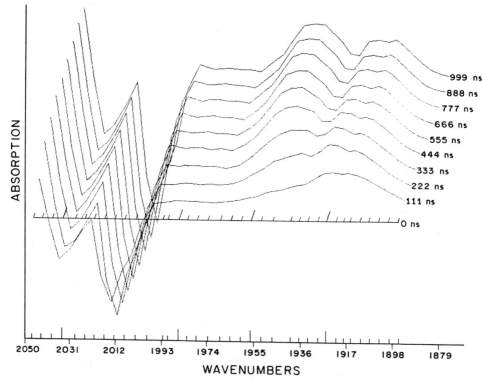

Figure 8.   Transient absorption spectrum resulting from KrF
photolysis of 20 mtorr of $Fe(CO)_5$ in 5 torr Ar.   (Reproduced with
permission from reference 3.   Copyright 1986 American Institute
of Physics.)

performed temperature dependent studies of the reactions in the
$Fe(CO)_x$ system over the limited temperature range of 10-55°C. Within
experimental error, over this temperature range, we do not see any
change in the rate of reaction of <u>any</u> of the $Fe(CO)_x$ species with CO
[8]. This implies that the activation energy for each reaction is
$\leq$2.5 kcal/mole. Thus the smaller preexponential causes the reaction
of $Fe(CO)_4$ to be slower than the reaction of the other coordinatively unsaturated iron carbonyls. Since the change of geometry of
$Fe(CO)_4$ on reaction with CO is not vastly different than the change
in geometry that occurs on reaction of $Cr(CO)_5$ with CO, it is
unlikely that this type of geometry change is a major factor in the
slowness of the reaction of $Fe(CO)_4$ with CO. Thus the effect of the
change in spin manifests itself in the preexponential.
    A further interesting difference in kinetic behavior can be
observed in the $Mn_2(CO)_{10}$ system. For this system the rate constant
for the reaction of $Mn_2(CO)_9$ with CO has been measured as (2.4 $\pm$
0.8)x 10^6 1 mole^{-1}s^{-1} which is very similar to the rate constant for
this reaction measured in solution [6,10,33a]. This is an additional order of magnitude slower than even the reaction of $Fe(CO)_4$
with CO. Why is this? $Mn_2(CO)_9$ has been studied in the matrix and
has been found to have a structure with a bridging CO located
between the two Mn metal centers. However, since the CO is not
symmetrically located relative to the two metal centers, it has been
designated as a "semi-bridging" CO group [33a]. Since this CO
shares electron density between the two metal centers, formally each
Mn atom can attain an 18 electron count. Thus the reaction of CO
with $Mn_2(CO)_9$ could better be viewed as a displacement reaction
rather than an addition reaction of CO to a coordinatively
unsaturated compound. Viewed in this light it is not surprising
that this reaction is much slower than addition reactions to
coordinatively unsaturated compounds.
    We have also measured the rate constant for the association
reaction of two $Mn(CO)_5$ radicals generated on photolysis of
$Mn_2(CO)_{10}$. With appropriate assumptions regarding the absorption
coefficient for $Mn(CO)_5$, the rate constant for this reaction was
determined to be (2.7 $\pm$ 0.6) x 10^{10} 1 mole^{-1} s^{-1} [6,10]. This is
compatible with the diffusion limited rate constant for this
reaction that has been measured in solution and is within an order
of magnitude of a gas kinetic rate constant as would be expected for
an essentially unactivated radical-radical association reaction
[33a].

Photophysics and Photochemistry. Figure 8 illustrates what has been
found to be a general feature in the production of coordinatively
unsaturated metal carbonyls in the gas phase: they are normally
formed with internal excitation. Typically, species that lose the
least ligands are formed with most internal excitation. As the
internal excitation relaxes the absorptions narrow and shift toward
higher frequency. Typically, the absorptions of the species formed
with the largest amount of internal excitation will take the longest
time to approach their final position and shape. This occurs since
the most highly excited species must lose the most internal energy
via collisional relaxation processes. This behavior has been
observed in all metal carbonyl systems studied to date with an

increasing degree of excitation observed in a given photofragment as the energy of the photolysis photon increases. This observation is compatible with other studies involving photodissociation of metal carbonyls [11,36].

Another interesting observation regarding the dissociation process is a general increase in the degree of unsaturation of the photoproducts as a function of the energy of the input photon. This behavior is apparent in either the iron or chromium system where the branching ratios for photoproducts changes dramatically with input energy. This is also observed in the $Mn_2(CO)_{10}$ system [10]. As shown in figure 1 , for the iron system, almost exclusively $Fe(CO)_3$ is formed for XeF laser photolysis while for KrF laser photolysis the product mix shifts toward $Fe(CO)_2$. For ArF laser photolysis almost exclusively $Fe(CO)_2$ is produced (not shown) [8]. Similar behavior is observed in the chromium system with XeF photolysis producing predominantly $Cr(CO)_5$, KrF photolysis predominantly $Cr(CO)_4$ and ArF photolysis a mix of products including $Cr(CO)_3$ and $Cr(CO)_2$ (see figure 4).

The behavior described in the two preceding paragraphs is qualitatively compatible with a straightforward mechanism for photodissociation which also reconciles observed differences in product distributions in the gas phase versus condensed phases. The initially absorbed photon initiates a photochemical event which results in loss of a CO ligand and the production of a photofragment which is highly internally excited. This excited molecule is rapidly relaxed in condensed phase due to the high density of surrounding collision partners. Thus the net result of photolysis in these and related systems in condensed phase is loss of one ligand. However, in the gas phase additional processes can occur. The energized photofragment can go on to further dissociate in an RRKM like process leading to multiple products [17]. Dissociative steps are terminated when the excited molecule can be collisionally stabilized on the timescale of the next possible dissociative event. We are currently working on further verifying this hypothesized photolysis mechanism by calculations and additional experiments.

This description must obviously be modified when there is the possibility for multiple initial photochemical events. Multiple initial photochemical events could occur in the $Fe(CO)_5$ and $Cr(CO)_6$ systems due to overlapping electron states and have been shown to occur in $Mn_2(CO)_{10}$, where either homolytic bond cleavage or decarbonylation occurs in photolysis [36-38]. For $Mn_2(CO)_{10}$, the nature of the initially excited electronic state, which can vary with wavelength, has now been convincingly shown to influence the branching ratio for these two paths [37]. This leads to a wavelength dependence for photolysis even in condensed phase. However, the general principles of the aforementioned dissociation mechanism can still valid once the initial photochemical event occurs. For the $Mn_2(CO)_{10}$ system in the gas phase a change in the ratio of the two initial photoproducts are observed as the energy of the input photon increases in going from XeF to KrF to ArF. Preliminary studies indicate that these dissociation products are dominated by dissociation of $Mn_2(CO)_9$ rather than further dissociation of $Mn(CO)_5$. This result is compatible with the observation originally made by Vaida that the Mn-Mn bond strength <u>increases</u> with loss of additional CO ligands [38].

Another interesting aspect of the photophysics of this system is revealed by inspection of the high energy region of figures 1 and 4. In these regions there are positive going absorptions (labeled e in figure 1) which are due to vibrationally and/or rotationally excited CO. Thus UV photolysis not only produces internally excited coordinatively unsaturated metal carbonyls but it also produces internally excited CO as the other photoproduct. Furthermore, as with the coordinatively unsaturated fragment, the internal energy of the CO increases with increasing photolysis energy. This can be readily observed in figure 1 by a comparison of the shape of the parent absorption as a function of photolysis energy. The high energy peak of the parent absorption appears weaker relative to the low energy peak as the photolysis energy increases. This is due to internally excited CO, which produces a positive absorption that is superimposed on the parent absorption. The effect is more pronounced for KrF than for XeF photolysis because the internal energy of the CO is greater leading it to absorb at lower frequencies. The effect is even more pronounced for ArF photolysis (not shown) where it actually causes part of the high frequency parent band to appear <u>above</u> the baseline [8].

<u>Spectroscopic Considerations</u>. Though spectroscopic considerations have not been emphasized in this manuscript, a few general comments are in order . As seen in figure 2, where the now partially resolved A_1, B_1 and B_2 bands are indicated, the gas phase spectrum of $Fe(CO)_4$ is compatible with a C_{2v} structure, as is the fact that gas phase $Fe(CO)_4$ has a triplet ground state [3,4,8]. This is the same structure that has been observed for matrix isolated $Fe(CO)_4$ [30,31]. Similarly, virtually all of the gas phase absorption features that we have observed for coordinatively unsaturated compounds of Fe, Cr and Mn are compatible with their reported matrix structures. This is an important observation in that it implies that matrix isolated coordinatively unsaturated metal carbonyls are not subject to "matrix effects" and the structure determined in the matrix is very likely to be that of the gas phase species. Some subtleties may modify this statement such as the effect of coordinated rare gas molecules or other coordinated matrix or glass substrate molecules on the structure. Nevertheless, this statement is likely to be accurate in a large majority of cases. The only possible exception to this statement that we have observed to date, deals with the difference in position of the semi-bridging CO band in $Mn_2(CO)_{10}$ in the matrix versus the gas phase. This band is observed to be at higher frequency in the matrix which is counter to typical behavior [10]. This could indicate a change in structure in the gas phase versus the matrix for this compound. However, until further studies are complete the previous statement should be regarded more as conjecture than proven fact.

Conclusions

Perhaps the best way to sum up the general conclusions regarding our studies of coordinatively unsaturated metal carbonyls is in a series of propensity rules with the understanding that these rules may be modified by future studies. The rules are:

1) Addition reactions to coordinatively unsaturated compounds are expected to be significantly faster than substitution reactions.
2) Spin conserving reactions are significantly faster than spin disallowed reactions.
3) Spin allowed addition reactions have rate constants near gas kinetic.
4) In the gas phase the degree of coordinative unsaturation increases with increasing photolysis energy.
5) The nature of the electronic state accessed can also influence branching ratios for products but within a given electronic state statement 4 will prevail.
6) Both the coordinatively unsaturated photofragment and the ejected CO tend to be produced with more internal excitation as the energy of the photolysis photon increases.
7) The structures of gas phase coordinatively unsaturated metal carbonyls are very similar to those of the matrix isolated species.

Acknowledgments

We acknowledge support of this work by the Air Force Office of Scientific Research under contract #83-0372, the National Science Foundation under grant #CHE 82-06976 and the donors of the Petroleum Research Fund administered by the American Chemical Society under grant #15163-AC3. We acknowledge many useful conversations with Dr. Martyn Poliakoff and Prof. J. J. Turner and thank NATO for a travel grant which facilitated these conversations. We also thank our coworkers in the field for their useful suggestions and comments.

Literature Cited

1. Poliakoff, M.; Weitz, E. Advances in Organometallic Chemistry, 1986, 25, 277.
2. Geoffroy, G. L.; Wrighton, M. S. Organometallic Photochemistry, Academic Press, N.Y., 1979.
3. Ouderkirk, A.; Weitz, E. J. Chem. Phys., 1983, 79, 1089.
4. Ouderkirk, A.; Wermer, P.; Schultz, N. L.; Weitz, E. J. Am. Chem. Soc., 1983, 105, 3354.
5. Seder, T. A.; Church, S. P.; Ouderkirk A. J.; Weitz, E. J. Am. Chem. Soc., 1985, 107, 1432.
6. Seder, T. A.; Church, S. P.; Weitz, E. J. Am. Chem. Soc., 1986, 108, 1084.
7. Ouderkirk, A. J.; Seder, T. A.; Weitz, E. Laser Applications to Industrial Chemistry SPIE; 1984; Vol. 458, p. 148.
8. Seder, T. A.; Ouderkirk, A. J.; Weitz, E. J. Chem. Phys., 1986, 85, 1977.

9.   Seder, T. A.; Church, S. P.; Weitz, E. J. Am. Chem. Soc., 1986, 108, 4721.

10. Seder, T. A.; Church, S. P.; Weitz, E. J. Am. Chem. Soc. - in press.

11. Bray, R. G.; Seidler, Jr., P. F.; Gethner, J. S.; Woodin, R. C. J. Am. Chem. Soc., 1986, 108, 1312.

12. Fletcher, T. R.; Rosenfeld, R. N. J. Am. Chem. Soc., 1985, 107, 2203.

13, Fletcher, T. R.; Rosenfeld, R. N. J. Am. Chem. Soc., 1986, 108, 1686.

14. Breckenridge, W. H.; Sinai, N. J. Phys. Chem., 1981, 85, 3557.

15. Breckenridge, W. H.; Stewart, G. M. J. Am. Chem. Soc., 1986, 108, 364.

16. Tumas, T.; Gitlan, B.; Rosan, A. M.; Yardley, J. T. J. Am. Chem. Soc., 1982, 104, 55.

17. Yardley, J. T.; Gitlan, B.; Nathanson G.; Rosan, A. M. J. Chem. Phys., 1981, 74, 361; ibid., 1981, 74, 370.

18. Rayner, D. N.; Nazran, A. S.; Drouin, M.; Hackett, P. B. J. Phys. Chem., 1986, 90, 2982.

19. Casey, C. P.; Cyr, C. R. J. Am. Chem. Soc., 1973, 95, 2248.

20. Whetten, R. L.; Fu, K. J.; Grant, E. R. J. Chem. Phys., 1982, 77, 3769.

21. Whetten, R. L.; Fu, K. J.; Grant, E. R. J. Am. Chem. Soc., 1982, 104, 4270.

22. Mitchener, J. C.; Wrighton, M. S. J. Am. Chem. Soc., 1981, 103, 975.

23. Miller, M. C.; Grant, E. R. SPIE, 1984, 458, 154.

24. Welch, J. A.; Peters, K. S.; Vaida, V. J. Phys. Chem., 1982, 86, 1941.

25. See, for example, Hepp, A. F.; Wrighton, M. S. J. Am. Chem. Soc., 1983, 105, 5934.

26. See, for example, Turner, J. J.; Simpson, M. B.; Poliakoff, M.; Maier II, W. B. J. Am. Chem. Soc., 1983, 105, 3898.

27. Callear, A. B.; Oldman, R. J. Nature, 1966, 210, 730; Trans. Faraday Soc., 1967, 63, 2888.

28. Kelly, J. M.; Bent, D. V.; Hermann, H.; Schulte-Frohlinde, D.; Koerner von Gustorf, E. J. Organomet. Chem., 1974, 69, 259.

29. Hermann, H.; Grevels, F. W.; Henne, A.; Schaffner, K. J. Phys. Chem.. 1982, 86, 5151; Moore, B. D.; Simpson, M. B.; Poliakoff, M.; Turner, J. J. J. Chem. Soc. Chem. Comm., 1984, 972.

30. Poliakoff, M. Chem. Soc. Rev., 1978, 7, 527 and references therein; Turner, J. J.; Burdett, J. K.; Perutz, R. N.; Poliakoff, M. Pure & Appl. Chem., 1977, 49, 271-285 and references therein.

31. Poliakoff, M. J. Chem. Soc. Dalton Trans., 1974, 210; Poliakoff, M.; Turner, J. J. ibid., 1974, 2276.

32. a) Perutz. R. N.; Turner, J. J. J. Am. Chem. Soc., 1975, 97, 4791.

     b) Perutz, R. N.; Turner, J. J. Inorg. Chem., 1975, 14, 262.

     c) Perutz, R. N.; Turner, J. J. J. Am. Chem. Soc., 1975, 97, 4800.

33.   a)   Church, S. P.; Hermann, H.; Grevels, F.; Schaffner, K. J. Chem. Soc. Chem. Comm., 1984, 785 and references therein.

      b)   Church, S. P.; Poliakoff, M.; Tinney, J. A.; Turner, J. J. J. Am. Chem. Soc., 1983, 103, 7515.

34.   Burdett, J. K. J. Chem. Soc. Faraday Trans. II, 1974, 70, 1599.

35.   Barton, T. J.; Grinter, R.; Thomson, A. J.; Davies, B.; Poliakoff, M. J. C. S. Chem. Comm., 1977 841.

36.   Freedman, A.; Bersohn, R. J. Am. Chem. Soc., 1978, 100, 4116.

37.   Kobayashi, T.; Yasufuku, K.; Iwai, J.; Yesaka, H.; Noda, H.; Ohtoni, H. Coord. Chem. Rev., 1985, 64, 1; Kobayashi, T.; Ohtani, H.; Noda, H.; Teratani, S.; Yamazaki, H.; Yasafuku, K Organometallics, 1986, 5, 110.

38.   Leopold, B. G.; Vaida, V. J. Am. Chem. Soc., 1984, 106, 3720.

39.   Graham, M. A.; Poliakoff, M.; Turner, J. J. J. Chem. Soc. A, 1971, 2939.

40.   Burdett, J. K.; Graham, M. A.; Perutz, R. N.; Poliakoff, M.; Rest, A. J.; Turner, J. J.; Turner, R. F. J. Am. Chem. Soc., 1975, 97, 4085.

RECEIVED November 3, 1986

Chapter 6

Primary and Secondary Processes in Organometallic Photochemistry

T. R. Fletcher and R. N. Rosenfeld

Department of Chemistry, University of California—Davis, Davis, CA 95616

Recent interest in the photochemistry of organometallic compounds is associated in part with the plethora of applications of this technology in chemical synthesis, homogeneous and heterogeneous catalysis and the deposition of metallic films. The field also provides fertile grounds for exploring some fundamental problems in polyatomic photochemistry, e.g. the relationship between electronic structure and reactivity, and the role of radiationless transitions in photochemistry. In fact, it is necessary to address problems of this type in order to develop a rational basis for utilizing the reactions of organometallic species in processes such as those noted. Here, we will discuss recent work from our laboratory on the photoactivated dissociation reactions of transition metal carbonyls. The use of time resolved spectroscopic methods in probing photodissociation mechanisms and the structure and reactivity of dissociation products will be described. Several research groups have made significant contributions to the current understanding of metal carbonyl photochemistry and an excellent review has been published.[2] The contributions of Turner, Poliakoff and co-workers[3] have been particularly noteworthy. Their studies of simple metal carbonyls [e.g. $Ni(CO)_4$, $Fe(CO)_5$, $Cr(CO)_6$, etc.] in cryogenic matrices demonstrate that the dominant chemical channel following photoactivation (in condensed phases) is cleavage of a single metal-CO bond, resulting in the formation of CO and a mono-unsaturated metal center. Moreover, they have exploited this observation in obtaining infrared absorption spectra of a variety of unsaturated metal carbonyls. The information obtained from these spectra has dramatically refined our understanding of bonding and electronic structure in the transition metal carbonyls. Studies reported by Grant and co-workers[4] have provided an impetus for much of the current research on gas-phase organometallic photochemistry. They have found that the products of the photolysis of $Fe(CO)_5$ are active catalysts for olefin hydrogenation and geometrical isomerization and that reaction rates for catalyzed processes in the vapor phase can be substantially larger than in condensed phases. Yardley and co-workers[5] have established that highly unsaturated metal centers can be prepared by the photolysis of $Fe(CO)_5$ and $Cr(CO)_6$ in the gas phase. This result is in marked contrast to those obtained in solution phase studies where only mono-unsaturated photoproducts are observed. Moreover, flash photolysis experiments by Breckenridge and co-workers[6] and others[7] indicate that unsaturated transition metal carbonyls can readily associate even

0097-6156/87/0333-0099$06.00/0

with relatively "inert" species, e.g. Xe and N_2. The findings noted here, as well as others, indicate several reasons why gas phase organometallic photochemical studies provide a useful complement to solution phase work:

(i) It is possible to prepare and study mono-, bi- and tri-unsaturated organometallics in the gas phase, while this is generally not possible in solution.

(ii) The intrinsic properties (i.e. spectra, kinetics) of unsaturated metal centers can be characterized in the vapor phase. Such properties can be perturbed by facile association with solvent molecules in condensed phases.

(iii) The rates of catalyzed reactions in the gas phase can exceed the corresponding solution phase rates. This suggests that, in some cases, it may be advantageous to carry out synthetic procedures in the gas phase.

The development of comprehensive models for transition metal carbonyl photochemistry requires that three types of data be obtained. First, information on the dynamics of the photochemical event is needed. Which reactant electronic states are involved? What is the role of radiationless transitions? Second, what are the primary photoproducts? Are they stable with respect to unimolecular decay? Can the unsaturated species produced by photolysis be spectroscopically characterized in the absence of solvent? Finally, we require thermochemical and kinetic data i.e. metal-ligand bond dissociation energies and association rate constants. We describe below how such data is being obtained in our laboratory.

Consider the photodissociation reaction, (1), where the asterisk denotes ro-vibrational excitation. We have previously shown how measurements of the ro-

$$M(CO)_6 \xrightarrow{h\nu} [M(CO)_5]^* + CO(v,J) \qquad (1)$$

vibrational energy distribution of the CO product can provide information on the dynamics of fragmentation reactions.[8] If the unsaturated product, $M(CO)_5$, can be spectroscopically observed, then one should be able to obtain both structural information on this species and data on its reaction kinetics. This can all be accomplished, in principle, by infrared absorption spectroscopy, as previously described.[8,9] Briefly, $M(CO)_6$ [0.01-0.03 torr] is contained in a one meter pyrex absorption cell where it may be mixed with an added gas, e.g. CO, NH_3, etc. at 0-5 torr, and an inert buffer gas (He, Ar at 0-50 torr). The mixture is irradiated with a pulse of ultraviolet (UV) light from an excimer laser (193-351 nm, 1-5 mJ/cm^2). The beam of a line tuneable, continuous wave CO laser is directed through the absorption cell, coaxially with respect to the UV laser beam, and then onto an InSb detector. The CO laser oscillates on $P_{v+1,v}(J)$ transitions, where v = 0-11 and J = 8-15. Thus, the region 2100-1830 cm^{-1} can be covered. This allows us to monitor $CO(v,J)$ by resonance absorption and various $M(CO)_n$ [n = 3-6] as a result of near coincidences between the CO laser lines and the carbonyl stretching vibrations of these species. The temporal response of the detection system is ca. 100 ns and is limited by the risetime of the InSb detector. Detection limits are approximately 10^{-5} torr for CO and $M(CO)_n$. The principal limitation of our instrumentation is associated with the use of a molecular, gas discharge laser as an infrared source. The CO laser is line tuneable; laser lines have widths of ca. 10^{-3}cm^{-1} and are spaced 3-4 cm^{-1} apart. Thus, spectra can only be recorded point-by-point, with an effective resolution of ca. 4 cm^{-1}. As a result, band maxima (e.g. in the carbonyl stretching region) cannot be located to better than 4 cm^{-1} and band shapes can be characterized only qualitatively. This is, nevertheless, sufficient to detect the various fragments, $M(CO)_n$, which can be generated by the photolysis of $M(CO)_6$. To date, we have studied the photochemistry of $W(CO)_6$ at 351 nm and $Cr(CO)_6$ at 248 and 351 nm. An overview of our findings is presented below.

Photochemistry of $W(CO)_6$ at 351 nm. The irradiation of $W(CO)_6$ at 351 nm results in the direct population of the $^3T_{1g}$ state.[10,11] Dissociation to form $W(CO)_5$ and CO may then occur from the $^3T_{1g}$ state or, following intersystem crossing, from the $^1A_{1g}$ ground state. CO can be observed by time resolved CO laser absorption spectroscopy in vibrational states, $v = 0$-2, following the 351 nm photoactivation of $W(CO)_6$ (see Figure 1). The vibrational energy distribution of the CO product is shown in Figure 2. Such data can be compared with distributions calculated using models for energy disposal in the fragmentation reaction.[12,13] In this way, some insight regarding the photophysics of $W(CO)_6$ and its dissociation dynamics can be obtained.

A simple phase space model can be used to compute the CO product vibrational energy distribution as a function of the available energy,[12-14] E_{av}. The maximum energy which can be partitioned among the products' degrees of freedom is the reaction exoergicity, $E_x = h\nu - DH^\circ[(CO)_5W\text{-}CO]$. For a 351 nm photolysis, $E_x \approx 35$ Kcal/mole.[15] We find that the CO product vibrational distribution calculated using the phase space model with $E_{av} = 35$-40 Kcal/mole is in good agreement with our experimental results (Figure 2). Thus, the measured CO vibrational distribution indicates that vibrational energy disposal to the photolysis products is determined at a point on the potential surface where the full reaction exoergicity is available. This suggests that the 351 nm excitation of $W(CO)_6$ results in the sequence of events, (2)-(4), where the asterisk denotes vibrational excitation.

$$W(CO)_6\,(\tilde{X}^1A_{1g}) \xrightarrow{\;351\text{ nm}\;} W(CO)_6\,(\tilde{a}^3T_{1g}) \qquad (2)$$

$$W(CO)_6\,(\tilde{a}^3T_{1g}) \dashrightarrow [W(CO)_6\,(\tilde{X}^1A_{1g})]^* \qquad (3)$$

$$[W(CO)_6\,(\tilde{X}^1A_{1g})]^* \longrightarrow W(CO)_5\,(\tilde{X}^1A_1) + CO(\tilde{X}^1\Sigma^+) \qquad (4)$$

Photochemistry of $Cr(CO)_6$ at 351 nm. The excitation of $Cr(CO)_6$ at 351 nm populates the $\tilde{a}^3T_{1g}$ state either directly, or via intersystem crossing from the $\tilde{A}^1T_{1g}$ state. The measured CO product vibrational energy distribution is shown in Figure 3, along with distributions calculated using the phase space model. In this case, the reaction exoergicity is $E_x \approx 44$ Kcal /mole.[17] However, the calculated CO product vibrational distribution, when the available energy equals the reaction

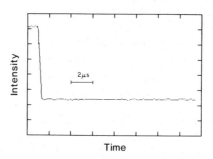

Figure 1. Time resolved absorption of the CO laser $P_{1,0}(10)$ line following the 351 nm photolysis of $W(CO)_6$. $[W(CO)_6] = 0.025$ torr, $[He] = 4.0$ torr.

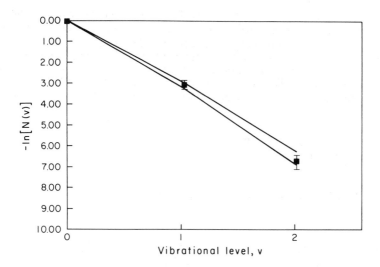

Figure 2. Vibrational energy distribution of the CO product formed via the 351 nm photolysis of $W(CO)_6$. Experimental data are indicated as ■. The lines correspond to results obtained by phase space calculations with an available energy of 40 and 35 Kcal/mole.

exoergicity, is significantly hotter than that observed in the laboratory. Agreement between the phase space model and experimental results can be obtained only if the available energy is reduced to ca. 25 Kcal/mole. This indicates that an energy of approximately 19 Kcal/mole is *not* available to the products' vibrational degrees of freedom. The calculated[19] ($\tilde{X}^1A_{1g}$ - $\tilde{a}^3T_{1g}$) energy splitting in $Cr(CO)_5$ is 17 Kcal/mole. On the basis of these findings, we can propose that the photodissociation of $Cr(CO)_6$ at 351 nm occurs via (5)-(6). This model is consistent with the results of solution phase quantum

$$Cr(CO)_6\,(\tilde{X}^1A_{1g}) \xrightarrow{\;351\ nm\;} Cr(CO)_6\,(\tilde{a}^3T_{1g}) \qquad (5)$$

$$Cr(CO)_6\,(\tilde{a}^3T_{1g}) \longrightarrow Cr(CO)_5\,(\tilde{a}^3E) + CO(\tilde{X}^1\Sigma^+) \qquad (6)$$

yield measurements for direct and triplet sensitized photosubstitution.[20] It differs from that proposed for $W(CO)_6$, (2)-(4), in that intersystem crossing, (3), is facile for $W(CO)_6$ relative to $Cr(CO)_6$. (The spin-orbit coupling constant for W exceeds that for Cr by at least an order of magnitude.) The $Cr(CO)_5$ ($\tilde{a}^3E$) formed via (6) apparently undergoes rapid relaxation to the ground electronic state, $\tilde{X}^1A_1$. The $Cr(CO)_5$ infrared absorption spectrum observed in our work is consistent with that reported by Turner and co-workers[21] in a cryogenic matrix for $Cr(CO)_5$ ($\tilde{X}^1A_1$). See Figure 4. Based on such data, we can conclude that the major organometallic product of the 351 nm photolysis of $Cr(CO)_6$ is $Cr(CO)_5$. The further fragmentation of this species, yielding $Cr(CO)_4$, occurs to a relatively minor (<10%) extent.

Photochemistry of $Cr(CO)_6$ at 248 nm. We have previously shown[8] that the 248 nm photolysis of $Cr(CO)_6$ yields $Cr(CO)_4$ as the principal organometallic product via (7)-(8). Time-resolved laser absorption methods can be used to record the

$$Cr(CO)_6 \xrightarrow{\;248\ nm\;} [Cr(CO)_5]^* + CO \qquad (7)$$

$$[Cr(CO)_5]^* \longrightarrow Cr(CO)_4 + CO \qquad (8)$$

infrared spectrum of $Cr(CO)_4$ following photolysis; see Figure 5. The gas phase spectrum observed here is consistent with the spectrum observed by Turner and co-workers[22] for $Cr(CO)_4$ in a cryogenic matrix. Once a spectrum, such as Figure 5, has been characterized, the kinetics of association reactions, (9), can be measured by

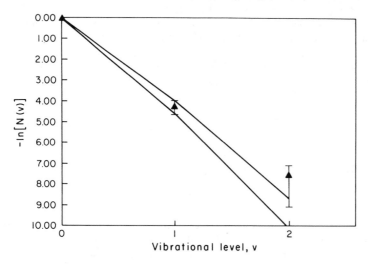

Figure 3. Vibrational energy distribution of the CO product formed via the 351 nm photolysis of $Cr(CO)_6$. Experimental data are indicated as ▲. The lines correspond to results obtained by phase space calculations with an available energy of 25 and 20 Kcal/mole.

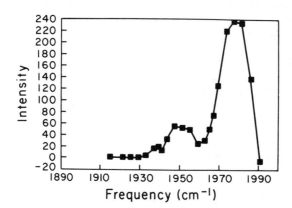

Figure 4. Transient infrared absorption spectrum obtained at 400 ns following the 351 nm photolysis of $Cr(CO)_6$. $[Cr(CO)_6] = 0.020$ torr, $[CO] = 0.400$ torr, $[He] = 20.0$ torr.

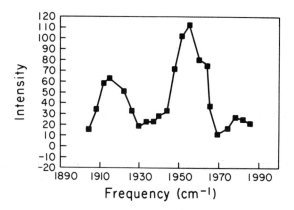

Figure 5. Transient infrared absorption spectrum obtained at 400 ns following the 248 nm photolysis of $Cr(CO)_6$. $[Cr(CO)_6] = 0.020$ torr, $[He] = 80.0$ torr.

$$Cr(CO)_4 + L \longrightarrow Cr(CO)_4 L \tag{9}$$

monitoring the time dependence of one of the $Cr(CO)_4$ absorption bands[23] in the presence of added reagent, L.   For example, when $L = NH_3$ and H_2, the recombination rate constant is $k_9 = 1.1 \times 10^7$ and 1.6×10^6 torr^{-1}s^{-1}, respectively.[23] Here we discuss the case $L = CO$. The measured rate of (9) for $L = CO$ depends on the pressure of added gas, e.g. He, showing a characteristic fall-off with decreasing pressure (see Figure 6).   Such observations indicate that a Lindemann-type recombination mechanism, (10)-(11), is operative, where the asterisk denotes vibrational excitation

$$Cr(CO)_4 + CO \; \underset{k_a}{\overset{k_c}{\rightleftarrows}} \; [Cr(CO)_5]^* \tag{10}$$

$$[Cr(CO)_5]^* + M \; \xrightarrow{k_s} \; Cr(CO)_5 + M \tag{11}$$

and M represents any third body. An expression for the observed rate constant for $Cr(CO)_4$ decay, (12), can be derived by making the steady state assumption,

$$k_{obs} = k_c k_s [M] / (k_a + k_s [M]) \tag{12}$$

$d[Cr(CO)_5]^*/dt = 0$. We have experimentally measured k_s for $M = He$ ($k_s{}^{He} = 2.6 \times 10^5$ torr^{-1}s^{-1}) by both absorption and infrared fluorescence methods. k_c can be determined from the measured value for k_{obs} in the high pressure limit[23] ($k_{obs,\infty} = k_c = 7.5 \times 10^6$ torr^{-1}s^{-1}). In principle, k_a can be calculated using RRKM theory. Then, we can compare the model and experimental results for k_{obs} vs. [M] as a means for evaluating some of the parameter values selected for the RRKM calculations. If the transition state frequencies and moments of inertia are chosen so as to reproduce the reported Arrhenius A-factor[16] ($A = 10^{15.7}$s^{-1}, $\Delta S^\ddagger = +9.9$ eu), then the only critical parameter in the calculations is the activation energy for $Cr(CO)_5 \rightarrow Cr(CO)_4 + CO$. The value that gives the best agreement with our experimental data is $E_a = 25 \pm 6$ Kcal/mole. Since the decarbonylation of $Cr(CO)_5$ occurs via a "loose" (or late) transition state, $E_a \approx DH°[(CO)_4Cr\text{-}CO]$. Thus, the kinetic data suggest $DH°[(CO)_4Cr\text{-}CO] \approx 25$ Kcal/mole. This value for the bond dissociation energy in $Cr(CO)_5$ is substantially less than the value proposed by Lewis et al,[16] $DH° \approx 40$ Kcal/mole, but is nevertheless consistent with trends found in other transition metal carbonyls, i.e. first bond dissociation energies are typically greater than second bond dissociation energies. See Table I. Note that the $DH°$ for

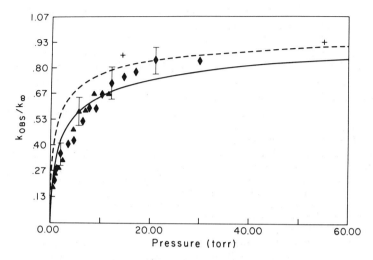

Figure 6. Rate constant for the recombination of $Cr(CO)_4$ with CO as a function of bath gas (helium) pressure. The various symbols correspond to data obtained using different CO laser lines. The lines, (———) and (------) represent RRKM calculations with E_0 = 23 and 25 Kcal/mole, respectively.

Table I. Bond Dissociation Energies for Transition Metal Carbonyls

Bond	DH°(Kcal/mole)	Reference
$(CO)_5Cr-CO$	37	14,16
$(CO)_4Cr-CO$	25	(a)
$(CO)_4Fe-CO$	55.4	24
$(CO)_3Fe-CO$	4.6	24
$(CO)_3Ni-CO$	25	25
$(CO)_2Ni-CO$	13	25
$[(CO)_4Mn-CO]^+$	20.2	26

(a) See text.

$[Mn(CO)_5]^+$, a species that is isoelectronic and isostructural with $Cr(CO)_5$, is comparable to the value of DH° for $Cr(CO)_5$ determined here.

Conclusions.   Time-resolved CO laser absorption spectroscopy can provide information useful in characterizing the primary photochemical channels in gas-phase transition metal carbonyls. We have found that product vibrational energy distributions indicate that $W(CO)_6$ and $Cr(CO)_6$ dissociate via different channels following photoactivation at 351 nm. $W(CO)_5$ ($\tilde{X}^1A_1$) is obtained in the former case while $Cr(CO)_5$ (a^3E) is obtained in the latter. The infrared absorption spectroscopy of coordinatively unsaturated metal centers can be studied using laser absorption methods. Spectra of $Cr(CO)_5$ and $Cr(CO)_4$ are reported here. These spectroscopic data provide a basis for investigating the association kinetics of unsaturated metal carbonyls with virtually any ligand. Results are described for the recombination reaction, $Cr(CO)_4 + CO \rightarrow Cr(CO)_5$. The pressure dependence of the recombination rate constant has been measured and compared with the behavior expected on the basis of RRKM theory. In this way, we find $DH°[(CO)4Cr-CO] \approx$ 25 Kcal/mole.

The potential of the techniques described here is just beginning to be realized. Our approach is clearly well-suited to studying the photochemistry of virtually any (volatile) transition metal carbonyl and the spectroscopy of the fragments thus obtained. Kinetic data on these fragments is similarly accessible. Important areas for further research appear to be:

(i) The systematic study of a *series* of transition metal carbonyls so that the structural basis for trends observed in reaction dynamics, spectroscopy and reactivity can be elucidated.

(ii) The study of substituted metal carbonyls, e.g. $C_5H_5Mn(CO)_3$ and $C_6H_6Cr(CO)_3$. Research here will be useful in assessing the influence of the organic substituent on the reactivity of unsaturated metal centers. Such data are particularly important in developing models for catalysis.

(iii) The application of continuously tuneable infrared laser sources in studying the spectroscopy of unsaturated metal carbonyls. The use of such sources will permit the structures of these species to be determined much more conclusively than is possible using line tuneable, molecular lasers.

(iv) The electronic spectroscopy of saturated metal carbonyls. Clearly, the prediction of likely photophysical and photochemical pathways depends on being able to identify the states that are prepared following optical excitation. Thus, detailed studies of the UV-visible spectroscopy of metal carbonyls are necessary.

Finally, it is worth noting that all of the research described here is greatly facilitated when accurate values are available for metal-ligand bond dissociation energies. Only limited data of this type is presently available and further work along these lines is certainly warranted.

Acknowledgment

Acknowledgment is made to the NSF for support of the research described here through grant CHE-8500713. One of us (RNR) also acknowledges the Alfred P. Sloan Foundation for the award of a fellowship (1985-1987).

Literature Cited

1.   Present address; JILA, University of Colorado, Boulder, Colorado 80309.
2.   M. Poliakoff and E. Weitz, to be published in Adv. Organomet. Chem, F.G.A. Stone (ed.), 1985.
3.   See for example (a) J.J. Turner, J.K. Burdett, R.N. Perutz and M. Poliakoff, Chem. Soc. Rev. 78, 527 (1978).
4.   (a) R.L. Whetten, K.J. Fu and E.R. Grant, J. Chem. Phys. 77, 3769 (1982); (b) K.J. Fu and E.R. Grant, J. Am. Chem. Soc. 104, 4270 (1982).

5.   (a) W.Tumas, B. Gitlin, A.M. Rosan and J.T. Yardley, J. Am. Chem. Soc. *104*, 55 (1982); (b) J.T. Yardley, B. Gitlin, G. Nathanson and A.M. Rosan, J. Chem. Phys. *74*, 361 (1981).

6.   (a) W.H. Breckenridge and N. Sinai, J. Phys. Chem. *85*, 3557 (1981); (b) W.H. Breckenridge and G.M. Stewart, J. Am. Chem. Soc. *108*, 364 (1986).

7.   (a) J.A. Welch, K.S. Peters and V. Vaida, J. Am. Chem. Soc. *86*, 1941 (1982); (b) J.D. Simon and K.S. Peters, Chem. Phys. Lett. *98*, 53 (1983).

8.   T.R. Fletcher and R.N. Rosenfeld, J. Am. Chem. Soc. *107*, 2203 (1985).

9.   Also see (a) A. Ouderkirk, P. Wermer, N.L. Schultz and E. Weitz, J. Am. Chem. Soc. *105*, 3354 (1983); (b) A.J. Ouderkirk and E. Weitz, J. Chem. Phys. *79*, 1089 (1983); (c) T.A. Seder, S.P. Church, A.J. Ouderkirk and E. Weitz, J. Am. Chem. Soc. *107*, 1432 (1985); (d) T.A. Seder, S.P. Church and E. Weitz, J. Am. Chem. Soc. *108*, 4721 (1986).

10.   S.R. Desjardins, Ph.D. dissertation, Massachusetts Institute of Technology (1981).

11.   N. Beach and H.B. Gray, J. Am. Chem. Soc. *90*, 5713 (1968).

12.   B.I. Sonobe, T.R. Fletcher and R.N. Rosenfeld, J. Am. Chem. Soc. *106*, 4352 (1984).

13.   B.I. Sonobe, T.R. Fletcher and R.N. Rosenfeld, J. Am. Chem. Soc. *106*, 5800 (1984).

14.   We have also investigated the application of Franck-Condon models in fitting the CO product vibrational energy distribution and find they are unable to do so. D. Anderson, J. Brown, R. Fletcher, K. Prather and R. Rosenfeld, manuscript in preparation.

15.   This follows from $DH°[(CO)_5W-CO] = 46$ Kcal/mole. See ref. 16.

16.   K.E. Lewis, D.M. Golden and G.P. Smith, J. Am. Chem. Soc. *106*, 3905 (1984).

17.   This follows from $DH°[(CO)_5Cr-CO] = 37$ Kcal/mole. See ref. 16 and 18.

18.   M. Bernstein, J.D. Simon and K.S. Peters, Chem. Phys. Lett. *100*, 241 (1983).

19.   P.J. Hay, J. Am. Chem. Soc. *100*, 2411 (1978).

20.   J. Nasielski and A. Colas, Inorg. Chem. *17*, 237 (1978).

21.   (a) M.A. Graham, M. Poliakoff and J.J. Turner, J. Chem. Soc. A, 2939 1971); (b) R.N. Perutz and J.J. Turner, Inorg. Chem. *14*, 262 (1975).

22.   (a) R.N. Perutz and J.J. Turner, J. Am. Chem. Soc. 97, 4800 (1975); (b) J.K. Burdett, M.A. Graham, R.N. Perutz, M. Pliakoff, A.J. Rest, J.J. Turner and R.F. Turner, J. Am. Chem. Soc. *97*, 4805 (1975).

23.   T.R. Fletcher and R.N. Rosenfeld, J. Am. Chem. Soc. *108*, 1686 (1986).

24.   P.C. Engelking and W.C. Lineberger, J. Am. Chem. Soc. *101*, 5569 (1979).

25.   A.E. Stevens, C.S. Feigerle and W.C. Lineberger, J.Am. Chem. Soc. *104*, 5026 (1982).

26.   F.E. Saalfeld, M.V. McDowell, J.J. DeCorpo, A.D. Berry and A.G. MacDiarmid, Inorg. Chem. *12*, 48 (1973).

RECEIVED November 13, 1986

Chapter 7

Infrared Spectroscopy of Organometallic Intermediates

James J. Turner, Michael A. Healy, and Martyn Poliakoff

Department of Chemistry, University of Nottingham, Nottingham, NG7 2RD, England

IR spectroscopy is a powerful spectroscopic technique
for examining the structure and behavior of intermediates
involved in organometallic photochemistry. Examples
are given of the combination of IR spectroscopy with
matrix isolation, with liquid noble gases as solvents,
and with flash generation, for probing novel transients
and intermediates.

It is important to know as much as possible about the intermediates
involved in photochemical reactions, particularly those of
relevance to catalysis (1,2). Over the past few years our approach
has involved three different techniques - matrix isolation,
liquefied noble gases as solvents, and time-resolved spectroscopy -
frequently in combination (3). In each method, intermediates are
generated by UV photolysis and particular use is made of IR
spectroscopy for detection, characterisation and kinetic
measurements. Such emphasis on IR spectroscopy might be thought
strange given the ubiquitous presence of X-ray crystallography and
nmr in organometallic chemistry. However, these intermediates are
generally unstable and frequently present in low concentrations;
moreover, IR spectroscopy can, with caution, be very revealing,
particularly if full use is made of isotopic substitution and,
where appropriate, force field calculations. In this article will
be described some recent examples which illustrate the application
of these techniques.

Matrix Isolation

The application of matrix isolation to organometallic chemistry has
been extensively described elsewhere (4,5,6,7). Two methods have
generally been employed. In the first, based on G.C. Pimentel's
original development, the solid matrix environment is a frozen
noble gas - usually Ar - at 10-20K and the unstable fragment is
generated either by photolysis of a parent molecule already trapped
in the matrix, or by cocondensation from the gas phase. In the

0097-6156/87/0333-0110$06.00/0

second, derived from G.N. Lewis and G. Porter, the matrix is a
frozen organic glass at 77K and the transient is generated by
photolysis of a parent which has been dissolved in the host solvent
prior to freezing. The two variants have advantages and
disadvantages, but when applied to organometallic photochemistry in
which the intermediates are detected by IR spectroscopy of intense
$\nu(C-O)$ bands, there is little to choose between them: the advantage
of frozen noble gases is that the IR bands are usually narrower and
hence the complex spectra produced on isotopic substitution are
more revealing; the disadvantage of the noble gas method is that
the parent species has to be sufficiently volatile to be deposited
via the gas phase without decomposition. Very many CO-containing
species have been investigated by these methods.

One specific advantage of the frozen noble gas is that it does
not absorb IR and thus it is possible, in principle, to examine the
spectrum of any coordinated ligand within the matrix. In practice,
it is more difficult because, although the intensity of $\nu(N-N)$ and
$\nu(N-O)$ bands are comparable with $\nu(C-O)$ bands, most of the
vibrations of other ligands give rather weak IR bands. However, as
an example, in recent important work, Perutz (8) has demonstrated
that photolysis of $CpRh(C_2H_4)_2$ in Ar matrices at 20K leads to
reversible loss of C_2H_4 to form $CpRh(C_2H_4)$ identified by IR
spectroscopy, whereas photolysis of $CpIr(C_2H_4)_2$ gives vinyl hydride
complexes.

In principle, IR spectroscopy also offers a method of probing
agostic interactions in which H atoms on alkyl groups interact with
empty coordination sites on a metal, and C-H activation, via the
behavior of the stretching vibration of CH_3 groups attached to
transition metals. Unfortunately, the picture is complicated by
coupling between CH bonds to give symmetric and antisymmetric modes
and by Fermi resonance between these two modes and overtones of
bending modes. However, McKean (9,10) has shown that these problems
are removed by observation of C-H stretching frequencies of the
CHD_2 group - the $\nu(C-H)_{is}$ method. A graph of $\nu(C-H)_{is}$ versus known
C-H bond length for a series of C/H compounds shows a remarkable
straight line with a gradient corresponding to a bond length change
of -0.0001023 $\overset{o}{A}$ per cm^{-1} shift in frequency [10]. In a striking
application to organometallic chemistry, McKean and colleagues (11)
have shown that in the gas phase the $\nu(C-H)_{is}$ frequencies of
$CHD_2Mn(CO)_5$ and $CHD_2Re(CO)_5$ are 2955.0 and 2934.6 cm^{-1}
respectively, implying C-H bond lengths of 1.095_9 and 1.098_0 $\overset{o}{A}$
respectively.

In collaboration with McKean and McQuillan we have been
applying the $\nu(C-H)_{is}$ technique to unstable species in matrices.
The complex photochemistry of $CH_3M(CO)_5$ (M = Mn, Re) has already
been unravelled by matrix isolation experiments, utilising both IR
($\nu(C-O)$ and $\nu(N-N)$ vibrations) and UV/visible spectroscopy (12).
The photochemistry pattern is illustrated in 1. Figure 1 shows the
$\nu(C-H)_{is}$ of some of these species. The first striking observation
is that each $\nu(C-H)_{is}$ is split, with some weaker features; this
behavior is related to the rotation of CHD_2 groups around the M-C
bond and will be discussed elsewhere (13). For present purposes
there are some important points to note. The mean positions of
the $\nu(C-H)_{is}$ bands of the parent compounds $CHD_2Mn(CO)_5$ and

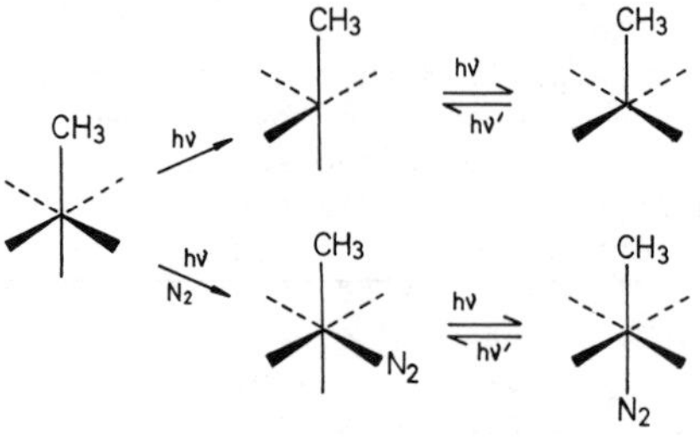

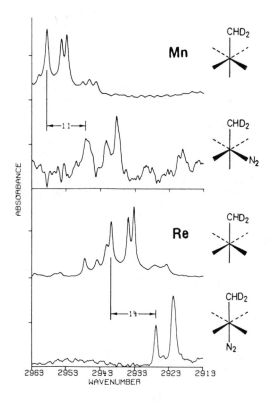

Scheme 1

Figure 1.   IR spectra in $\nu(C-H)_{is}$ region of the parent molecules $CHD_2Mn(CO)_5$ and $CHD_2Re(CO)_5$ in N_2 matrices at 20K and of two of the N_2-containing photoproducts.

$CHD_2Re(CO)_5$ are very close to the gas phase positions and the shift
$(Mn$ to Re$)$ is almost identical in gas phase and matrix. Thus,
$\nu(C-H)_{is}$ in the matrix will provide information on the C-H bond
length. On generation of the unstable N_2 complexes, there is a
downward shift in $\nu(C-H)_{is}$ of 11-14 cm^{-1} implying a bond
lengthening of .0011-.0015 Å. This is clearly a very sensitive
probe of the effect of slight electronic perturbation of the metal
centre and is likely to be useful for other systems.

Liquefied Noble Gases as Solvents

Liquefied Xe and Kr have two features which make them particularly
attractive as low-temperature solvents for examining the spectra of
unstable organometallics. These solvents are inert, and this is
important as the significant role of even innocuous solvents such
as cyclohexane is more readily recognised; in contrast to
conventional solvents for IR spectroscopy they have no absorptions
over a wide range of the spectrum, which also permits the use of
special long path cells to overcome problems of low solubility.

We have used such cells (14) to generate and measure the
kinetic stability over a wide temperature range of a wide variety
of unstable species. The easiest experiments - and the first
performed - were simply the photolysis of a metal carbonyl in
liquid xenon doped with dissolved N_2 and hence replacement of CO by
N_2. The $\nu(N-N)$ IR bands are little weaker than $\nu(C-O)$ bands and
hence detection and characterization are straightforward,
particularly when use is made of previous matrix studies. Species
examined include $Cr(CO)_{6-x}(N_2)_x$ (x = 1-5) (15), $Ni(CO)_3(N_2)$ (16).
With $\nu(N-O)$ bands it is equally straightforward to characterize,
for instance, $Fe(CO)_{2-x}(NO)_2(N_2)_x$ (x = 1,2) (17).

However, as in solid noble gas matrices, the real value of
liquefied noble gases becomes apparent when other and more
interesting ligands are attached to the transition metal.
Following Kubas's (18) original synthesis of $(OC)_3Mo(PCy_3)_2(H_2)$
there has been a great flurry of activity in compounds in which H_2
is sideways bonded, undissociated, to a transition metal (19-22).
There is considerable theoretical interest (23-27) and it is likely
that a whole range of compounds will be made of which many will be
unstable. A convenient general route to such molecules is similar
to the N_2 synthesis method described above but replacing the N_2 gas
by H_2 in the high pressure cell. We can characterize the H_2 unit
by IR spectrocopy. For instance, Figure 2 shows the $\nu(H-H)$ bands
of $M(CO)_5H_2$ (M = Cr, Mo, W) in liquid xenon at -70°C (28); the
simplest theoretical argument would relate the shift in $\nu(H-H)$
frequency between free H_2 and coordinated H_2, to the degree of
interaction of H_2 with the transition metal. It is gratifying that
the stability of these three species is in the order W>Cr>Mo, the
order of increasing $\nu(H-H)$ frequencies. It should be noted that in
a number of matrix experiments (29,30) involving H_2-doped matrices,
spectroscopic evidence for H_2 complexes has been obtained but such
studies have had to rely on $\nu(C-O)$ bands since it is not possible
to build up a sufficient concentration of H_2 complex in the matrix
to observe IR bands associated with the $M-H_2$ moiety.

It is possible under high H_2 pressure to photolyse $Cr(CO)_5(H_2)$ further and to generate $Cr(CO)_4(H_2)_2$ and then to observe exchange and substitution reactions (28). These reactions are summarised in Scheme 2 where the most significant process is the H_2/D_2 exchange.

One of the reasons for the current interest in H_2 complexes is connected with catalysis. Thus it is important to examine compounds in which both H_2 and alkenes (or dienes) are coordinated to the same metal centre; in real catalytic situations such intermediates are likely to be very reactive. As a first step we have investigated some unstable ethylene, 1-butene and butadiene complexes generated by photolysis of parent molecules such as $Cr(CO)_6$, $Fe(CO)_2(NO)_2$, $Co(CO)_3(NO)$, dissolved in liquid xenon doped with the appropriate organic molecule (31,32). Again, IR spectroscopy of the organic ligand has been crucial in assigning the structure of the photoproduct; for instance, Figure 3 shows the derivation of the bands assigned to coordinated 1-butene in $Co(CO)_2(NO)(1\text{-butene})$ produced on photolysis of $Co(CO)_3(NO)$ in 1-butene doped liquid xenon.

Since $Cr(CO)_6$ is a well known hydrogenation photocatalyst (2), an interesting experiment is to characterize a $Cr(alkene)(H_2)$ complex starting from an alkene complex parent. Until recently there were no stable monoalkene complexes of Cr, but Grevels (33) has synthesised $(OC)_5Cr(trans\text{-cyclooctene})$ ($\equiv (OC)_5Cr(ol)$) which is stabilised because of relief of ring strain. When this compound was dissolved in liquid Xe and photolysed under high D_2 pressure, nothing happened; however, in liquid Kr at $-120°C$ there was strong evidence for the generation of $(OC)_4Cr(ol)(D_2)$, with a half-life of about an hour, as shown in Figure 4 (34). Given the instability of this species it was not possible to generate a sufficient concentration to observe a $\nu(D-D)$ band but further evidence for the correct assignment comes from observing its reaction with dissolved N_2. An interesting question is, why does coordination of the olefin so weaken the metal/H_2 interaction to make $Cr(CO)_4(ol)(D_2)$ much less stable than $Cr(CO)_5(H_2)$?

Time-Resolved IR Spectroscopy

Although matrix isolation and low-temperature solvents have great potential for identifying intermediates, and for obtaining structural information, and - in the case of the solvents - for obtaining kinetic data, it is necessary to relate these observations to more ordinary conditions, i.e. conventional solvents at room temperature.

UV/visible spectroscopy of organometallic transients has been extremely important for kinetic measurements on previously identified species, but the spectra are less valuable for structural identification since most spectra show broad and featureless bands. One way round this problem has been to utilise matrix IR spectroscopy for characterisation and to use the data obtained from the matrix UV/visible spectrum to monitor the room-temperature kinetics. A more satisfactory method is to record the IR spectra of transients directly and there has been much activity in both gas phase and solution organometallic chemistry; this field has been recently reviewed (35). In our laboratory,

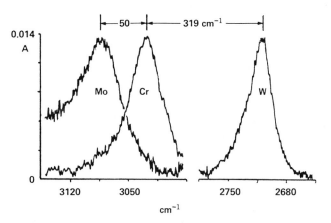

Figure 2.   IR spectra showing the substantial wavenumber shift between bands assigned to the $\nu(H-H)$ vibration of coordinated molecular hydrogen in $Mo(CO)_5(H_2)$, $Cr(CO)_5(H_2)$, and $W(CO)_5(H)_2$ in liquid Xe at $-70°C$.   Note that the three traces were recorded in separate experiments and that the absorbance scale refers to the Mo trace [28].

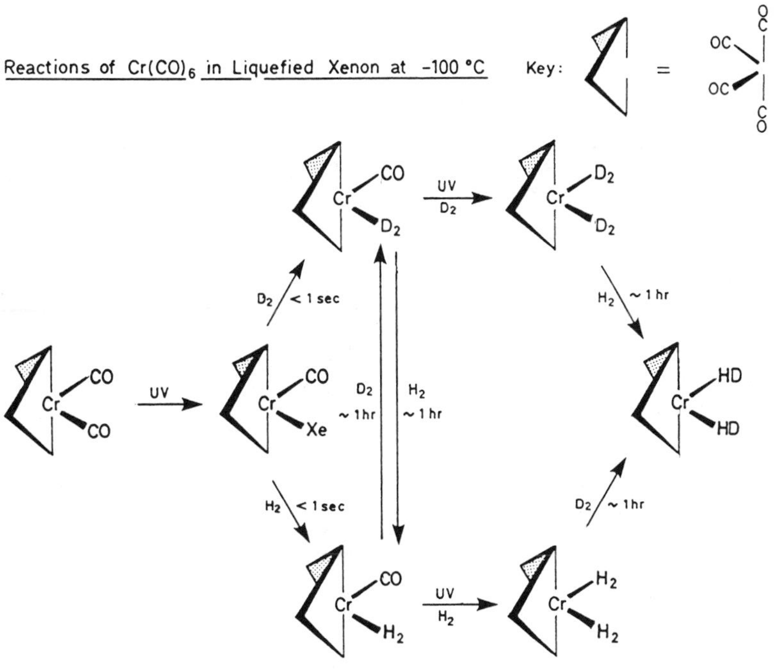

Reactions of $Cr(CO)_6$ in Liquefied Xenon at $-100 °C$

Scheme 2

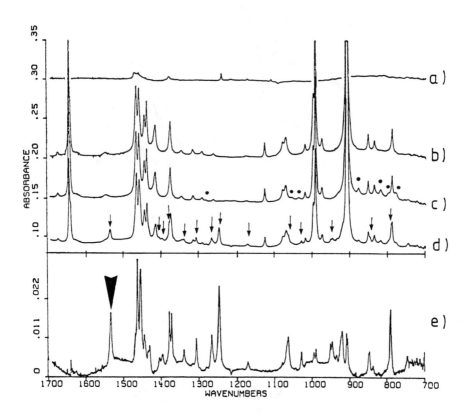

Figure 3. Series of IR spectra illustrating how the absorptions (spectrum e) of coordinated 1-butene were identified: (a) spectrum of LXe before addition of reactants (weak bands due to trace impurities); (b) spectrum after addition of 1-butene; (c) spectrum after addition of Co(CO)$_3$NO to the butene/LXe mixture (bands due to Co(CO)$_3$NO marked with o); (d) spectrum obtained after UV photolysis (bands due to Co(CO)$_2$(NO)(η^2-1-btn) marked with ↓); (e) spectrum of coordinated 1-butene in Co(CO)$_2$(NO)(η^2-1-butene) obtained from spectrum d by computer subtraction of the bands of Co(CO)$_3$NO and 1-butene (large arrow marks the ν(C=C) band) [31].

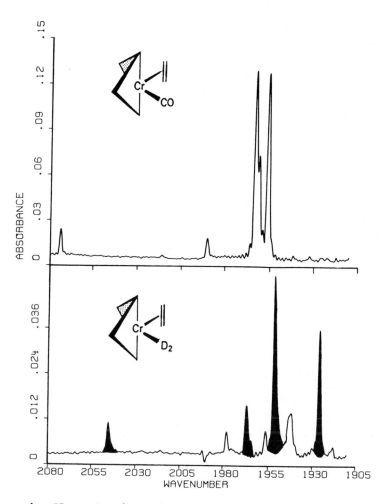

Figure 4. IR spectra in D_2-doped liquid Kr at $-120°C$ of $(OC)_5Cr(ol)$ and of the photolysis product $(OC)_5Cr(ol)(D_2)$ [ol = trans-cyclooctene]. D_2 was used rather than H_2 to avoid potential spectral overlap of any $\nu(H-H)$ bands with bands due to the hydrocarbon ligand.

transients are generated by flash photolysis using either a xenon flash lamp or an XeCl excimer laser; the rapidly changing IR signal is monitored at a particular IR frequency by a line-tunable CO laser. The experiment is repeated over a whole series of CO laser lines separated by approximately 4 cm^{-1}. Fortunately, there is a good overlap between the ν(C-O) absorption bands of organometallic compounds and the output of a CO laser. Full details of the apparatus have been described elsewhere (35,36).

A simple application of time-resolved IR spectroscopy (TRIR) is distinguishing between CO-bridged and non-bridged dinuclear metal carbonyls in solution (37). For example, photolysis of MM'(CO)$_{10}$ (M,M' = Mn, Re) in solution involves two pathways, believed to be the radical M(CO)$_5$ and the CO-loss species MM'(CO)$_9$. A combination of matrix isolation and TRIR shows that Mn$_2$(CO)$_9$ and MnRe(CO)$_9$ (38-42) have CO bridges but that Re$_2$(CO)$_9$ (42-44) does not and that the rate of reaction of CO with MnRe(CO)$_9$ is greater than with Re$_2$(CO)$_9$. Extensive use has been made of this method in identifying and monitoring the behavior of the two phototransients from [CpFe(CO)$_2$]$_2$, i.e. the radical CpFe(CO)$_2$ and the novel triply-bridged species CpFe(μ-CO)$_3$FeCp (45,46), the structure of which has been firmly identified by matrix experiments (47,48). When [CpFe(CO)$_2$]$_2$ is photolysed in CH$_3$CN solution, the major product is (CO)CpFe(μ-CO)$_2$FeCpCH$_3$CN. An interesting question arises: Which of the two intermediates is responsible for the overall photochemistry? Figure 5 shows the TRIR spectra from a flash experiment involving [CpFe(CO)$_2$]$_2$ and CH$_3$CN in cyclohexane solution (46). From these spectra it is clear that the product grows in at the same rate as the triply bridged intermediate decays; the radical disappears much faster. Hence the mechanism of this reaction involves the reaction of the triply bridged species with CH$_3$CN.

In recent work in collaboration with G.R. Dobson, we have examined the flash photolysis of W(CO)$_4$(piperidine)L (L = PPh$_3$, P(OEt)$_3$, P(O^{i}Pr)$_3$) in n-heptane solution. There was previous evidence for the production of two intermediates tentatively identified as W(CO)$_4$L(S) (S = solvent), with L and S cis and trans to each other (49,50). It has now proved possible to obtain firm identification of these two species in room temperature solution from their ν(C-O) absorption spectra. It could be proved that their structures were indeed cis and trans W(CO)$_4$L by generating exactly the same species in low-temperature matrices. One noteworthy feature of the experiments is that, even with such a weakly coordinating solvent as heptane, the cis and trans intermediates do not interconvert on a millesecond timescale [53].

It will be clear that in these examples the short-lived transients have been detected by their ν(C-O) vibrations. Church (51) has managed to observe ν(N-N) of Cr(CO)$_5$(N$_2$) and Weitz has observed ν(N-O) of transients produced in the reaction of Fe(CO)$_5$ with NO (52). However, at this stage it looks as though detection of organometallic intermediates via, for example, ν(C-H) or ν(C=C) absorptions, will require extensive improvements in sensitivity.

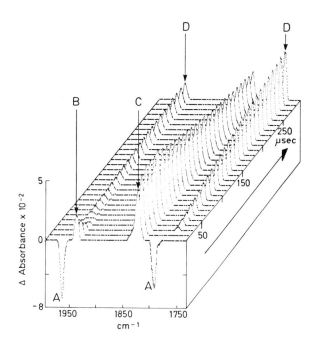

Figure 5. Time-resolved IR spectra obtained after UV flash photolysis of [CpFe(CO)$_2$]$_2$, A, (6x10^{-4}M) and CH$_3$CN (6x10^{-3}M) in cyclohexane solution at 25°C. The bands are labelled B [CpFe(CO)$_2$], C [Cp$_2$Fe$_2$(μ-CO)$_3$] and D [Cp$_2$Fe$_2$(CO)$_3$(CH$_3$CN). The spectra have been reconstituted from ~70 kinetic traces recorded at intervals of ~4 cm^{-1} across the wavenumber region illustrated. The spectra are drawn by interpolation between these discrete points. The first three spectra correspond to the duration of the firing of the UV flash lamp and subsequent spectra are shown at intervals of 10 μsec. The negative peaks in the first spectrum are due to material destroyed by the flash. These negative peaks have been omitted from the subsequent traces to avoid undue confusion in the figure. Reproduced with permission from Ref. 46. Copyright 1986, The Chemical Society.

Conclusions

These examples show that IR spectroscopy continues to be a powerful technique for the characterisation of unstable intermediates important in organometallic photochemistry.

Acknowledgments

Work at Nottingham has been supported by the SERC, the Donors of the Petroleum Research Fund adminstered by the American Chemical Society, the Paul Instrument Fund of the Royal Society and the Scientific Stimulation Programme of the EEC. We also gratefully acknowledge the help of many colleagues both at Nottingham and elsewhere.

Literature Cited

1. Geoffroy, G.L.; Wrighton, M.S. 'Organometallic Photochemistry', Academic Press: New York, 1979.
2. Moggi, L.; Juris, A.; Sandrini, D.; Manfrin, M.F. Rev. Chem. Intermed. 1981, 4, 171.
3. For further background discussion see e.g. Turner, J.J.; Poliakoff, M. 'Photochemical Intermediates' in Chisholm, M.A. Ed. 'Inorganic Chemistry Towards the 21st Century', American Chemical Society, 1983: Fresenius Z. Anal. Chem. 1986, 324, 819.
4. Hitam, R.B.; Mahmoud, K.A.; Rest, A.J. Coord. Chem. Rev. 1984, 55, 1.
5. Burdett, J.K. Coord. Chem. Rev. 1978, 27, 1.
6. Wrighton, M.S.; Graff, J.L.; Kazlauskas, R.J.; Mitchener, J.C.; Reichel, C.L. Pure Appl. Chem. 1982, 54, 161.
7. Perutz, R.N. Chem. Rev. 1985, 85, 77, 97.
8. Haddleton, D.M.; Perutz, R.N. J. Chem. Soc. Chem. Commun. 1985, 1372, and in press.
9. McKean, D.C. Chem. Soc. Rev. 1978, 7, 399.
10. McKean, D.C. J. Mol. Struct. 1984, 113, 251.
11. Long, C.; Morrison, A.R.; McKean, D.C.; McQuillan, G.P. J. Am. Chem. Soc. 1984, 106, 7418.
12. Horton-Mastin, A.; Poliakoff, M.; Turner, J.J. Organometallics 1986, 5, 405.
13. Horton-Mastin, A.; McKean, D.C.; McQuillan, G.P.; Poliakoff, M.; Turner, J.J. to be published.
14. Maier, W.B. II; Freund, S.M.; Holland, R.F.; Beattie, W.H. J. Chem. Phys. 1978, 69, 1961.
15. Turner, J.J.; Poliakoff, M.; Maier, W.B.; Simpson, M.B. Inorg. Chem. 1983, 22, 911.
16. Turner, J.J.; Simpson, M.B.; Poliakoff, M.; Maier, W.B. J. Am. Chem. Soc. 1983, 105, 3898.
17. Gadd, G.E.; Poliakoff, M.; Turner, J.J. Inorg. Chem. 1984, 23, 630.
18. Kubas, G.J.; Ryan, R.R.; Swanson, B.I.; Vergamini, P.J.; Wasseman, J.J. J. Am. Chem. Soc. 1984, 106, 451.
19. Morris, R.H.; Sawyer, J.F.; Shiralian, M.; Zubkowski, J.D. J. Am. Chem. Soc. 1985, 107, 5381.

20. Crabtree, R.H.; Lavin, M. J. Chem. Soc. Chem. Commun. 1985, 794: 1985, 1661.
21. Crabtree, R.H.; Hamilton, D.G. J. Am. Chem. Soc. 1986, 108, 3124.
22. Crabtree, R.H.; Lavin, M.; Bonneviot, L. J. Am. Chem. Soc. 1986, 108, 4032.
23. Hay, P.J. Chem. Phys. Lettrs. 1984, 103, 466.
24. Saillard, J.Y.; Hoffmann, R. J. Am. Chem. Soc. 1984, 106, 2006.
25. Jean, Y.; Lledos, A. Nouv. J. Chim. in press.
26. Jean, Y.; Eisentein, O.; Volatron, F.; Maouche, B.; Sefta, F. J. Am. Chem. Soc. in press.
27. Burdett, J.K. private communication.
28. Upmacis, R.K.; Poliakoff, M.; Turner, J.J. J. Am. Chem. Soc. 1986, 108, 3645.
29. Perutz, R.N. private communication.
30. e.g. Sweany, R.L., J. Am. Chem. Soc. 1985, 107, 2374: private communication.
31. Gadd, G.E.; Poliakoff, M.; Turner, J.J. Inorg. Chem. 1986, 3604.
32. Gregory, M.F.; Jackson, S.A.; Poliakoff, M.; Turner, J.J. J. Chem. Soc. Chem. Commun. 1986, 1175.
33. Grevels, F-W.; Skibbe, V. J. Chem. Soc. Chem. Commun. 1984, 681.
34. S.A. Jackson, unpublished observations.
35. Poliakoff, M.; Weitz, E. 'Detection of Transient Organometallic Species by Fast Time-Resolved IR Spectroscopy', in Stone, F.G.A.; R. West, Eds 'Advances in Organometallic Chemistry', 1986, 25, 277.
36. Dixon, A.J.; Healy, M.A.; Hodges, P.M.; Moore, B.D.; Poliakoff, M.; Simpson, M.B.; Turner, J.J.; West, M.A. J. Chem. Soc. Faraday Trans. 2 in press.
37. For an excellent review of the photochemistry of binuclear metal species see Meyer, T.J.; Caspar, J.V. Chem.Rev. 1985, 85, 187.
38. Hepp, A.F.; Wrighton, M.S. J. Am. Chem. Soc. 1983, 105, 5934.
39. Dunkin, I.R.; Härter, P.; Shields, C.S. J. Am. Chem. Soc. 1984, 106, 7248.
40. Church, S.P.; Hermann, H.; Grevels, F-W.; Schaffner, K. J. Chem. Soc. Chem. Commun. 1984, 785.
41. Sedar, T.A.; Church, S.P.; Weitz, E. J. Am. Chem. Soc. 1986, 108, 1084.
42. Firth, S.; Hodges, P.M.; Poliakoff, M.; Turner, J.J. Inorg. Chem. in press.
43. Church, S.P.; Grevels, F-W.; Schaffner, K. Organometallics in press.
44. Church, S.P., private communication.
45. Moore, B.D.; Simpson, M.B.; Poliakoff, M.; Turner, J.J. J. Chem. Soc. Chem. Commun. 1984, 972.
46. Dixon, A.J.; Healy, M.A.; Poliakoff, M.; Turner, J.J. J. Chem. Soc. Chem. Commun. 1986, 994.
47. Hepp, A.F.; Blaha, J.P.; Lewis, C.; Wrighton, M.S. Organometallics 1984, 3, 174.

48. Hooker, R.H.; Mahmoud, K.A.; Rest, A.J. J. Chem. Soc. Chem. Commun. 1983, 1022.
49. Dobson, G.R.; Bernal, I.; Reisner, M.G.; Dobson, C.B.; Mansour, S.E. J. Am. Chem. Soc. 1985, 107, 525.
50. Dobson, G.R.; Dobson, C.B.; Mansour, S.E. Inorg. Chim. Acta 1985, 100, C7.
51. Church, S.P.; Grevels, F-W.; Hermann, H.; Schaffner, K. Inorg. Chem. 1984, 23, 3830.
52. Weitz, E. private communication.
53. Dobson, G.R.; Hodges, P.M.; Firth, S.; Healy, M.A.; Poliakoff, M.; Turner, J.J.; Asali, K.J. J. Am. Chem. Soc., submitted.

RECEIVED November 3, 1986

Chapter 8

Reactive Intermediates in the Thermal and Photochemical Reactions of Trinuclear Ruthenium Carbonyl Clusters

Peter C. Ford, Alan E. Friedman, and Douglas J. Taube

Department of Chemistry, University of California—Santa Barbara, Santa Barbara, CA 93106

Summarized are a series of investigations using both thermal and photochemical techniques to probe the reaction dynamics of intermediates formed in various reactions of triruthenium cluster complexes.

The chemistry of metal carbonyl clusters has been largely dominated by synthetic chemists who have prepared a remarkable array of structures involving wide varieties of metal cage structures and modes of coordination of even simple ligands. Interest in such systems has ranged from simple curiosity in what types of systems can indeed be constructed and whether these species will display unique chemical properties to attempts to use clusters as models for ligand interactions with metal surfaces and metal particles. Perhaps with the exception of ligand fluxionality processes, quantitative mechanistic investigations of cluster reactions have lagged behind the synthetic advances. However, in recent years there has been increasing attention to mechanistic details, stimulated in part by interest in the possible role of various clusters in the homogeneous catalytic activation of carbon monoxide, dihydrogen and other small molecules. Our own interest in the reaction mechanisms of triangular and tetrahedral clusters arose initially from the discovery that such species may be active components of homogeneous catalysts for the water gas shift reaction ([1,2]), but this has expanded to other chemical properties including the transformations stimulated by photoexcitation. The goal of the present manuscript is to review our investigations of several thermal and photochemical reactions of triangular triruthenium carbonyl complexes with an emphasis on attempts to characterize the quantitative reactivities of high energy, reactive intermediates along the trajectories of these chemical transformations.

Photoreactions of $Ru_3(CO)_{12}$

The photochemistry of $Ru_3(CO)_{12}$ has been investigated in our laboratory ([3-5]) and others ([6-11]) and has been shown to involve both photofragmentation of the cluster (Equations 1 and 2) and photolabilization of carbonyls to give substituted trinuclear clusters $Ru_3(CO)_{11}L$ (Equation 3).

0097-6156/87/0333-0123$06.00/0
© 1987 American Chemical Society

$$Ru_3(CO)_{12} + 3\ CO \xrightarrow{h\nu} 3\ Ru(CO)_5 \qquad (1)$$

$$Ru_3(CO)_{12} + 3\ L \xrightarrow{h\nu} 3\ Ru(CO)_4L \qquad (2)$$

$$Ru_3(CO)_{12} + L \xrightarrow{h\nu} Ru_3(CO)_{11}L + CO \qquad (3)$$

The electronic spectrum of $Ru_3(CO)_{12}$ (Figure 1) is dominated by an intense absorption band centered at 392 nm (ϵ_{max} = 7.7 × 10^3 $M^{-1}cm^{-1}$ in cyclohexane solution). Photofragmentation is indicated by a decrease in this band's intensity without a shift in the λ_{max}, while photosubstitution by L is indicated by shifts in this band to longer wavelengths. Photolysis at 405 nm in the presence of $P(OCH_3)_3$ led to photofragmentation only, while photolysis at shorter wavelengths gave spectral shifts indicative of the formation of substituted clusters. Quantum yields for photofragmentation Φ_f and photosubstitution Φ_s in octane solutions containing 0.012 M $P(OCH_3)_3$ are illustrated in Figure 1.

Photofragmentation Mechanisms. Photolysis (λ_{irr} 405 nm) of $Ru_3(CO)_{12}$ in hydrocarbon solvents under CO gave $Ru(CO)_5$ as the sole product (Equation 1). The quantum yield Φ_f proved markedly dependent on P_{CO} and on the solvent (Table I), with donor solvents such as THF giving much smaller values. Photofragmentation in octane (λ_{irr} 405 nm, P_{CO} = 1 atm) with various concentrations of cosolvents added led to significant quenching of Φ_f by donor solvents and gave linear Stern-Volmer type plots (e.g., $\Phi_f°/\Phi_f$ versus [THF]) with slopes (K_{SV}) of 34 ± 1, 26 ± 1 and 16 ± 1 M^{-1} for THF, diglyme and cyclohexene, respectively. In contrast, photolysis in 2,5-dimethyltetrahydrofuran led to quantum yields comparable to those observed in hydrocarbon solutions, an observation which reinforces the view that the ability to coordinate may be important to the quenching process.

Given the well documented role of homolytic cleavage of metal-metal bonds in the photoreactions of dimeric complexes (7), a logical hypothesis would be for the photofragmentations of trinuclear complexes to follow a similar path, e.g.,

One diagnostic test for such homolytic photofragmentation has been the trapping of the metal radicals M· by chlorocarbons to give the respective chlorides M-Cl. Photolysis (405 nm) of $Ru_3(CO)_{12}$ in a 1.0 M CCl_4 solution in octane under CO (1.0 atm) did indeed give a different product than in the absence of CCl_4, and this was identified as a mixture of two isomeric chloro complexes $Ru_2(CO)_6Cl_4$ (3). However, the addition of CCl_4 to octane solutions of $Ru_3(CO)_{12}$ had little influence on Φ_f values measured under CO (1.0 atm) (Table I), and Φ_f measured under argon in CCl_4/octane was found to be several orders of magnitude smaller than that measured under CO in the same mixed solvent, even though CO is not required in the stoichiometry for the formation of chloro carbonyl ruthenium products.

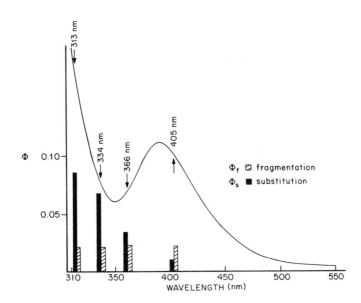

Figure 1.   Spectrum of $Ru_3(CO)_{12}$ in octane.   Quantum yields for photosubstitution and photofragmentation in 25°C argon flushed octane in the presence in 0.012 M $P(OCH_3)_3$ represented as a function of irradiation wavelength (from reference 5).

Table I.   Photofragmentation Quantum Yields for the 405 nm Photolysis of $Ru_3(CO)_{12}$ in Different Solutions (25°C)[a]

solvent	P_{CO}[b]	other additives	$10^3 \times \Phi_f$
Octane	1.0	—	28. ± 4
	0.0	—	< 0.1
	1.0	0.5 M THF	1.7 ± 0.0
	1.0	1.0 M CCl_4	24. ± 4[c]
	0.0	1.0 M CCl_4	0.2 [c]
Cyclohexane	1.0	—	18. ± 1
	0.25	—	4.4 ± 1.3
	0.10	—	2.2 ± 0.1
	0.0	0.001 M $P(OCH_3)_3$	3.2 ± 0.6[d]
	0.0	0.005 M $P(OCH_3)_3$	11.0 ± 1.4[d]
	0.0	0.010 M $P(OCH_3)_3$	18.3 ± 0.6[d]
	0.0	0.10 M $P(OCH_3)_3$	42.0 ± 1.7[d]
THF	1.0	—	3.5 ± 0.7
Diglyme	1.0	—	0.7 ± 0.1
CCl_4	1.0	—	13. ± 3[c]
	0.0	—	0.7 ± 0.1[c]
2,5-Me_2THF	1.0	—	20 ± 2

[a] Product is $Ru(CO)_5$ except where noted.
[b] Total P is 1.0 atm, balance being N_2 or Ar.
[c] Product is $Ru_z(CO)_xCl_y$.
[d] Product is $Ru(CO)_4(P(OCH_3)_3)$.

These results were explained by the discovery that the chloro-
ruthenium complexes are <u>not</u> the primary photoproducts under CO in 1.0
M CCl_4/octane solution. Instead $Ru(CO)_5$ proved to be the initial
product even after nearly complete photofragmentation of the starting
material, and the chlorocarbonyl ruthenium products to be the result
of a secondary, dark reaction between the $Ru(CO)_5$ and CCl_4 (<u>3</u>):

$$Ru(CO)_5 + CCl_4 \longrightarrow Ru_z(CO)_xCl_y + ? \qquad (4)$$

<u>It is therefore clear that a diradical sufficiently long-lived to be
trappable in the manner seen in metal dimer systems is not formed
with $Ru_3(CO)_{12}$</u>. Thus, the dominant feature of the photofragmentation
pathways is the role of two electron donors. Ligands which are
π-acids such as CO, ethylene or phosphorous donors PR_3, each gave
photofragmentation while little photoactivity occurs for longer λ_{irr}
when L is a harder donor such as cyclohexene, THF, diglyme or
2-methyltetrahydrofuran.
       Such characteristics led to the proposal (<u>3</u>,<u>8</u>) that the mechanism
for the fragmentation pathway must involve the formation of a reactive intermediate, an isomer of $Ru_3(CO)_{12}$ capable of first order
return to the initial cluster or of capture by a two electron donor.
Scheme 1 illustrates the proposed mechanism for photofragmentation.

$$Ru_3(CO)_{12} \xrightleftharpoons{h\nu} [Ru_3(CO)_{12}{}^*] \longrightarrow \underline{I} \qquad (5)$$

$$\underline{I} \xrightarrow{k_1} Ru_3(CO)_{12} \qquad (6)$$

$$\underline{I} + L \underset{k_3}{\overset{k_2}{\rightleftharpoons}} \underset{\underline{I}'}{Ru_3(CO)_{12}L} \qquad (7)$$

$$\underline{I}' \xrightarrow{k_4} Ru(CO)_4L + Ru_2(CO)_8 \xrightarrow{+2L, fast} 3\ Ru(CO)_4L \qquad (8)$$

$$\underline{I}' \xrightarrow{k_5} Ru_3(CO)_{12} + L \qquad (9)$$

SCHEME 1

A possible formulation for $\underline{I}$ is illustrated below. This could be
formed by the heterolytic cleavage of a Ru-Ru bond an corresponding
movement of a carbonyl from a terminal site to a bridging one to
maintain the charge neutrality of both Ru atoms. The result would be
to leave one ruthenium atom electron deficient (a 16 electron
species) and capable of coordinating a two electron donor to give
another intermediate $\underline{I}'$.

$$
\begin{array}{cc}
\text{Ru(CO)}_4 & \text{Ru(CO)}_4 \\
\diagup\ \diagdown & \diagup\ \diagdown \\
\text{(CO)}_4\text{Ru}\qquad\text{Ru(CO)}_3 & \text{(CO)}_4\text{Ru}\qquad\text{Ru(CO)}_3\text{L} \\
\diagdown\ \text{C}\ \diagup & \diagdown\ \text{C}\ \diagup \\
\text{O} & \text{O} \\
\underline{I} & \underline{I}'
\end{array}
$$

Flash photolysis studies were therefore conducted with the goal of probing for the presence of such intermediates in the photofragmentation ($\underline{4},\underline{5}$). Flash photolysis ($\lambda_{irr} > 395$ nm) of $Ru_3(CO)_{12}$ in cyclohexane solution under CO (1.0 atm) showed some net photoreaction, but no transients were detectable with lifetimes > 30 μs. Neither observable transients nor net photochemistry resulted from a similar flash experiment under argon. In contrast, a CO equilibrated cyclohexane solution of $Ru_3(CO)_{12}$ containing THF (1.0 M) displayed transient bleaching in the spectral region 380-460 nm which decayed exponentially ($k_d = 20 \pm 5$ s^{-1}) to give a final absorbance consistent with a small amount of net photoreaction. The same transient behavior with an identical k_d value was noted for an analogous THF/cyclohexane solution under argon with the exceptions that considerably more bleaching was apparent immediately after the flash and that no net photochemistry was seen. These results are consistent with formation of a species such as $\underline{I}'$ which decays largely back to $Ru_3(CO)_{12}$.

Similar transient bleaching at 390 nm followed by exponential decay to a final absorbance was seen in argon equilibrated cyclohexane solutions containing cyclohexene, PPh_3, or $P(OCH_3)_3$. Flash photolysis with added cyclohexene led to just small net photoreaction, but photolysis with added PPh_3 or $P(OCH_3)_3$ gave net cluster fragmentation. For these ligands, the flash photolysis kinetics were more conveniently investigated at 480 nm, where transient absorbance increases were seen (Figure 2). For $P(OCH_3)_3$, variation of this ligand's concentration (0.005 to 0.05 M) did not affect k_d but did affect the amount of transient formed and the extent of net photoreaction. The k_d values determined for the various donor ligands follow the order THF < cyclohexene < PPh_3 < $P(OCH_3)_3$ (Table II).

If $k_3 \ll k_4 + k_5$, the Φ_f is determined by three pairs of competitive processes. The first is the formation of $\underline{I}$ from $Ru_3(CO)_{12}{}^*$ in competition with decay to $Ru_3(CO)_{12}$ and occurs with an efficiency Φ_I. The second is the competition between decay of $\underline{I}$ back to $Ru_3(CO)_{12}$ (rate constant k_1) and capture of $\underline{I}$ by L to give $\underline{I}'$ (k_2). The third is the competition between Equation 9 to reform $Ru_3(CO)_{12}$ and fragmentation via Equation 8 to give products. Analysis of the various quantum yield ($\underline{5},\underline{8}$) and flash photolysis ($\underline{5}$) experiments in terms of Scheme I have led to the following conclusions: 1) The limiting quantum yield for photofragmentation (λ_{irr} 405 nm) in hydrocarbon solutions would be Φ_I, which was determined to be about 0.05 moles/einstein. 2) Trapping of $\underline{I}$ to give $\underline{I}'$ is relatively insensitive to the nature of L, relative values of k_2 being 1.6, 1.1 and 1.0 for CO, $P(OCH_3)_3$ and PPh_3, respectively. These values are consistent with the nature proposed for $\underline{I}$, i.e., a coordinatively unsaturated species which is essentially unselective in reacting with available ligands. 3) The apparent values of k_4 (fragmentation of $\underline{I}'$ to products of lower nuclearity) fall into the sequence: CO, $CH_2=CH_2 \gg P(OCH_3)_3 > PPh_3 \gg$ cyclohexene > THF, an

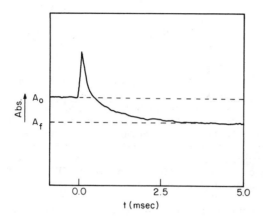

Figure 2. Absorbance (λ_{mon} = 480 nm) vs time trace for the longer wavelength flash photolysis (λ_{irr} > 395 nm) of a cyclohexane solution of $Ru_3(CO)_{12}$ plus $P(OCH_3)_3$ (0.010 M) (from reference 5).

Table II. First Order Rate Constants for Decay of Transients Seen by Longer Wavelength (λ_{irr} > 390 nm) Flash Photolysis of $Ru_3(CO)_{12}$ in Cyclohexane Solutions with Various Added Ligands [a]

Ligand	Concentration(M)	k_d (s^{-1})	Comments
none	—	> 5×10^4 [b]	no net photoreaction
CO	0.0084 M[c]	> 5×10^4 [b]	net photofragmentation
$H_2C=CH_2$	d	> 5×10^4 [b]	net photofragmentation
$P(OCH_3)_3$	0.005-0.05 M	900 ± 30	net photofragmentation
PPh_3	0.005-0.01 M	200 ± 30	net photofragmentation
cyclohexene	0.01 M	59 ± 10	no net photoreaction
THF/CO	[THF] = 1.0 M	20 ± 5	small net photofrag.
THF	1.0 M	20 ± 5[e]	no net photoreaction

[a] T = 25°C, Table from reference 5.
[b] No transient seen, estimated rate is lower limit.
[c] P_{CO} = 1.0 atm.
[d] $P_{C_2H_4}$ = 1.0 atm.

[e] Identical behavior noted in neat THF solution.

order which qualitatively parallels the π-acidity of L. A possible explanation is that the activation barrier for initial fragmentation (to give $Ru(CO)_4L$ plus $Ru_2(CO)_8$?), which would involve the bridging $CO \longrightarrow$ terminal CO transformation, may be lower for a π-acid L owing to the more electron withdrawing nature of the bridging CO.

Photosubstitution mechanisms. Continuous photolysis of this cluster in the presence of PPh_3 or $P(OCH_3)_3$ and at wavelengths shorter than 405 nm led to spectral changes indicating formation of substituted clusters (4). The marked wavelength dependence of the photosubstitution quantum yields is consistent with the direct reaction from an upper level excited state prior to internal conversion to the state(s) responsible for fragmentation. Unlike the fragmentation pathway, the photosubstitution quantum yields were little affected by solvent; therefore, Φ_s/Φ_f ratios were much higher in THF solutions than in hydrocarbon solutions.

Flash photolyses of $Ru_3(CO)_{12}$ at the shorter wavelengths ($\lambda_{irr} >$ 315 nm) were carried out both in THF and cyclohexane solutions. No transients were noted in the latter solvent, but in THF under excess CO, transient absorbance in the wavelength range 480 to 550 nm, which decayed exponentially back to the starting spectrum with a [CO] dependent k_{obs}, was observed. Similar flash photolysis of $Ru_3(CO)_{12}$ in argon flushed THF solution with excess PPh_3 or $P(OCH_3)_3$ also gave initial transient absorptions at these monitoring wavelengths similar to those noted under CO. However, in these cases, the system was shown to undergo further absorbance increases exponentially to a final product spectrum consistent with net reaction to give, principally, the substituted clusters $Ru_3(CO)_{11}L$. Plots of k_{obs} vs [L] or [CO] were curved, but the double reciprocal plots (k_{obs}^{-1} vs $[L]^{-1}$) were linear in each case (Figure 3).

These data are interpretable in terms of a reaction scheme where the primary photoreaction is the dissociation of CO to give, first a $Ru_3(CO)_{11}$ intermediate (II), then the solvated species $Ru_3(CO)_{11}S$ (II').

$$Ru_3(CO)_{12} \xrightarrow{\ h\nu\ } Ru_3(CO)_{11} + CO \qquad (10)$$

$$II$$

$$II + S \underset{k_{-s}}{\overset{k_s}{\rightleftharpoons}} Ru_3(CO)_{11}S \qquad (11)$$

$$II'$$

$$II + CO \xrightarrow{\ k_{CO}\ } Ru_3(CO)_{12} \qquad (12)$$

$$II + L \xrightarrow{\ k_L\ } Ru_3(CO)_{11}L \qquad (13)$$

SCHEME 2

The relative solvent independence of Φ_s supports the view that the first step is CO dissociation rather than an associative displacement by solvent or another ligand.

The transient seen by flash photolysis in THF is proposed to be $\underline{II}'$ (S = THF), since no transient precursor to substitution with a lifetime > 30 μs was seen in cyclohexane despite the comparable Φ_s in both solvents. According to Scheme 2, when L = CO or [L] >> [CO] the following relationship would hold true:

$$k_{obs} = \frac{k_L k_{-s}[L]}{k_s + k_L[L]} \tag{14}$$

A plot of k_{obs}^{-1} vs $[L]^{-1}$ is thus predicted to be linear with slope = $k_s/k_{-s}k_L$ and a nonzero intercept k_{-s}^{-1} as seen in Figure 3. From the average intercept of $(0.9 \pm 0.4) \times 10^{-3}$ s, a k_{-s} value of about 1.1×10^3 s^{-1} can be estimated for the dissociation of THF from $\underline{II}'$. The relative values of k_L can be determined from ratios of the slopes given that k_s/k_{-s} should be ligand independent. These relative values are 8, 1.5 and 1.0 for CO, $P(OCH_3)_3$ and PPh_3, respectively.

It is interesting to note that the photosubstitution intermediate $\underline{II}$ appears to be significantly more selective toward reaction with various two electron donor substrates than is the photofragmentation intermediate $\underline{I}$. One speculative rationalization of this is that the $Ru_3(CO)_{11}$ intermediate has the opportunity to "delocalize" its unsaturation by having one CO bridge an edge of the metal triangle with concomitant formation of a multiple metal-metal bond. A similar rearrangement is not accessible to $\underline{I}$.

Summary. Figure 4 illustrates the proposed photofragmentation and photosubstitution mechanisms $Ru_3(CO)_{12}$ ($\underline{5}$). Quantum yields for the latter process are markedly wavelength dependent, very small or undetectable at λ_{irr} > 400 nm but dominant for UV excitation. The photofragmentation pathway, which is dominant for longer λ_{irr}, is quenched by donor ligands such as THF, but Φ_f values are little affected by the presence of CCl_4. Thus, it is proposed that fragmentation occurs via a nonradical isomer of the starting cluster having an unsaturated ruthenium center which reacts rather nonselectively with two electron donors L to give $Ru_3(CO)_{12}L$, the precursor to the photofragmentation. The photosubstitution pathway is proposed to proceed by CO dissociation to give the intermediate $Ru_3(CO)_{11}$, which is trapped by THF to give another transient, $Ru_3(CO)_{11}S$. Analyses of both CW quantum yield and kinetic flash photolysis data lead to the conclusion that $\underline{II}$ is significantly more selective than is $\underline{I}$ toward reactions with ligands, the greater selectivity of $\underline{II}$ suggested to be a consequence of a greater ability to delocalize the unsaturation.

Recent low temperature photochemical studies by Bentsen and Wrighton ($\underline{12}$), who used FTIR to characterize intermediates and products, appear to confirm key qualitative features of the models proposed for the photofragmentation and photosubstitution mechanisms in Schemes I and II and in Figure 4 ($\underline{5}$). Short wavelength excitation (313 nm) of $Ru_3(CO)_{12}$ in 90 K alkane glass was shown to give first a $Ru_3(CO)_{11}$ species with only terminal CO's which then rearranged to an isomeric form of $Ru_3(CO)_{11}$ having a bridging CO. In the presence of

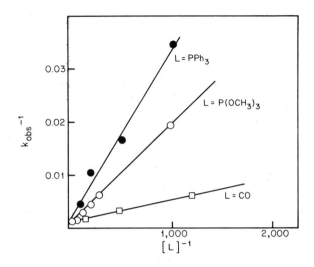

Figure 3. Double reciprocal plots of the kinetics data obtained
for the decay of the transients seen for the short wavelength
flash photolysis ($\lambda_{irr} > 315$ nm) of THF solutions of $Ru_3(CO)_{12}$
in the presence of various ligands L (from reference 5).

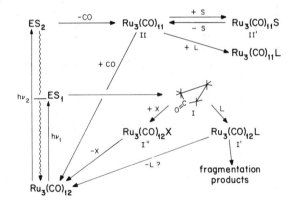

Figure 4. Qualitative model for the photoreactions of $Ru_3(CO)_{12}$
in solution.

various L, these intermediates reacted to form the $Ru_3(CO)_{11}L$ photosubstitution product. Longer wavelength excitation ($\lambda_{irr} > 420$ nm) of $Ru_3(CO)_{12}$ in 195 K alkane solution containing excess CO was shown to give $Ru(CO)_5$ and $Ru_2(CO)_9$ as the initial products. Similar reaction in the presence of PPh_3 gave an intermediate formulated as $Ru_3(CO)_{12}(PPh_3)$, which displayed an IR band (1791 cm^{-1}) characteristic of a bridging CO as proposed above for $\underline{I}'$. However, under these conditions (195 K), this intermediate did not fragment but underwent CO or PPh_3 loss to give $Ru_3(CO)_{11}(PPh_3)$ or $Ru_3(CO)_{12}$.

Lastly, it is appropriate to comment on the relationships between the intermediates seen in photochemical studies and possible reactive intermediates along the reaction coordinates of related thermal transformations. Earlier kinetics studies ($\underline{13}$) of the reactions of $Ru_3(CO)_{12}$ with various phosphorous ligands PR_3 have found evidence for both first order and second order pathways leading to substitution plus some cluster fragmentation. The first order path was proposed to proceed via reversible CO dissociation to give an intermediate analogous to $\underline{II}$.

$$Ru_3(CO)_{12} \underset{k_{-6}}{\overset{k_6}{\rightleftharpoons}} Ru_3(CO)_{11} + CO \qquad (15)$$

$$Ru_3(CO)_{11} + L \xrightarrow{k_7} Ru_3(CO)_{11}L \qquad (16)$$

The k_6 value was determined to be about 6.9×10^{-5} s^{-1} independent of the nature of L in 50°C decalin ($\Delta H^{\ddagger} = 31.8$ kcal mol^{-1}; $\Delta S^{\ddagger} = +20.2$ cal mol^{-1} K^{-1}). Competition ratios k_{-6}/k_7 equal to 3 and 5 were determined for L = $P(OPh_3)_3$ and PPh_3, respectively under the same conditions. The second order pathway was proposed to occur via nucleophilic attack of L on the cluster, and an intermediate with a formulation the same as $\underline{II}'$ was suggested, without supporting evidence of its existence, as a possible initial product of this nucleophilic attack. However, since fragmentation was only a minor side reaction of the substitution reactions with L = PPh_3, it is quite unlikely that the photofragmentation and second order thermal substitution reactions occur via a common intermediate.

Photoisomerization of $HRu_3(CO)_{10}(\mu-\eta^1\text{-}COCH_3)$

Photolysis of the methylidyne cluster $HRu_3(CO)_{10}(\mu-\eta^1\text{-}COCH_3)$ ($\underline{A}$) ($\underline{14}$) in cyclohexane solution leads to an unprecedented oxygen-to-carbon alkyl migration to form the bridging acyl complex $HRu_3(CO)_{10}(\mu-\eta^2\text{-}C(O)CH_3)$ ($\underline{B}$):

$$(17)$$

This transformation was demonstrated ($\underline{14}$) by evaluating changes in the UV, IR and NMR spectra and comparing these to the spectra of authentic samples of each cluster ($\underline{15},\underline{16}$). Quantum yields for the photoisomerization depicted in Equation 17 were found to be notably dependent both on the CO concentration and on the λ_{irr}. Although the resulting optical changes were the same for different λ_{irr}, the quantum yields in CO saturated cyclohexane ranged from $< 10^{-5}$ at 405 nm to 4.9×10^{-2} at 313 nm. Furthermore, Φ_2 varied linearly from 1.2×10^{-4} at $P_{CO} = 0.0$ to 4.9×10^{-2} at $P_{CO} = 1.0$ atm for 313 nm photolysis in cyclohexane.

No cluster fragmentation was observed initially, although long term photolysis (313 nm) of $\underline{B}$ in CO saturated cyclohexane eventually did lead to fragmentation to $Ru(CO)_5$ plus acetaldehyde:

$$(CO)_3Ru\underset{H-Ru(CO)_3}{\overset{CH_3}{\diagdown}}Ru(CO)_4 \;+\; 5\;CO \;\xrightarrow[\Phi_3]{h\nu}\; 3\;Ru(CO)_5 \;+\; CH_3-\overset{O}{\overset{\|}{C}}\diagdown_H \qquad (18)$$

This photoreaction was studied quantitatively using authentic samples of $HRu_3(CO)_{10}(\mu\text{-}\eta^2\text{-}C(O)CH_3)$, and a quantum yield of 1.1×10^{-3} moles/einstein was determined.

Any proposed mechanism for the unprecedented transformation described by Equation 18 must account for the promotion of this photoisomerization by CO, although CO is not required by the stoichiometry. A possible initial step would be similar to that for the $Ru_3(CO)_{12}$ fragmentation (Scheme 1). In this a Ru-Ru bond is broken concomitant with the movement of a CO from a terminal to a bridging site to form an unsaturated intermediate analogous to $\underline{I}$. A speculative proposal along these lines is presented in Figure 5. The key feature of this proposal would be the formation of $\underline{III}$ with one unsaturated ruthenium, which could be captured by CO to promote the subsequent steps leading from the $\mu\text{-}\eta^1$-methylidyne to the $\mu\text{-}\eta^2$-acyl complex.

If such a scheme indeed were responsible for the above isomerization then Equation 17 should also be facilitated by other two-electron donors capable of capturing $\underline{III}$. In this context, it is notable that use of THF rather than cyclohexane as the solvent (under argon) gives a much larger quantum yield 1.4×10^{-3}. Thus, unlike the photofragmentation of $Ru_3(CO)_{12}$, which is quenched by the donor solvent, the ligand isomerization is promoted by THF, probably because CO is not required in the overall stoichiometry of the latter transformation. Preliminary flash photolysis experiments are also consistent with this view. Flash photolysis ($\lambda_{irr} > 313$ nm) of $\underline{A}$ in argon deaerated THF gave a long lived ($\tau > 1$ s) transient while reproducible transients with lifetimes greater than 30 μs could not be observed when analogous experiments were carried out in cyclohexane. Thus we conclude that a key intermediate in the photoisomerization of $\underline{A}$ is an unsaturated cluster such as $\underline{III}$ which can be trapped by the two electron donor THF. Although preliminary results suggest that subsequent rearrangement to $\underline{B}$ by the THF adduct may be less efficient than from the proposed CO adduct, the former apparently can function in this manner.

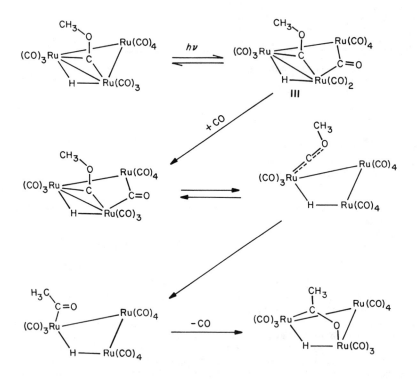

Figure 5. Proposed scheme for the photoisomerization of $HRu_3(CO)_{10}(\mu-\eta^1-COCH_3)$ to $HRu_3(CO)_{10}(\mu-\eta^2-C(O)CH_3)$.

Substitution Reactions of the Hydride Cluster $HRu_3(CO)_{11}^-$

The trinuclear ruthenium hydride ion $HRu_3(CO)_{11}^-$ has drawn considerable recent attention as a prominent species in homogeneous catalysts
for the water gas shift reaction (1) and for the hydrogenation,
hydroformylation, and hydrosilation of alkenes (17,18). In the
course of investigating the reactivity of $HRu_3(CO)_{11}^-$ and its
potential roles in such catalytic cycles, this ion was found (19) to
be remarkably more labile toward ligand substitution (Equation 19)
than the parent neutral carbonyl $Ru_3(CO)_{12}$ which was described above
as being rather slow to undergo substitutions.

$$HRu_3(CO)_{11}^- + PPh_3 \rightleftharpoons HRu_3(CO)_{10}(PPh_3)^- + CO \qquad (19)$$

A systematic investigation (19) of this reaction in THF has shown
the kinetics to be consistent with the following scheme:

$$HRu_3(CO)_{11}^- \underset{k_{-8}}{\overset{k_8}{\rightleftharpoons}} HRu_3(CO)_{10}^- + CO \qquad (20)$$
$$\underline{IV}$$

$$\underline{IV} + PPh_3 \underset{k_{-9}}{\overset{k_9}{\rightleftharpoons}} HRu_3(CO)_{10}(PPh_3)^- \qquad (21)$$

This mechanism predicts that (at low [CO]) a plot of k_{obs} vs $[PPh_3]$
will approach the condition where $k_9[PPh_3] \gg k_{-8}[CO]$ and k_{obs}
reaches a limiting value equal to k_8, the rate of CO dissociation
from $HRu_3(CO)_{11}^-$. This has been shown to be the case for several
different [CO] (Figure 6), and k_{obs}(limiting) = 2.1 s^{-1} ± 0.1 has
been determined at 25°C and ambient pressure (19). Thus, the rate
data are consistent with this model and argue against reaction of
PPh_3 with $HRu_3(CO)_{11}^-$ in an associative or interchange pathway to
displace CO.

However, an alternative mechanism by which CO is displaced by the
nucleophilic attack of solvent was not excluded by these kinetics
results, especially given that the activation parameters $\Delta H^{\ddagger}$ = 16.0 ±
1.7 kcal/mol and $\Delta S^{\ddagger}$ = -1.9 ± 3.0 cal $mol^{-1} K^{-1}$ for k_{obs}(limiting)
(19) would indeed appear to be more consistent with an associative
type mechanism than with the dissociative path described above. This
problem was addressed by measuring the stopped-flow kinetics of Equation 19 under limiting conditions (k_{obs} = k_8) at various pressures
(20). A plot of $\ln(k_{obs})$ vs P (Figure 7) gave the activation volume
$\Delta V^{\ddagger}_{ave}$ = +21.2 ± 1.4 $cm^3 mol^{-1}$). Quantitative prediction of the
expected $\Delta V^{\ddagger}$'s for these models is restricted by the absence of
partial molar volume data in THF for the various reactants and intermediates. However, one may assume that $HRu_3(CO)_{11}^-$ and the dissoc-
iated intermediate $\underline{IV}$ have similar $\bar{V}$'s and that $\bar{V}(CO)$ in this
solvent is close to that of liquid CO (about 23 cm^3/mol). Thus, the
measured $\Delta V^{\ddagger}$ value strongly supports the concept of a limiting
dissociative mechanism. Further consistent with this view is the
observation (19) that k_{obs} is essentially independent of whether the
solvent is THF or the more sterically demanding 2,5-dimethyl-THF.

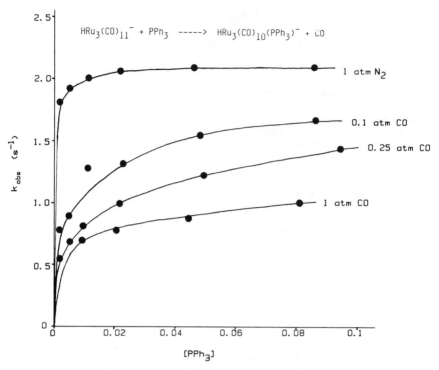

Figure 6. Plots of k_{obs} vs [PPh₃] for the reaction of HRu₃(CO)₁₁⁻ plus PPh₃ in THF under varied P_{CO} at 25°C (curves drawn for illustrative purposes only) (from reference 19).

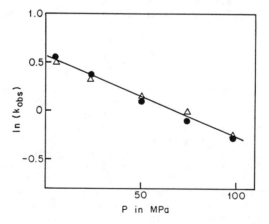

Figure 7. Plots of (k_{obs}) vs pressure for the reaction

$$HRu_3(CO)_{11}^- + PPh_3 \longrightarrow HRu_3(CO)_{10}(PPh_3)^- + CO$$

Reaction run in 25°, N₂ flushed THF with limiting concentrations of PPh₃ (O = 0.053 M PPh₃; △ = 0.086 M PPh₃).

The above kinetics studies of the thermal reactions provide powerful indirect evidence for the operation of a limiting dissociative
mechanism in this solvent and for the formation of a reactive
intermediate such as $\underline{IV}$. Such studies also allow one to evaluate the
relative reactivities of that intermediate with different substrates.
For example, k_{-8}/k_9, the ratio of the rate constants for reaction of
$\underline{IV}$ with CO or PPh$_3$ in 25° THF, was determined to have the value 15 ±
4 by analysis of the rate data presented in Figure 7. However, under
favorable conditions, it should be possible to use flash photolysis
to observe the reactive intermediate directly and to measure absolute
rate constants for its various reactions. In this context the photochemistry of the HRu$_3$(CO)$_{11}^-$ cluster anion has been briefly explored
in these laboratories. Continuous photolysis (λ_{irr}) of a solution of
[Bu$_4$N][HRu$_3$(CO)$_{11}$] (λ_{max} 387 nm, ϵ = 6,900 L mol^{-1} cm^{-1}) in CO
saturated or argon flushed THF led to no observable photochemistry;
i.e. no fragmentation was seen. (Photosubstitution with CO would be
undetectable and thermal reactions with other ligands are too rapid
for convenient investigation of the photochemical analogs.) However,
flash photolysis (λ_{irr} > 315 nm) of this salt (1 × 10^{-5} mol L^{-1}) in
argon flushed THF led to observable transient bleaching in the
370-440 nm wavelength region and transient absorption in the 450-540
nm region indicating the formation of new species. The absorbance
changes at all wavelengths detectable decayed to the starting spectrum via second order kinetics and with the same lifetime. When a
small but known concentration of CO (1.3 × 10^{-4} mol L^{-1}) was
introduced to the reaction solution, the same transient was observed,
but the decay kinetics became first order. This observation clearly
suggests that the transient formed in this experiment is the result
of CO photodissociation.

$$HRu_3(CO)_{11}^- \xrightarrow{h\nu} HRu_3(CO)_{10}^- + CO \qquad (22)$$

The rate constant for the exponential relaxation of the latter system
to the starting system was calculated to be 1.4 × 10^3 s^{-1}. From this
value, an approximate second order rate constant of 1.0 × 10^7 L mol^{-1}
s^{-1} was calculated for the reaction between $\underline{IV}$ and CO. Given the
above determination of the limiting rate constant for CO dissociation
of 2 s^{-1}, the equilibrium constant for thermal CO dissociation from
HRu$_3$(CO)$_{11}^-$ in THF to give $\underline{IV}$ can be calculated from the ratio of the
forward and back rate constants (k_8/k_{-8}) to be 2 × 10^{-7} mol L^{-1}.

Concluding remarks

In this article we have summarized the use of both photochemical and
more classical thermal kinetics techniques to deduce the nature of
intermediates in the ambient temperature, fluid solution chemistry of
several triruthenium clusters. In some cases the photochemically
generated intermediates appear to be the same as those proposed to
be formed along thermal reaction coordinates, while in other cases
unique pathways are the results of electronic excitation. The use of
pulse photolysis methodology allows direct observation, and the measurement of the reaction dynamics of such transients and provides
quantitative evaluation of the absolute reactivities of these
species. In some cases, detailed complementary information regarding

photoreaction intermediates can be deduced also by trapping these
species at low temperatures and characterizing their spectroscopic
properties (12). The examples described here illustrate the power of
a comprehensive approach using all of the above techniques to inves-
tigate the chemistries of these high energy species.

Acknowledgments

This work was supported by grants from the National Science Foundation (INT83-04030; CHE-8419283) and the US Department of Energy,
Office of Basic Energy Sciences (DE-FG03-85ER13317). Key aspects of
the experimental studies described here were carried out in these
laboratories by Marc Desrosiers and David Wink, whose intellectual
contributions are greatly appreciated.

Literature Cited

1. Ford, P. C. Accounts of Chem. Research 1981, 14, 31-37.
2. Gross, D. C.; Ford, P. C. J. Am. Chem. Soc. 1985, 107, 585-593.
3. Desrosiers, M. F.; Ford, P. C. Organometallics 1982, 1,
    1715-1716.
4. Desrosiers, M. F.; Wink, D. A.; Ford, P. C. Inorg. Chem 1985,
    24, 1-2.
5. Desrosiers, M. F.; Wink, D. A.; Trautman, R.; Friedman, A. E.;
    Ford, P. C. J. Am. Chem. Soc. 1986, 108, 1917-1927.
6. Johnson, B. F. G.; Lewis, J.; Twigg, M. V. J. Organomet. Chem.
    1974, 67, C75-76.
7. Geoffroy, G. L.; Wrighton, M. S. "Organometallic Photochemis-
    try"; Academic Press: New York, 1979.
8. Malito, J.; Markiewicz, S.; Poë, A. Inorg. Chem. 1982, 21,
    4335-4338.
9. Burke, M. R.; Takats, J.; Grevels, F.-W.; Reuvers, J. G. A. J.
    Am. Chem. Soc. 1983, 105, 4092-4093.
10. Leopold, D. G.; Vaida, V.; J. Am. Chem. Soc. 1983, 105,
    6809-6811.
11. Austin, R. G.; Paonessa, R. S.; Giordano, P. J.; Wrighton, M. S.
    ACS Adv. Chem. Ser. 1978, 168, 189-214.
12. Bentson, J. G.; Wrighton, M. S., submitted for publication, pri-
    vate communication from M. S. Wrighton.
13. Poë, A.; Twigg, M. V. J. Chem. Soc., Dalton Trans. 1974,
    1860-1866.
14. Friedman, A. E.; Ford, P. C. J. Am. Chem. Soc., in press.
15. Keister, J. B.; Payne, M. W.; Muscatella, M. J. Organometallics
    1983, 2, 219.
16. Boag, N. M.; Kampe, C. E.; Lin, Y. C.; Kaesz, H. D. Inorg. Chem.
    1982, 21, 1704-1708.
17. Süss-Fink, G. Angew. Chem. Int. Ed. Engl. 1982, 21, 73-74.
18. Süss-Fink, G.; Reiner, J. J. Mol. Catal. 1982, 16, 231-242.
19. Taube, D. J.; Ford, P. C. Organometallics 1986, 5, 99-104.
20. Taube, D. J.; van Eldik, R.; Ford, P. C. Organometallics, in
    press.

RECEIVED November 12, 1986

Chapter 9

Photochemical Reactions Between Transition Metal Complexes and Gases at High Pressures

Manfred J. Mirbach

BBC-Brown, Boveri & Co. Ltd., Central Laboratory RLC, Baden, Switzerland

Applications of photochemical techniques to reactions between coordination compounds in solution and gases at high pressures are reviewed. The main result of increasing the pressure in gas-liquid phase systems is to increase the solubility of the gas in the liquid. The high pressure photolysis technique has been used (1) to synthesize complexes containing H_2, N_2, CO, or CH_4 ligands, (2) to initiate homogeneous catalytic reactions such as hydrogenation, hydroformylation, and the water gas shift reaction, and (3) to study the mechanism of thermal reactions by the photochemical preparation of possible intermediates.

This account summarizes our own results and the reports of other authors regarding the photochemical reactions between transition metal complexes and gases at high pressures. The reactions usually take place in a liquid solvent between dissolved substrates, metal complexes, and dissolved gases which are in equilibrium with a gas phase reservoir.

In such two-phase systems, a higher pressure increases the solubility of the gas in the liquid phase. The changes in the solution itself are similar to those which occur when a low-density liquid is mixed with a high-density solvent: the volume of the solution increases and its density and internal pressure decrease. /1/ (see Figure 1a). These effects are opposite to those observed in condensed singlephase systems, in which a pressure increase causes a volume decrease and a density increase /2,3/ (see Figure 1 b).

The fundamental differences between the chemical reactions when compressing a single liquid phase and when compressing a gas-liquid two-phase system are best explained by comparing the rate and equilibrium equations (Equation (1) and (2), respectively) of the reaction

$$a A + b B \rightleftharpoons c C$$

$$d[C]/dt = k [A]^m [B]^n \qquad (1)$$

$$K = [C]^c/[A]^a [B]^b \qquad (2)$$

If one of the reactants, e.g. A, is a gas, and the reaction takes place in solution, a pressure increase leads to a higher concentration of A in the solution: $([A] = f(P))$. The reaction rate may increase or decrease, depending on the reaction order, but the rate constant k changes little when increasing the pressure in gas liquid systems to 200 or 300 atm. Similar arguments apply to chemical equilibria (Equation 2). A higher pressure in gas A leads to a higher concentration of A in the solution, thereby shifting the equilibrium to the right.

If A and B are miscible liquids, increasing the pressure will change the rate constant k according to Equation 3. The rate may increase or decrease, depending on the sign of the activation volume $\Delta V^{\neq}$.

$$\ln k/dP = - \Delta V^{\neq}/RT \qquad (3)$$

In a single-phase liquid system, pressure can influence an equilibrium only by altering equilibrium constant K. This requires very high pressures ($\gg$1000 atm). The pressures applied in gas-liquid systems are much lower. They range from 1 atm to a few hundred atm. Higher pressures are usually not necessary, since they would not change the concentration of the gas in the liquid phase by very much.

High-Pressure Photoreactors

A reaction vessel for high pressure gas-liquid photochemical reactions must fulfill the following basic requirements:

1.) It must be safe, since compressed gases store considerable amounts of energy.
2.) It must be transparent to the radiation used.
3.) It must permit the efficient mixing of gas and liquid phases, so as to maintain equilibrium conditions.
4.) It should provide a means for monitoring the reaction.

A variety of designs has been used [1]. For small volumes and low pressures, quartz or glass tubes may suffice. Preparative scale experiments require autoclaves, which may have either external or internal light sources [4-8]. Figure 2 shows an autoclave having an external light source and a magnetic stirrer [7]. Similar devices with windows transparent to IR allow for in situ monitoring of the reaction. They are well suited for mechanistic studies and for small-scale preparations [7, 8].

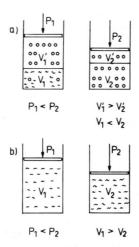

Figure 1 : a) The effect of high pressure in gas—liquid two-phase systems, b) The effect of high pressure in a liquid one-phase system, (o) = gas molecules; (~) = liquid molecules

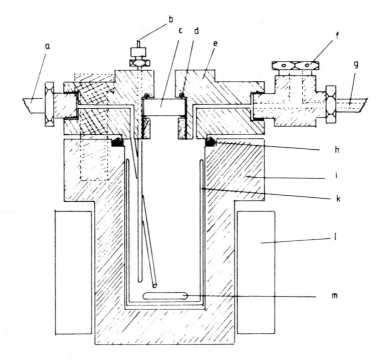

Figure 2: Schematic diagram of a 100-mL UV-autoclave. a = gas and sampling valve, b = thermocouple, c = quartz window, d = Teflon O-rings, e = autoclave lid, f = rupture disc, g = valve and pressure gauge, h = gaskets, i = autoclave body, k = glass insert, l = temperature control, m = stirring bar. (P_{max} = 300 bar, T_{max} = 150 °C) (Reproduced with permission from Ref. 7. Copyright 1983 Elsevier Sequoia.)

For large-scale preparations, an internal light source utilizes the light better. The autoclave shown in Figure 3 consists of a stainless steel body with two caps flanged to it. The bottom cap contains the UV lamp, which is protected from the pressurized solution by a thick (e.g. 6 mm) quartz dome. A second, smaller dome of either quartz or pyrex provides space for a liquid to cool the lamp, and for filters. The top cap contains the valves, controls, and magnetic stirrer /4/. For our mechanistic studies, an autoclave of this type was connected to a system in which part of the reaction mixture was pumped through high-pressure UV and IR spectroscopic cells, thereby permitting continuous monitoring of the reaction /9/.

Synthesis of Metal Complexes

The synthesis of metal complexes can benefit in several ways from the application of high pressure photochemical methods. Compounds requiring high pressures and temperatures for synthesis by conventional methods can be prepared under milder conditions when the reactants are irradiated. A typical example is the synthesis of cobalt hydrocarbonyl from cobalt acetate and CO/H_2 in a methanol solution. This reaction normally requires temperatures above $150°C$ and pressures of 200 - 300 atm., and a promotor is required to reduce the long induction time. Under photochemical conditions the reduction takes place smoothly at $80°C$ and 90 atm (syngas pressure) without a promotor. In this case, $Co(MeOH)_6^{++}$ ions are the light-absorbing species and molecular hydrogen is the reducing agent /10/. A similar reaction is also possible, however, with CO alone, although it is somewhat slower. At present it is not certain whether CO or methanol is the reducing agent in the absence of H_2 /11/.

$$Co(OAc)_2 \quad \xrightarrow[MeOH]{hv} \quad \begin{array}{l} \xrightarrow{+\ CO} Co(MeOH)_6^{++} + Co(CO)_4^- \\ \xrightarrow{+\ CO/H_2} H^+ + Co(CO)_4^- + HOAc \end{array} \quad (4)$$

Photochemical reductions can be carried out with a variety of complexes. An interesting example is the ionic complex 1, (Equation 5), because of the multitude of reactions observed /12/. When this complex is irradiated with light >300nm, only the cation is excited. In the primary photolysis step, the cation loses a ligand, either CO or L. In the absence of any other potential ligand, the coordinatively unsaturated species 2a and 2b react with the anion to give the neutral complexes 3 and 4. Under hydrogen at 50 atm, however, the intermediates 2a and 2b are trapped by hydrogen to yield the hydrides 3 and 4. If synthesis gas is used instead of hydrogen, the CO inhibits the formation of 6, and the monosubstituted cobalt carbonyl hydride 5 is the major product (>90 %). These examples demonstrate how the presence of gaseous reactants can change the product distribution completely. Indeed, many more reactions of this type have been observed or can be envisioned.

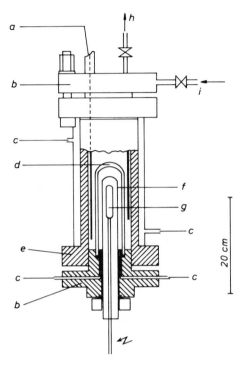

Figure 3: UV autoclave (after Gascard and Saus /4/).
a = magnetic stirrer, b = autoclave lid, c = connection to
thermostat, d = pressure-stable quartz tube, e = autoclave
housing, f = quartz or Pyrex tube, g = lamp, h = gas entrance
and sampling valve, i = gas outlet.
(Reproduced with permission from Ref. 1)

$$\left[Co(CO)_3L_2\right]\left[Co(CO)_4\right] \underset{1}{\overset{MeOH}{\rightleftharpoons}} \underset{1a}{Co(CO)_3L_2^+} + \underset{1b}{Co(CO)_4^-} \qquad (5)$$

$$\underset{1a}{Co(CO)_3L_2^+} \quad \xrightarrow{h\nu} \quad \begin{cases} \underset{2a}{Co(CO)_3L^+} + L \\ \underset{2b}{Co(CO)_2L_2^+} + CO \end{cases} \qquad (6)$$

$$\underset{2a}{Co(CO)_3L^+} \quad \begin{cases} \xrightarrow{+H_2} H^+ + \underset{5}{HCo(CO)_3L} \\ \xrightarrow{+Co(CO)_4^-} \underset{3}{Co_2(CO)_7L} \end{cases} \qquad (7)$$

$$\underset{2b}{Co(CO)_2L_2^+} \quad \begin{cases} \xrightarrow{+H_2} H^+ + \underset{6}{HCo(CO)_2L_2} \\ \xrightarrow{+Co(CO)_4^-} \underset{4}{Co_2(CO)_6L_2} \end{cases} \qquad (8)$$

$L = P(n\text{-}Bu)_3$

The application of high-pressure photochemistry is not limited to the formation of stable hydrido-carbonyls. P. Krusic prepared the hydrido iron carbonyl radical $HFe_2(CO)_8$ by irradiation of iron pentacarbonyl under hydrogen at 30 atm, and studied its ESR spectrum in situ by means of an ingeniously designed experiment /13/. He suggested that the dinuclear radical is formed through a reaction of the mononuclear analogue with unconverted $Fe(CO)_5$. In the absence of hydrogen, stable $Fe_2(CO)_9$ is the only product.

Another promising technique is to use liquid xenon as a solvent, as it uniquely inert and is transparent to radiation. High pressure extends the temperature range in which xenon is liquid, and permits the addition of gaseous reactants like hydrogen or nitrogen /14, 15/. Using this technique, R.K. Upmacis et al prepared the chromium pentacarbonyl - molecular hydrogen complex. An alternative method of preparation is by the irradiation of chromium hexacarbonyl in heptane, under hydrogen at 100 atm /16/.

$$Cr(CO)_6 + H_2 \; \underset{367 \text{ nm}}{\overset{228 \text{ nm}}{\rightleftharpoons}} \; Cr(CO)_5(H_2) + CO \qquad (9)$$

The liquid xenon technique was also used for preparing the analogous nitrogen complex /17/.

$$Cr(CO)_6 + N_2 \; \underset{367 \text{ nm}}{\overset{UV}{\rightleftharpoons}} \; Cr(CO)_5(N_2) + CO \qquad (10)$$

The nitrogen complex had already been synthesized in a solid matrix, but its decomposition kinetics and its further photolysis could be studied only in solution. The liquid noble gas technique is superior to the solid matrix technique, especially for the synthesis of multiple substituted chromium carbonyl nitrogen complexes. Their IR spectra were extremely complex in matrices, due to "site splittings" which arise when different molecules are trapped in different matrix environments /18/.

$$Cr(CO)_5(N_2) + N_2 \longrightarrow Cr(CO)_4(N_2)_2 \quad (cis) + CO \qquad (11)$$

$$Cr(CO)_4(N_2)_2 + N_2 \longrightarrow fac\text{-}Cr(CO)_3(N_2)_3 + mer\text{-}Cr(CO)_3(N_2)_3 + CO \qquad (12)$$

$$Cr(CO)_3(N_2)_3 + N_2 \longrightarrow Cr(CO)_2(N_2)_4 + CO \qquad (13)$$

$$Cr(CO)_2(N_2)_4 + N_2 \longrightarrow Cr(CO)(N_2)_5 + CO \qquad (14)$$

The more nitrogen the complexes contain, the less stable they are. The last member of the series, chromium hexanitrogen, was not detected, probably because it is too unstable at the temperature of liquid xenon. Other complexes and reactions were also studied, using liquid xenon or liquid krypton as the solvent. A list is given in Table 1.

Actually, a similar approach was used in studying the oxidative addition of methane to an iridium complex. Hydrocarbon solvents would have reacted faster than methane with the photochemically produced unsaturated iridium species, therefore J.K. Hoyano et al chose perfluorinated hexane as being an inert solvent. The elevated pressure was necessary in order to increase the concentration of the methane in the solution sufficiently to shift equilibrium (15) to the right /20/.

$$MeCpIr(CO)_2 + CH_4 \; \underset{}{\overset{h\nu, \ 8 \text{ atm}}{\rightleftharpoons}} \; MeCpIr{\overset{H}{\underset{CO}{\diagdown}}}CH_3 + CO \qquad (15)$$

The photocatalyzed reduction of carbon dioxide at elevated pressure was also investigated. Porous glass beads were used to obtain efficient gas-liquid contact. With isopropanol as the solvent and 2-propyl formate as the reducing agent,the reaction products were carbon monoxide and hydrogen. The catalyst, chloro(tetraphenyl-porphinato)rhodium(III), was irradiated with visible light /21/.

Table I: Photochemical reactions in liquefied noble gases

Starting material	Product	Ref.
$CpRh(C_2H_4)_2 + N_2$	$CpRh(C_2H_4)N_2$	47
$CpRh(C_2H_4)N_2 + CO$	$CpRh(C_2H_4)CO$	47
$Fe(CO)_2(NO)_2 + H_2$	$Fe(CO)(NO)_2(H_2)$	48
$Co(CO)_3(NO) + H_2$	$Co(CO)_2(NO)(H_2)$	48
$Fe(CO)_2(NO) + N_2$	$Fe(CO)(NO)_2(N_2)$	19
$Fe(CO)_2(NO)(H_2) + N_2$	$Fe(NO)_2(H_2)_2$	19
$Ni(CO)_4 + N_2$	$Ni(CO)_3(N_2)$	15
$Fe(CO)_2(NO)_2 + butene$	$Fe(CO)_2(NO)(^2\text{-butene})$	49
$Co(CO)_3(NO) + butene$	$Co(CO)_2(NO)(^2\text{-butene})$	49
$M(CO)_6 + H_2$ (M=Cr, Mo, W)	$M(CO)_5(H_2) + M(CO)_4(H_2)_2$	50
$Cr(CO)_6 + Xe$	$Cr(CO)_5Xe$	51
$Cr(CO)_6 + H_2$	$Cr(CO)_5(H_2)$	16
$Cr(CO)_6 + C_2H_4$	$Cr(CO)_5(C_2H_4)$	52
$Cr(CO)_6 + n\ N_2$	$Cr(CO)_{6-n}(N_2)_n$	17
$Ru(CO)_5 + n\ ^{13}CO$	$Ru(CO)_{5-n}(^{13}CO)_n$	53
$Fe(CO)_2(NO)_2 + butadiene$	$Fe(CO)(NO)_2(butadiene)$	54[a]
$Co(CO)_3(NO) + butadiene$	$Co(CO)_2(NO)(butadiene)$	54[a]

a = other substitution products are also formed.

$$2\ CO_2 + 4\ CH_3\underset{OH}{CH}\text{-}CH_3 \xrightarrow[Rh(III)]{h\nu} HCOOC_3H_7 + CO + 3\ CH_3\text{-}\underset{O}{C}\text{-}CH_3 + H_2 + 2\ H_2O \tag{16}$$

This may be of significance in connetion with the biosynthesis of acetate from carbon dioxide, because the next step, the fixation of carbon monoxide, was demonstrated by B. Kräutler. He irradiated methyl cobalamin under Co at 30 atm and obtained the acyl cobalamin as the product. Interestingly, a radical mechanism was iproposed, involving the reaction of methyl radicals with CO to give acyl radicals, which then recombine with the cobalt complex /55/.

$$CH_3\text{-}Co(III) \xrightarrow{h\nu} \cdot CH_3 + Co(II) \xrightarrow{+\ CO} CH_3CO\text{-}Co(III) \tag{16a}$$

Photocatalysis

The potentially most promising application of high pressure photo-chemistry is in catalysis. Most industrial processes are catalytic, and many of these require high temperatures and pressures. Activation of the catalysts by light can lead to higher activity and selectivity or to novel reaction paths which yield products not obtained under conventional thermal conditions.

To our knowledge, the first published report of a photocatalytic reaction at elevated pressure was W. Strohmeyer's hydrogenation of 1,3-cyclohexadiene under hydrogen at 10 atm /22/. On photolysis, the iridium complex 8 formed a very active catalyst, probably by dissociation of a phosphine ligand (Equation 17). At $70^{\circ}C$, with hydrogen at 10 atm, and a catalyst/substrate ratio of 1/100,000, the activity was 196 per minute and the turnover number was 96,000 mol of product/mol catalyst.

$$\underset{8}{IrCl(CO)L_2} \;\overset{h\nu}{\rightleftharpoons}\; IrCl(CO)L + L \qquad\qquad (17)$$

One of the most selective hydroformylation catalysts was obtained when cobalt acetate was irradiated in the presence of an excess of a phosphine, with synthesis gas at 80 atm, in methanol as the solvent. Propylene was hydroformylated with this catalyst to give butyraldehyde with an n/i ratio of more than 99/1 /10/. In the absence of phosphine, the cobalt acetate forms a more active catalyst which is, however, less selective for straight chain products /23/.

$$CH_3-CH=CH_2 + CO + H_2 \;\xrightarrow[\substack{80^{\circ}C,80\ atm \\ in\ methanol \\ +\ P(n-Bu)_3}]{h\nu,\quad Co(OAc)_2}\; \begin{array}{l} CH_3-CH_2-CH_2-CHO \quad\ 99\ \% \\[6pt] +\ CH_3-\underset{CHO}{CH}-CH_3 \quad 1\ \% \end{array} \qquad (18)$$

The mechanism of this reaction was studied in detail, using high-pressure UV and IR spectroscopy. The first step is a fast thermal reaction of cobalt acetate with syn-gas and phosphine to from the ionic complex 7. The yellow cation is the photoactive species. On irradiation it is converted to the cobalt carbonyl hydride 8, as discussed above. The hydride then reacts with the olefin in a reaction sequence containing only thermal steps /10/.

$$Co^{++} + 2\ OAc^- + 2\ L + 3\ CO + 1/2\ H_2 \longrightarrow \underset{7}{Co(CO)_3L_2^+} + OAc^- + HOAc$$

$$(19)$$

$$\underset{7}{Co(CO)_3L_2}^+ + H_2 \xrightarrow[CO]{h\nu} \underset{8}{HCo(CO)_3L} + H^+ + L \qquad (20)$$

When irradiated in the presence of norbornadiene and high-pressure synthesis gas, rhodium chloride is converted to a catalyst which is active for a variety of reactions. /24/. The salt is probably converted photochemically to the rhodium norbornadiene complex 9. This dimer may undergo a consecutive photoreaction to give the monomeric hydrido complex 10, which is the actual catalyst for polymerisation, hydrogenation, and hydroformylation reactions.

$$2\ RhCl_3 x3H_2O + 2\ NBD \xrightarrow[-HCl]{h\nu} \underset{9}{(NBD)Rh(Cl_2)Rh(NBD)} \qquad (21)$$

$$\xrightarrow[\substack{h\nu \\ -HCl}]{+\ H_2} 2\ \underset{10}{RhH(NBD)_2}$$

Polymerisation is the dominating reaction when norbornadiene is used as the only organic reactant. Hydrogenation dominates when aromatic olefins such as styrene are added. Particularly noteworthy are the reactions of 1-octene in the presence of the norbornadiene rhodium system. At low syn-gas pressures (<50 atm) the major products are the methyl nonanoates resulting from a hydroesterfication reaction (Equation 22a). At higher syn-gas pressures (>50 atm), the hydroformylation dominates, leading to the formation of dimethyl acetals of the nonanals (Equation 22b).

$$R-CH = CH_2 \xrightarrow[20^{\circ}C,\ CO/H_2]{h\nu,\ RhCl_3 + NBD} \begin{cases} R-CH(OMe)_2 & (22a) \\ \text{major product, >50 atm} \\ \\ R'-COOMe & (22b) \\ \text{major product, 10-50 atm} \end{cases}$$

The rhodium catalyst is rather unselective with respect to the isomeric distribution of the products.

Rhodium and cobalt carbonyls have long been known as thermally active hydroformylation catalysts. With thermal activation alone, however, they require higher temperatures and pressures than in the photocatalytic reaction. Iron carbonyl, on the other hand, is a poor hydroformylation catalyst at all temperatures under thermal activation. When irradiated under synthesis gas at 100 atm, the iron carbonyl catalyzes the hydroformylation of terminal olefins even at room temperatures, as was first discovered by P. Krusic. ESR studies suggested the formation of $HFe_2(CO)_8$ radicals as the active catalyst, /25, 26/. Our own results support this idea, /27,28/. Light is necessary to start the hydroformylation of 1-octene with the iron carbonyl catalyst. Once initiated, the reaction proceeds even in the

dark, and additional irradiation leads to only a minor enhancement
of the conversion. This is consistent with the idea that iron pentacarbonyl photoreacts to $H_2Fe(CO)_4$, which in turn forms the hydrido
radical, the thermally active catalyst of the hydroformylation
(Equations 24-27).

$$R-CH=CH_2 \xrightarrow[\text{Fe(CO)}_5]{\text{hv,CO/H}_2} R-CH_2-CH_2-CHO + R-CH-CHO \tag{23}$$

$$\underset{3}{} \quad / \quad \overset{CH_3}{\underset{1}{}}$$

$$Fe(CO)_5 \underset{}{\overset{\text{hv}}{\rightleftharpoons}} Fe(CO)_4 + CO \tag{24}$$

$$Fe(CO)_4 + H_2 \rightleftharpoons H_2Fe(CO)_4 \tag{25}$$

$$H_2Fe(CO)_4 + Fe(CO)_4 \longrightarrow 2\ HFe(CO)_4 \tag{26}$$

$$HFe(CO)_4 + Fe(CO)_5 \rightleftharpoons HFe_2(CO)_8 + CO \tag{27}$$

Mechanistic Studies

High-pressure photochemistry has been used very successfully in
studying the mechanisms of catalytic reactions. Irradiation of a
suitable precursor permits in-situ preparation of reactive intermediates such as coordinatively unsaturated complexes or radicals. It
is thus possible to check whether these species are involved in the
catalytic cycle.

A typical example of this is the dicobalt octacarbonyl catalyzed hydroformylation of olefins to yield aldehydes. According to the
classical mechanism proposed by Heck and Breslow /29/ (Equations
28-31), the cobalt carbonyl reacts with hydrogen to form hydrido
cobalt tetracarbonyl, which is in equilibrium with the coordinatively unsaturated $HCo(CO)_3$. The tricarbonyl coordinates the olefin,
and rearranges to form the alkyl cobalt carbonyl.

$$Co(CO)_8 + H_2 \rightleftharpoons 2\ HCo(CO)_4 \tag{28}$$

$$HCo(CO)_4 \rightleftharpoons HCo(CO)_3 + CO \tag{29}$$

$$HCo(CO)_3 + \text{olefin} \rightleftharpoons (\pi\text{-olefin})Co(CO)_3 \tag{30}$$

$$(\pi\text{-olefin})Co(CO)_3 \rightleftharpoons R\text{-}Co(CO)_3 \tag{31}$$

The alkyl complex reacts further with carbon monoxide to give the
acyl complex, which in turn yields the aldehyde by hydrogenolysis.

The Heck and Breslow mechanism was widely accepted until evidence was found that $Co(CO)_4$ radicals may be involved in the hydroformylation /30, 31/ (Equations 32-34).

$$Co_2(CO)_8 \rightleftharpoons 2 \cdot Co(CO)_4 \qquad (32)$$

$$\cdot Co(CO)_4 + olefin \rightleftharpoons \cdot Co(CO)_3(\pi\text{-olefin}) + CO \qquad (33)$$

$$\cdot Co(CO)_3(\pi\text{-olefin}) + HCo(CO)_4 \longrightarrow HCo(CO)_3(\pi\text{-olefin} + \cdot Co(CO)_3$$

$$(34)$$

The hypothesis that the cobalt carbonyl radicals are the carriers of catalytic activity was disproved by a high pressure photochemistry experiment /32/, in which the $Co(CO)_4$ radical was prepared under hydroformylation conditions by photolysis of dicobalt octacarbonyl in hydrocarbon solvents. The catalytic reaction was not enhanced by the irradiation, as would be expected if the radicals were the active catalyst. On the contrary, the $Co(CO)_4$ radicals were found to inhibit the hydroformylation. They initiate the decomposition of the real active catalyst, $HCo(CO)_4$, in a radical chain process /32, 33/.

A similar set of experiments was carried out with the phosphine substituted derivative $Co_2(CO)_6L_2$ (L = PBu_3). Again it was shown that $Co(CO)_3L$ radicals were not involved in the hydroformylation of aliphatic olefins /32/. It is quite clear now, however, that such radicals may play a role in the hydrogenation of aromatic olefins, like styrene, stilbene, etc. /34, 35/.

The question as to the existence of 17 versus 16 electron intermediates was also raised in the example of the photocatalytic hydrogenation of olefins using iron pentacarbonyl as the catalyst precursor (Equation 35). Schroeder and Wrighton studied this reaction at normal pressure, and they suggested $H_2Fe(CO)_4$ and $H_2Fe(CO)_3$, respectively, as the active catalysts /36/.

$$R\text{-CH=CH}_2 + H_2 \xrightarrow[Fe(CO)_5]{hv} R\text{-CH}_2\text{-CH}_3 \qquad (35)$$

On the other hand, the previously mentioned ESR study by P. Krusic showed that $HFe(CO)_4$ and $HFe_2(CO)_8$ radicals are formed during photolysis of iron pentacarbonyl in the presence of hydrogen /13/. The ESR detection of radicals, however, does not prove that they are involved in the catalytic cycle.

$$Fe(CO)_5 + H_2 \xrightarrow{hv} H_2Fe(CO)_4 \text{ or } HFe_2(CO)_8 \quad ? \qquad (36)$$

Therefore we studied this reaction at various hydrogen pressures, using IR and UV spectroscopy /37/, and found that the rate of the $H_2Fe(CO)_4$ formation increased with the hydrogen pressure, but that the rate of hydrogenation decreased with pressure. This inversely proportional relation proved that $H_2Fe(CO)_4$ cannot be the active hydrogenation catalyst, and that a mechanism via Krusic's radical is more likely.

The high-pressure photochemistry technique also contributed to clarifying the mechanism of the chromium carbonyl catalyzed water gas shift reaction (Equation 37) /38/.

$$CO + H_2O \xrightarrow[Cr(CO)_6]{} CO_2 + H_2 \tag{37}$$

Two alternatives had been proposed earlier. The first assumed the dissociation of a CO ligand as the first step (Equation 38), and the coordination of a formate ion as the second step (Equations 39-40) /39/.

$$Cr(CO)_6 \underset{hv}{\rightleftarrows} Cr(CO)_5 + CO \tag{38}$$

$$CO + OH^- \rightleftarrows HCOO^- \tag{39}$$

$$Cr(CO)_5 + HCOO^- \longrightarrow Cr(CO)_5(OOCH)^- \tag{40}$$

The second alternative assumed that chromium hexacarbonyl reacts with hydroxyl ions to give a formate complex, without any preceding dissociation (Equation 41) /40/.

$$Cr(CO)_6 + OH^- \rightleftarrows Cr(CO)_5(COOH)^- \tag{41}$$

The dissociation process should be enhanced by UV irradiation and inhibited by high CO pressure, whereas a reaction through an associative pathway should be unaffected by either. We carried out the corresponding experiments /38/, and found that the chromium carbonyl catalyzed water gas shift reaction is much faster with irradiation than without, and that it is inhibited by high CO pressure. Therefore it seems quite clear that this reaction takes place through a dissociative mechanism.

We made a similar set of experiments for the iron pentacarbonyl catalyzed water gas shift reaction. In this case, the activity of the catalyst is unaffected by UV light, showing that here an associative mechanism is in operation here /41/.

Another subject of dispute was the mechanism of the photochemical, chromium carbonyl catalyzed hydrogenation of dienes /42/. The question here was whether the catalytic reaction is started by the dissociation of CO (Equation 42) or by the dissociation of the coordinated diene (Equation 43) /42, 43/.

$$Cr(CO)_4(\eta^4\text{-diene}) \xrightarrow{h\nu} \begin{cases} Cr(CO)_3(\eta^4\text{-diene}) + CO & (42) \\ Cr(CO)_4(\eta^2\text{-diene}) & (43) \end{cases}$$

11 12 13

In a series of experiments using various pressures of CO and H_2O, with norbornadiene as the diene /44, 45/, we were able to verify an earlier suggestion /43/ that the 1,2 addition product norbornene is formed through a diene dissociation (Equation 43) and the 1,5 addition product nortricyclene through a CO dissociation (Equation 42).

$$(44)$$

$$(45)$$

At high hydrogen pressures, norbornane is formed by hydrogenation of a nortricyclene precursor /45/.

With conjugated dienes, it is mainly 1,4-hydrogenation which is observed. The product ratio, however, does not reflect the ratio of the initial photoprocesses, since many thermal catalytic cycles follow each primary step. These thermal cycles take place mainly through intermediate 12 /45/.

By using hydrogen at high pressure, M.A. Green et al. were able to show that the first step in the photolysis of OsH_4L_3 (L=PMe$_2$Ph) is the reductive elimination of H_2. The 16-electron intermediate can react with excess phosphine, or can dimerize, or can exchange hydrogen with the benzene solvent /46/.

$$OsH_4L_3 \underset{}{\overset{h\nu}{\rightleftharpoons}} H_2 + OsH_2L_3 \qquad (46)$$

Acknowledgements : I would like to extend my sincere thanks and appreciation to the co-authors of our papers listed in the references. Our work was carried out partly at the Technische Hochschule Aachen, and at the University of Duisburg, both in Germany, and partly at the University of Petroleum and Minerals in Dhahran, Saudi Arabia.

References

1 Mirbach,M.F.; Mirbach,M.J., Saus,A. Chem. Rev. 1981, 82, 59
2 Palmer,D.A.; Kelm,H. Coord.Chem.Rev. 1981, 36, 89
3 Dibenedetto,J.; Ford,P.C. Coord.Chem.Rev. 1985, 64, 361
4 Gascard,T.; Saus,A. J.Phys.E 1982, 15, 627
5 Shekhtman,I.R.; Rybakova,I.A.; Shakhovskoi,G.P.; Lutsek,V.P. Khim.Vys.Energ. 1985, 19, 349
6 Cirjak,L.M. U.S.Patent 4,517,063 (14.5.1985)
7 Saus,A.; Phu,T.N.; Mirbach,M.J.; Mirbach,M.F. J.Mol. Catal. 1983, 18, 117
8 Rigby,W.; Whyman,R.; Wilding,K.; J.Phys. E 1970, 3, 572
9 Mirbach,M.F.; Mirbach,M.J.; Saus,A. Chemiker Ztg. 1982, 106, 335
10 Mirbach,M.J.; Mirbach,M.F.; Saus,A.; Topalsavoglou,N.; Phu,T.N. J.Am.Chem.Soc. 1981, 103,7594; Angew. Chem. 1981, 93, 391
11 Mirbach,M.J.; Mirbach,M.F. Unpublished results.
12 Mirbach,M.F.; Mirbach,M.J.; Wegmann,R.W. Organometallics 1984, 3, 900
13 Krusic,P.J. Am.Chem.Soc. 1981, 103, 2131
14 Beattie,W.H.; Maier II,W.B.; Holland,R.F.; Freund,S.M.; Steward,B. Proc.Soc.Photo-Opt.Instrum.Eng. 1978, 158, 113
15 Turner,J.J.; Simpson,M.B.; Poliakoff,M.; Maier II,W.B. J.Am.Chem.Soc. 1983, 105, 3898
16 Upmacis,R.K.; Gadd,G.E.; Poliakoff,M.; Simpson,M.B.; Turner,J.J.; Whyman,R.; Simpson,A.F. J.Chem.Soc.,Chem. Comm. 1985, 27
17 Maier II,W.B.; Poliakoff,M.; Simpson,M.B.; Turner,J.J. J.C.S. Chem.Comm. 1980, 587
18 Turner,J.J.; Simpson,M.B.; Poliakoff,M.; Maier II,W.B.; Graham,M.A. Inorg.Chem. 1983, 22, 911
19 Gadd,G.E.; Poliakoff,M.; Turner,J.J. Inorg.Chem. 1984, 23, 630
20 Hoyano,J.K.; McMaster,A.D.; Graham,W.A.G. J.Am.Chem.Soc. 1983, 105, 7190
21 Saito,Y.; Li,X.; Kurahashi,K.; Shinoda,S. Sci.Pap. Inst.Phys.Chem.Res.(Jap.) 1984, 78, 150; from Chem. Abstr. 1984, 103,150747h
22 Strohmeier,W.; Steigerwald,H. J.Organomet.Chem. 1977,125,C37
23 Mirbach,M.J.; Topalsavoglou,N.; Phu,T.N.; Mirbach,M.F.; Saus,A. Chem.Ber. 1983, 116, 1422
24 Saus,A.; Phu,T.N.; Mirbach,M.J.; Mirbach,M.F. J.Mol.Catal. 1983, 18, 117
25 Krusic,P. 2nd Int. Symp.Homog.Catal. Düsseldorf 1980
26 Cote,W.J.; Krusic,P.J. X Int.Conf.Organometal. Chem. Toronto 1981, Book of Abstracts, p. 48
27 Phu,T.N. Ph.D.Thesis, University of Duisburg, 1981
28 Mirbach,M.J. Habilitationsschrift, University of Duisburg 1982
29 Heck,F.; Breslow,D.S. J.Am.Chem.Soc. 1961, 83 4023

30 Markò,L. in "Fundamental Research in Homogeneous Catalysis"
    Vol.4, Ed. Graziani,M. and Giongo,M., Plenum Press,
    New York, 1984, p. 1-18
31 Ungvàry,F.; Markò,L. J. Organometal.Chem. 1980, 219, 397
32 Mirbach,M.J.; Mirbach,M.F.; Saus,A.; Topalsavoglou,N.;
    Phu,T.N. J.Am.Chem.Soc. 1981, 103, 7590
33 Wegmann,R.W.; Brown,T.L. J.Am.Chem.Soc. 1980, 102, 2494
34 Nalesnik,T.e.; Orchin,M. Organometallics 1982, 1, 223
35 Ungvàry,F.; Markò,L. J. Organometal. Chem. 1983, 249, 441
36 Schroeder,M.; Wrighton,M.S. J.Am.Chem.Soc. 1976, 98, 551
37 Nagorski,H.; Mirbach,M.J. J.Organometal. Chem. 1985, 291,
    199
38 Nagorski,H.; Mirbach,M.J.; Mirbach,M.F. J.Organometal.
    Chem. 1985, 297, 171
39 King,Jr.,A.D.; King,R.B.; Yang,D.B. J.Am.Chem.Soc. 1981,
    103, 2699
40 Darensbourg,D.J.; Rokicki,R. Organometallics 1982, 1, 1685
41 Nagorski,H. Ph.D.Thesis, University of Duisburg, 1984
42 Nasielski,J.; Kirsch,P.; Wilputte-Steinert,L.
    J.Organometal.Chem. 1971, 27, C13
43 Darensbourg,D.J.; Nelson III,H.H.; Murphy,M.A. J.Am.Chem.
    Soc. 1977, 99, 896
44 Mirbach,M.J.; Steinmetz,D.; Saus,A. J.Organometal. Chem.
    1979, 168, C13
45 Mirbach,M.J.; Phu,T.N.; Saus,A. J.Organometal.Chem. 1982,
    236, 309
46 Green,M.A.; Huffman,J.C.; Caulton,K.G. J.Organometal.Chem.
    1983, 243, C78
47 Haddleton,D.M.; Perutz,R.N.; Jackson,S.A.; Upmacis,R.K.;
    Poliakoff,M. J. Organometal Chem., in press
48 Gadd,G.E.; Upmacis,R.K.; Poliakoff,M.; Turner,J.J.
    J.Am.Chem.Soc. 1986, 108, 2547
49 Gadd,G.E.; Poliakoff,M.; Turner,J.J. Inorg.Chem., in press
50 Upmacis,R.K.; Poliakoff,M.; Turner,J.J. J.Am.Chem.Soc. 1986,
    108, 3645
51 Simpson,M.B.; Poliakoff,M.; Turner,J.J.; MaierII,W.B.;
    McLaughling,J.G. JCS, Chem.Comm. 1983, 1355
52 Gregory,M.F.; Jackson,S.A.; Poliakoff,M.; Turner,J.J.
    JCS, Chem.Comm., in press
53 Gregory,M.F.; Poliakoff,M.; Turner,J.J. J.Molecular Struct.
    1985, 127, 247
54 Gadd,G.E.; Poliakoff,M.; Turner,J.J. Organometallics, in press
55 Kräutler,B. Helv. Chim.Acta 1984, 67, 1053

RECEIVED November 3, 1986

Chapter 10

Electrochemiluminescence of Organometallics and Other Transition Metal Complexes

A. Vogler and H. Kunkely

Universität Regensburg, Institut für Anorganische Chemie, D-8400 Regensburg, Federal Republic of Germany

A variety of transition metal complexes including organometallics was subjected to an ac electrolysis in a simple undivided electrochemical cell, containing only two current-carrying platinum electrodes. The compounds (A) are reduced and oxidized at the same electrode. If the excitation energy of these compounds is smaller than the potential difference of the reduced (A^-) and oxidized (A^+) forms, back electron transfer may regenerate the complexes in an electronically excited state ($A^+ + A^- \rightarrow A^* + A$). Under favorable conditions an electrochemiluminescence (ecl) is then observed ($A^* \rightarrow A + h\nu$). A weak ecl appeared upon electrolysis of the following complexes: Ir(III)-(2-phenylpyridine-C^2,N^1)$_3$; [Cu(I)(pyridine)I]$_4$, Pt(II)(8-quinolinolate)$_2$, Tb(III)(TTFA)$_3$(o-phen) with TTFA = thenoyltrifluoroacetonate and o-phen = 1.10-phenanthroline, Tb(III)(TTFA)$_4^-$, and Eu(III)(TTFA)$_3$-(o-phen). An ecl of Re(o-phen)(CO)$_3$Cl occured during the electrolysis of tetralin hydroperoxide in the presence of the rhenium compound. The mechanism of these electrochemical reactions is discussed.

An electrolysis can be considered to be a high-energy process if reduction and oxidation take place at large potential differences. The primary redox products thus formed can participate in a variety of competing processes (1). They may be kinetically labile and undergo fragmentation or substitution reactions. As an alternative a rapid back electron transfer should take place due to the large driving force. This leads to the regeneration of the starting compounds either in the ground state or in an electronically excited state if the potential difference exceeds the excitation energy. At very large driving forces excited state generation is expected to be favored over ground state formation in many cases according to the Marcus theory (2). Various experimental methods are available to gain more insight into such high-energy electron transfer reactions. The detection of excited products is possible if they are luminescent. Chemiluminescence resulting from electron trans-

0097-6156/87/0333-0155$06.00/0
© 1987 American Chemical Society

fer reactions of transition metal complexes has been observed in a
few cases (3-14). However, the electrochemical generation of an
appropriate redox pair in situ offers various advantages. Under
suitable conditions an electrolysis will then be accompanied by
light emission (electrochemiluminescence or electrogenerated chemiluminescence, ecl) (15). By application of an alternating current
a redox pair is generated at the same electrode.

$$A - e^- \rightarrow A^+ \qquad \text{anodic cycle}$$
$$A + e^- \rightarrow A^- \qquad \text{cathodic cycle}$$
$$A^+ + A^- \rightarrow A* + A \qquad \text{annihilation}$$
$$A* \rightarrow A + h\nu \qquad \text{emission}$$

Back electron transfer takes place from the electrogenerated reduc-
tant to the oxidant near the electrode surface. At a sufficient
potential difference this annihilation leads to the formation of
excited (*) products which may emit light (ecl) or react "photochemically" without light (1,16). Redox pairs of limited stability
can be investigated by ac electrolysis. The frequency of the ac
current must be adjusted to the lifetime of the more labile redox
partner. Many organic compounds have been shown to undergo ecl
(17-19). Much less is known about transition metal complexes
despite the fact that they participate in many redox reactions.
Most observations of ecl involve $Ru(bipy)_3^{2+}$ (bipy = 2,2'-bipy-
ridyl) and related complexes which possess emissive charge transfer
(CT) metal to ligand (ML) excited states (11,20-33). The organometallic compound $Re(o-phen)(CO)_3Cl$ (o-phen = 1,10-phenanthroline)
is a further example of this category (34). Palladium and platinum
porphyrins with emitting intraligand (IL) excited states are also
ecl active (35). Under suitable conditions ecl was observed for
$Cr(bipy)_3^{3+}$. In this case the emission originates from a
ligand field (LF) excited state (27). Finally, it has been shown
that the electrolysis of $Pt_2(pop)_4^{4-}$ (36) (pop^{2-} = diphosphonate) or Mo_6Cl_{12} (37) is also accompanied by light emission. The redox processes as well as the subsequent excited state
formation involve the metal-metal bonding of these polynuclear complexes.

The present investigation was carried out in order to extend
ecl to other types of transition metal compounds including organometallics. In addition to the search for new systems the modification of a well-known ecl was used to learn more about the reaction
mechanism.

The choice of new complexes was guided by some simple considerations. The overall ecl efficiency of any compound is the product
of the photoluminescence quantum yield and the efficiency of excited state formation. This latter parameter is difficult to evaluate. It may be very small depending on many factors. An irreversible decomposition of the primary redox pair can compete with back
electron transfer. This back electron transfer could favor the
formation of ground state products even if excited state formation
is energy sufficient (13,14,38,39). Taking into account these
possibilities we selected complexes which show an intense photoluminescence ($\Phi > 0.01$) in order to increase the probability for
detection of ecl. In addition, the choice of suitable complexes
was also based on the expectation that reduction and oxidation
would occur in an appropriate potential range.

Experimental Section

Materials. The compounds Re(o-phen)(CO)$_3$Cl (40), Ir(2-phenylpy-
ridine-C^2,N^1)$_3$ (41), [Cu(pyridine)I]$_4$ (42,43), Pt(8-quino-
linolate)$_2$ (44), Tb(TTFA)$_3$(o-phen) (45,46) with TTFA = thenoyltrifluoroacetonate, [NH(C$_2$H$_5$)$_3$][Tb(TTFA)$_4$] (45), Eu(TTFA)$_3$-
(o-phen) (45), and tetralin hydroperoxide (47) were prepared according to published procedures. For the electrochemical experiments
acetonitrile and CH$_2$Cl$_2$ were triple vacuum line distilled from
P$_4$O$_{10}$ and degassed by several freeze-thaw cycles. The supporting electrolyte Bu$_4$NBF$_4$ was crystallized from dry acetone
several times and dried in vacuo.

Equipment and Methods. The ac electrolyses were carried out under
argon in 1-cm quartz spectrophotometer cells which were equipped
with two platinum foil electrodes directly connected to a Kröncke
Model 1246 sine wave generator as an ac voltage source. Ecl was
detected and spectrally analyzed by several procedures. The first
detection was achieved by connecting the spectrophotometer cell
directly with a photomultiplier (Hamamatsu 1 P21). This arrangement was also used to obtain maximum ecl intensity by variation of
the terminal ac voltage and the ac frequency. A crude spectral
analysis of the ecl was accomplished by placing appropriate broad-
band interference filters and cut-off filters between the cell and
the photomultiplier. The interference filters (Balzer) K3, K4, K5,
and K7 transmitted maximum intensity at λ = 510, 565, 610, and 700
nm. The Schott cut-off filter KV 550 transmitted light of λ > 530
nm. Ecl spectra were recorded on a Hitachi 850 Fluorescence
Spectrophotometer.

Results

As reported previously ac electrolyses were carried out in a simple
undivided electrochemical cell containing only the two current-
carrying electrodes (16). Most compounds investigated in the present study showed only very weak ecl intensities. First experiments were carried out by placing the ecl cell directly in front of
a photomultiplier. By this simple procedure the lowest light intensities could be detected. Under comparable experimental conditions integrated ecl intensities were detected which were roughly
by a factor of 10^{-4} lower than that of Ru(bipy)$_3$$^{2+}$. These
measurements were used to adjust the experimental parameters such
as ac voltage and frequency to maximum ecl intensity. A qualitative analysis of the spectral distribution of ecl was achieved by
inserting broad-band interference filters and cut-off glass filters
between the cell and the photomultiplier. Finally, complete ecl
spectra were recorded on a luminescence spectrometer. Measurements
by cyclic voltammetry were carried out by A. Haimerl (48). For the
individual compounds the following experimental details of the ecl
experiments are given: solvent, concentration of the supporting
electrolyte, concentration of the compound subjected to electrolysis, terminal ac voltage, ac frequency, current, integrated ecl
intensity in arbitrary units not corrected for photomultiplier
response, and wavelength of maximum light intensity, transmitted by
appropriate filters.

$\underline{Re(o\text{-}phen)(CO)_3Cl}$. CH_3CN, 0.1 M Bu_4NBF_4, 3×10^{-4} M complex, 2 V, 30 Hz, 1.1 mA, 40 units, λ_{max} = 610 nm. Upon addition of 3×10^{-4} M tetralin hydroperoxide the ecl intensity increased to 120 units.

$\underline{Ir(ppy)_3 \text{ with ppy} = 2\text{-phenylpyridine-}C^2,N^1}$. CH_3CN, 0.05 M Bu_4NBF_4, 10^{-4} M complex, 4 V, 10 Hz, 9 mA, 4 units, λ_{max} = 510 nm.

$\underline{[Cu(py)I]_4 \text{ with py} = \text{pyridine}}$. CH_2Cl_2, 0.05 M Bu_4NBF_4, 2×10^{-4} M complex, 5 V, 1 Hz, 10 mA, 20 units, λ_{max} = 700 nm.

$\underline{Pt(QO)_2 \text{ with QU} = 8\text{-quinolinolate}}$. CH_3CN, 0.005 M Bu_4NBF_4, 3×10^{-4} M complex, 4 V, 30 Hz, 8 mA, 10 units, λ_{max} > 530 nm; at 6 V, 30 Hz, and 21 mA the ecl intensity increased to 200 units.

$\underline{Tb(TTFA)_3(o\text{-phen}) \text{ with TTFA} = \text{thenoyltrifluoroacetonate}}$. CH_3CN, 0.05 m Bu_4NBF_4, 2.9×10^{-4} M complex, 4 V, 300 Hz, 20 mA, 50 units, λ_{max} = 565 nm. During electrolysis a solid separates and covers the electrodes.

$\underline{[NH(C_2H_5)_3]Tb(TTFA)_4}$. CH_3CN, 0.05 M Bu_4NBF_4, 3×10^{-4} M complex, 4 V, 300 Hz, 20 mA, 2 units, λ_{max} = 565 nm.

$\underline{Eu(TTFA)_3(o\text{-phen})}$. CH_3CN, 0.005 M Bu_4NBF_4, 2.7×10^{-4} M complex, 4 V, 30 Hz, 8.7 mA, 60 units, λ_{max} = 610 nm. Electrodes are covered by a solid during the electrolysis.

Discussion

The ac electrolyses of this work were carried out in an undivided electrochemical cell containing only the two current-carrying electrodes. This simple apparatus has certainly its limitations but was appropriate for the detection of new ecl-active compounds. The ecl intensity of most systems studied here was only very small. Generally, there may be several explanations for this observation. In some cases the reduced and oxidized species formed at the electrodes are not very stable as revealed by cyclic voltammetry. Only a small fraction of these reactive molecules may undergo the desired annihilation reaction competing with an irreversible decay. Moreover, the back electron transfer could favor the formation of ground state products even if excited state generation is energy sufficient (13,14,38,39). Finally, for some complexes it is difficult to obtain the materials free of impurities. In other cases the complexes are thermally not completely stable and dissolution is accompanied by a small degree of decomposition. These impurities may either interfere with the desired electrode process or act as quenchers for the excited molecules undergoing ecl.

The main goal of the present study was to discover new ecl-active complexes. But the first example may demonstrate that complexes known to show ecl can serve to gain more insight into the mechanism of electron transfer processes.

Re(o-phen)(CO)$_3$Cl and Tetraline Hydroperoxide

In 1978 Wrighton and his group showed that the complex Re(o-phen)-(CO)$_3$Cl undergoes ecl from its lowest excited state which lies about +2.3 eV above the ground state (34). The annihilation is energy sufficient. The oxidation of the neutral complex occurs at $E_{1/2}$ = 1.3 V vs. SCE while the reduction takes place at -1.3 V. In 1981 we found that Re(o-phen)(CO)$_3$Cl shows an intense chemiluminescence during the catalytic decomposition of tetralin hydroperoxide (THPO) in boiling tetraline (12).

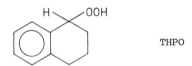

THPO

It was suggested that the mechanism of this reaction can be explained on the basis of a "chemically initiated electron exchange luminescence (CIEEL)" (49,50) according to the following scheme:

$$Re(o\text{-phen})(CO)_3Cl + THPO \rightarrow Re(o\text{-phen})(CO)_3Cl^+ + THPO^-$$
$$THPO^- \rightarrow \alpha\text{-tetralone}^- + H_2O$$
$$Re(o\text{-phen})(CO)_3Cl^+ + \alpha\text{-tetralone}^- \rightarrow Re(o\text{-phen})(CO)_3Cl^* + \alpha\text{-tetralone}$$
$$Re(o\text{-phen})(CO)_3Cl^* \rightarrow Re(o\text{-phen})(CO)_3Cl + h\nu$$

The first step is an activated electron transfer which takes place only at higher temperatures (T > 400 K). In the second step the reduced hydroperoxide is converted to the tetralone anion by elimination of water. This ketyl radical anion is strongly reducing ($E_{1/2}$ = -1.12 V) vs. SCE) (51). Electron transfer to the complex cation provides enough energy ($\approx$ 2.4 eV) to generate Re(o-phen)(CO)$_3$Cl in the emitting excited state. The overall process can be described as a catalyzed decomposition of tetralin hydroperoxide. The rhenium complex serves as an electron transfer catalyst which finally takes up the decomposition energy of the peroxide.

Apparently the same reaction sequence takes place when THPO and Re(o-phen)(CO)$_3$Cl are electrolyzed in acetonitrile at room temperature. The electrolysis replaces only the first activated electron transfer step of the CIEEL mechanism.

At an ac frequency of 30 Hz and a voltage larger than 2.6 V the ecl of Re(o-phen)(CO)$_3$Cl was very intense. If the voltage dropped below 2.6 V the efficiency of the electrolysis decreased. At 2 V the ecl was very weak. Upon addition of equimolar amounts of THPO the ecl intensity increased by a factor of $\approx$ 3. The hydroperoxide which is known to undergo an irreversible reduction at $E_{1/2}$ = -0.73 V vs. SCE (52) is apparently reduced during the cathodic cycle while the complex is oxidized during the anodic cycle. The subsequent reactions are assumed to be the same as those of the CIEEL mechanism. The overall reaction is an electro-catalyzed decomposition of THPO. The complex acts as an electro-catalyst.

Ir(2-phenylpyridine-c^2,N^1)$_3$

With regard to transition metal complexes the majority of ecl studies have been carried out with Ru(bipy)$_3^{2+}$ and its deriva-tives (11,20-33). Recently, King, Spellane, and Watts reported on the emission properties of Ir(ppy)$_3$ with ppy = 2-phenylpyridine-c^2,N^1 (41) which can be considered to be an organometallic counterpart of Ru(bipy)$_3^{2+}$.

Ir(ppy)$_3$

The lowest excited state ($\approx$ 2.5 eV) of the iridium complex which is also of the MLCT type undergoes an efficient emission. The quantum yield was about 0.4 in deoxygenated toluene at room temperature. The complex can be oxidized at $E_{1/2}$ = +0.7 V vs. SCE. The reduction was not reported but can be estimated to occur at $E_{1/2}$ = -1.9 V. This potential was obtained for the reduction of Pt(ppy)$_2$ (53). In this case the reduction was also assumed to take place at the ortho-metalated ppy$^-$ ligand.

The potential difference for reduction and oxidation ($\Delta E \approx$ 2.6 V) provides sufficient energy to generate an excited Ir complex in the annihilation reaction. At an ac voltage of 4 V and 10 Hz we observed a weak ecl of Ir(ppy)$_3$ in acetonitrile. The following reaction sequence may explain this observation:

$$Ir(ppy)_3 - e^- \rightarrow Ir(ppy)_3^+ \qquad \text{anodic}$$
$$Ir(ppy)_3 + e^- \rightarrow Ir(ppy)_3^- \qquad \text{cathodic}$$
$$Ir(ppy)_3^+ + Ir(ppy)_3^- \rightarrow Ir(ppy)_3^* + Ir(ppy)_3 \qquad \text{annihilation}$$
$$Ir(ppy)_3^* \rightarrow Ir(ppy)_3 + h\nu \qquad \text{emission}$$

Compared to the efficient ecl of Ru(bipy)$_3^{2+}$ the low ecl inten-sity of the Ir complex is rather surprising.

[Cu(pyridine)I]$_4$

Binuclear and polynuclear compounds with direct metal–metal interaction constitute a large class of transition metal complexes which
play an important role also in organometallic chemistry. Generally,
the frontier orbitals of these compounds are engaged in metal–metal
bonding. Consequently, redox processes affect the metal–metal
interaction. The same is true for the luminescence of such complexes since it involves also the frontier orbitals. The binuclear
complex Pt$_2$(pop)$_4^{4-}$ (36) (pop^{2-} = diphosphonate) and
the cluster Mo$_6$Cl$_{12}^{2-}$ (37) are rare examples of compounds
which contain metal–metal bonds and show photoluminescence at
ambient conditions. Both complexes are also ecl active. In the
present study the tetrameric complex [Cu(py)I]$_4$ with py =
pyridine as another compound of this type was investigated.

The colorless tetramer [Cu(py)I]$_4$ is fairly stable only in
non- or weakly coordination solvents such as benzene, CH$_2$Cl$_2$,
or acetone. At room temperature in solution this copper complex
shows an intense red photoluminescence ($\Phi \approx 0.04$) at λ_{max} = 698 nm
(54). The emitting state is a metal-centered 3d^{9}4s^1 excited
state which is strongly modified by Cu(I)–Cu(I) interaction in the
tetramer. This consists of a (CuI)$_4$ cubane core.

At a terminal voltage of 5 V and a frequency of 1 Hz the complex [Cu(py)I]$_4$ in CH$_2$Cl$_2$ showed a weak ecl which was clearly
identified as the red emission originating from the lowest excited
state of the complex.

It seems feasible that the ecl occurs according to the usual
mechanism. Reduction and oxidation of the complex is followed by
the annihilation and luminescence. However, there must be an
efficient competition by other processes since the ecl intensity is
rather low compared to the photoluminescence. As indicated by CV
measurements the reduction at $E_{1/2}$ = -0.7 V and -1.6 V and
oxidation at +0.8 V vs. SCE are largely associated with irreversible reactions. Hence, the reduced and oxidized forms of the
complex seem to be not stable. The ecl intensity is then low
because only a small fraction of the electrogenerated redox pair
escapes an irreversible decay and undergoes an annihilation. It is
also possible that the back electron transfer is not quite energy
sufficient for the formation of excited [Cu(py)I]$_4$ ($E \approx 2$ eV)
since the potential difference between first reduction and
oxidation is only 1.5 V. Finally, the large voltage of 5 V
required for the observation of the ecl could also indicate that
the solvent CH$_2$Cl$_2$ which is reduced at $E_{1/2}$ = -2.33 (55)
participates in the electrolysis and generation of excited
[Cu(py)I]$_4$.

Pt(8-quinolinolate)$_2$

The excited states which are responsible for the ecl of the previous examples are of the CTML type or involved in metal–metal
bonding of polynuclear complexes. Photoluminescence, or ecl in our
case, can also originate from intraligand (IL) excited states provided these states are the lowest excited states of such complexes. IL emissions are characteristic for many transition metal

porphyrins due to the low energy of the $\pi\pi*$ transitions of the porphyrin ligand (56). In 1974 Tokel-Takvoryan and Bard observed ecl from porphyrin IL excited states of Pd(II) and Pt(II) tetra-phenylporphyrin (35). In the present study we investigated the ecl of $Pt(QO)_2$ (QO^- = 8-quinolinolate) which is associated with an IL state of this Pt(II) chelate.

$Pt(QO)_2$

In 1978 Scandola and his group observed that solutions of $Pt(QO)_2$ exhibit an intense photoluminescence ($\Phi \approx 0.01$) at λ_{max} = 650 nm under ambient conditions (44). More details on the photo-physics and photochemistry of this compound were reported later (57-59). The emitting excited state was assigned to the lowest-energy IL triplet of the chelate ligand.

At a terminal ac voltage of 4 V and a frequency of 30 Hz we observed a weak ecl which was clearly identified as the IL emission of $Pt(QO)_2$. It is assumed that the ac electrolysis generates a redox pair $Pt(QO)_2^+Pt(QO)_2^-$. The subsequent annihilation leads to the formation of electronically excited $Pt(QO)_2$. The low ecl intensity may be associated with the observation that the electrochemical oxidation and reduction of $Pt(QO)_2$ is largely irreversible. CV measurements revealed an oxidation at $E_{1/2}$ = +0.9 and a reduction at -1.7 V vs. SCE. These redox reactions are probably ligand-based processes. The redox pair generated in the ac electrolysis decays irreversibly to a large extent. Only a small fraction undergoes the annihilation. The potential diffe-rence between $Pt(QO)_2^+$ and $Pt(QO)_2^-$ is 2.6 V. This is certainly sufficient to populate the emitting IL state which lies around 2.0 V above the ground state.

Terbium and Europium Complexes

The previous examples of ecl were interpreted on the basis of a relatively simple mechanism. In these cases the back electron transfer generates directly the emitting excited state (annihila-tion). However, in more complicated systems back electron trans-fer and formation of an emitting state may be separate processes

(17-19). The back electron transfer leads to a non-emitting excited state which undergoes energy transfer to a luminescent state.
This state does not participate in the redox reaction. The energy
transfer can occur as an intra- or intermolecular process. An
intermolecular sensitization of this type involving a metal complex
was studied by Bard and his group (60). An europium chelate served
as emitting energy acceptor. In the present work we studied ecl
which involves intramolecular energy transfer. The back electron
transfer leads to a non-emitting excited state at a certain part of
a molecule. Energy transfer populates an emitting excited state at
a different part. We selected rare earth metal chelates for this
study since they are well known to undergo intramolecular energy
transfer from excited IL states to emitting metal-centered f-levels
upon IL light absorption (61-64). Moreover, at least Tb(III) is
rather redox inert and does certainly not participate in electron
transfer processes at moderate potentials.

Tb(thenoyltrifluoroacetonate)$_3$(o-phen)

The complex Tb(TTFA)$_3$(o-phen) with TTFA = thenoyltrifluoroacetonate is a octacoordinate rare earth chelate which contains one
o-phen and three TTFA ligands (45-46). The latter are related to
acetylacetonate.

Tb(TTFA)$_3$(o-phen)

The longest wavelength absorption band of Tb(TTFA)$_3$(o-phen)
appears at λ_{max} = 336 nm. This intense ($\epsilon \approx 10^4$) and broad band
is assigned to an IL transition of TTFA (65). The metal centered
f-f bands are all narrow and of low intensity. Light absorption by
this IL band caused the typical green emission of Tb(III). The
main feature of this structured emmission spectrum at λ_{max} = 543
nm is assigned to the $^5D_4 \rightarrow {}^7F_5$ transition of the Tb^{3+}
ion (63,64). In analogy to many other rare earth chelates an
energy transfer occurs from the ligand to the emitting $^5D_4$
state of Tb^{3+} at 20500 cm^{-1} or 2.54 eV. The donor state for
energy transfer is the lowest TTFA triplet at 20660 cm^{-1} or 2.56
eV (63,64).

At a terminal voltage of 4 V and an ac frequency of 300 Hz $Tb(TTFA)_3$(o-phen) showed a weak ecl which was definitely identified as the emission from the $^5D_4$ state of Tb(III). The following scheme may describe the ecl mechanism.

$$Tb(III)(TTFA)_3(o\text{-phen}) - e^- \rightarrow Tb(III)(TTFA)_3(o\text{-phen})^+$$

anodic

$$Tb(III)(TTFA)_3(o\text{-phen}) + e^- \rightarrow Tb(III)(TTFA)_3(o\text{-phen})^-$$

cathodic

$$Tb(III)(TTFA)_3(o\text{-phen})^+ + Tb(III)(TTFA)_3(o\text{-phen})^-$$
$$\rightarrow Tb(III)[(TTFA)_3(o\text{-phen})]* + Tb(TTFA)_3(o\text{-phen})$$

back electron transfer

$$Tb(III)[(TTFA)_3(o\text{-phen})]* \rightarrow Tb(III)*(TTFA)_3(o\text{-phen})$$

intramolecular energy transfer

$$Tb(III)*(TTFA)_3(o\text{-phen}) \rightarrow Tb(III)(TTFA)_3(o\text{-phen}) + h\nu$$

emission

The complex $Tb(TTFA)_3$(o-phen) underwent a reduction at $E_{1/2}$ = -1.5 V vs. SCE which was partially reversible. An oxidation was not observed below +2 V. All redox reactions should be ligand-based processes. The potential difference of $\Delta E > 3.5$ V is energy sufficient to generate the IL triplet at 2.56 eV. The low ecl intensity could be due to a competing irreversible decay of the primary redox pair.

Tb(thenoyltrifluoroacetonate)$_4^-$

The absorption (λ_{max} = 335 nm) and emission (λ_{max} = 543 nm) spectrum of $Tb(TTFA)_4^-$ are very similar to those of $Tb(TTFA)_3$(o-phen). The IL excitation of $Tb(TTFA)_4^-$ is certainly also followed by energy transfer to the emitting $^5D_4$ f-level of Tb(III).

At a terminal voltage of 4 V and a frequency of 300 Hz the anion $Tb(TTFA)_4^-$ shows a very weak ecl. It was identified by a broad-band interference filter to appear around 565 nm. The intensity was too low to record the ecl spectrum on the emission spectrometer.

The ecl mechanism is assumed to be the same as that of $Tb(TTFA)_3$(o-phen). However, the very low ecl intensity of $Tb(TTFA)_4^-$ requires an explanation. The first reduction of this anion takes place at $E_{1/2}$ = -1.75 V vs. SCE and is partially

reversible. However, an oxidation at +0.85 V could be due to the
free ligand. In solution rare earth complexes are not very stable
with regard to loss of ligands. The free ligand interferes then
with the redox processes of the complex in the ac electrolysis.

Eu(thenoyltrifluoroacetonate)$_3$(o-phen)

In analogy to Tb(III) similar Eu(III) complexes show an intense
metal-centered photoluminescence involving the f-levels (61-65).
An ecl of an europium(III) chelate was reported before (60). The
complex was not involved in the electrolysis. Excited organic
compounds formed electrochemically underwent an intermolecular
energy transfer to the emitting Eu compound. Interestingly, in the
absence of the redox-active organic compounds an ecl of the
europium chelate was not observed.
    While Tb^{3+} is redox inert Eu^{3+} can be reduced to Eu^{2+} at
rather low potentials ($\approx$ -0.5 V) (66,67). This adds a further
complication to any possible ecl mechanism involving Eu(III)
complexes. Furthermore, the back electron transfer from Eu^{2+} to
an oxidizing ligand radical generated in the electrolysis is a
spin-allowed process if Eu^{3+} is formed in its ground state (68).
However, the formation of the emitting excited state of Eu^{3+} is
spin-forbidden. It follows that the generation of Eu^{3+} in the
ground state could be favored even if the excited state formation
is an energy-sufficient process. But it was pointed out that the
spin-selection rule may not be important due to the heavy atom
effect of europium.
    While the absorption spectrum of Eu(TTFA)$_3$(o-phen) is nearly
identical to that of the corresponding Tb complex, the intense red
emission is characteristic for Eu^{3+}. The main band of the
structured spectrum appears at λ_{max} = 613 nm and is assigned to
the metal-centered $^5D_o \rightarrow {}^7F_2$ transition (63-65). The
emitting ^5D_o state has an energy of 17150 cm^{-1} or 2.13 eV
while the lowest IL excited state which is a triplet of the TTFA
ligand occurs at 20450 cm^{-1} or 2.54 eV. The initially excited
ligand undergoes an efficient intramolecular energy transfer to the
emitting ^5D_o state of Eu^{3+}.
    At a terminal ac voltage of 4 V and a frequency of 30 Hz
Eu(TTFA)$_3$(o-phen) shows a weak ecl which was identified as the
typical emission of Eu^{3+} at λ_{max} = 613 nm. Without extensive
speculation it is difficult to propose an ecl mechanism due to the
complications discussed above.
    Eu(TTFA)$_3$(o-phen) was not oxidized below +2 V vs. SCE as
indicated by CV measurements. The reduction of Eu^{3+} to Eu^{2+} is
expected to occur around -0.5 V. A clear reduction wave was not
observed in this region. Another related Eu(III) chelate (60) was
also not reduced near this potential while some Eu(III) cryptates
(66,67) undergo a reversible reduction in this range. The complex
Eu(TTFA)$_3$(o-phen) showed two irreversible reductions at -1.3 and
-1.63 V. These are certainly ligand-based processes. The irrever-
sibility may be due to a direct decomposition of the reduced complex. As an alternative the reduced ligand could rapidly transfer
an electron to Eu^{3+}. The Eu(II) complex may undergo a facile
ligand displacement. All these complications at various stages of

the ac electrolysis can contribute to the low ecl intensity of $Eu(TTFA)_3(o-phen)$.

Conclusion

For transition metal complexes an intense ecl as it was observed for $Ru(bipy)_3^{2+}$ seems to be rather an exception. It is certainly difficult to draw definite mechanistic conclusions based on small ecl efficiencies because ecl may originate from side reactions in these cases. However, our results do show that electron transfer reactions with large driving forces can generate electronically excited transition metal complexes as a rather general phenomenon.

Acknowledgments

We thank A. Merz and A. Haimerl for measurements and the discussion of electroanalytical data. Financial support of this work by the Deutsche Forschungsgemeinschaft and the Fonds der Chemischen Industrie is gratefully acknowledged.

Literature Cited

1. Vogler, A.; Kunkely, H.; Schäffl, S. In "The Chemistry of Excited States and Reactive Intermediates"; Lever, A. B. P., Ed.; ACS Symposium Series No. 307, American Chemical Society: Washington, D. C., 1986, p. 120.
2. Siders, P.; Marcus, R. A. J. Am. Chem. Soc. 1981, 103, 748.
3. Lytle, F. E.; Hercules, D. M. Photochem. Photobiol. 1971, 13, 123.
4. Martin, J. E.; Hart, E. J.; Adamson, A. W.; Halpern, J. J. Am. Chem. Soc. 1972, 94, 9238.
5. Gafney, H. D.; Adamson, A. W. J. Chem. Ed. 1975, 52, 480.
6. Jonah, C. D.; Matheson, M. S.; Meisel, D. J. Am. Chem. Soc. 1978, 100, 1449.
7. Bolletta, F.; Rossi, A.; Balzani, V. Inorg. Chim. Acta 1981, 53, L 23.
8. Vogler, A.; El-Sayed, L.; Jones, R. G.; Namuath, J.; Adamson, A. W. Inorg. Chim. Acta 1981, 53, L 35.
9. Balzani, V.; Bolletta, F. J. Photochem. 1981, 17, 479.
10. Bolletta, F.; Balzani, V. J. Am. Chem. Soc. 1982, 104, 4250.
11. Rubinstein, I.; Bard, A. J. J. Am. Chem. Soc. 1981, 103, 512.
12. Vogler, A.; Kunkely, H. Angew. Chem. Int. Ed. Engl. 1981, 20, 469.
13. Ghosh, P. K.; Brunschwig, B. S.; Chou, M.; Creutz, C.; Sutin, N. J. Am. Chem. Soc. 1984, 106, 4772.
14. Liu, D. K.; Brunschwig, B. S.; Creutz, C.; Sutin, N. J. Am. Chem. Soc. 1986, 108, 1749.
15. Faulkner, L. R.; Bard, A. J. In "Electroanalytical Chemistry"; Bard, A. J., Ed.; Marcel Dekker Inc.: New York, 1977; Vol. 10, p. 1.
16. Kunkely, H.; Merz, A.; Vogler, A. J. Am. Chem. Soc. 1983, 105, 7241.

17. Faulkner, L. R.; Glass, R. S. In "Chemical and Biological Generation of Excited States"; Adam, W.; Cilento, G., Eds.; Academic Press: New York, 1982; chapter 6.
18. Park, S.-M.; Tryk, D. A. Rev. Chem. Intermediates 1981, 4, 43.
19. Pragst, F. Z. Chem. 1978, 18, 41.
20. Tokel, N. E.; Bard, A. J. J. Am. Chem. Soc. 1972, 94, 2862.
21. Tokel-Takvoryan, N. E.; Hemingway, R. E.; Bard, A. J. J. Am. Chem. Soc. 1973, 95, 6582.
22. Chang, M. M.; Saji, T.; Bard, A. J. J. Am. Chem. Soc. 1977, 99, 5399.
22. Chang, M. M.; Saji, T.; Bard, A. J. J. Am. Chem. Soc. 1977, 99, 5399.
23. Wallace, W. L.; Bard, A. J. J. Phys. Chem. 1979, 83, 1350.
24. Rubinstein, I.; Bard, A. J. J. Am. Chem. Soc. 1980, 102, 6641.
25. Luttmer, J. D.; Bard, A. J. J. Phys. Chem. 1981, 85, 1155.
26. Rubinstein, I.; Bard, A. J. J. Am. Chem. Soc. 1981, 103, 5007.
27. Bolletta, F.; Ciano, M.; Balzani, V.; Serpone, N. Inorg. Chim. Acta 1982, 62, 207.
28. Glass, R. S.; Faulkner, L. R. J. Phys. Chem. 1981, 85, 1160.
29. Itoh, K.; Honda, K. Chem. Lett. 1979, 99.
30. Abruna, H. D.; Bard, A. J. J. Am. Chem. Soc. 1982, 104, 2641.
31. Gonzales-Velasco, J.; Rubinstein, I.; Crutchley, R. J.; Lever, A. B. P.; Bard, A. J. Inorg. Chem. 1983, 22, 822.
32. Abruna, H. D. J. Electrochem. Soc. 1985, 132, 842.
33. Abruna, H. D. J. Electroanal. Chem. 1984, 175, 321.
34. Luong, J. C.; Nadjo, L.; Wrighton, M. S. J. Am. Chem. Soc. 1978, 100, 5790.
35. Tokel-Takvoryan, N. E.; Bard, A. J. Chem. Phys. Lett. 1974, 25, 235.
36. Vogler, A.; Kunkely, H. Angew. Chem. Int. Ed. Engl. 1984, 23, 316.
37. Nocera, D. G.; Gray, H. B. J. Am. Chem. Soc. 1984, 106, 824.
38. Sutin, N. Prog. Inorg. Chem. 1983, 30, 441.
39. Indelli, M. T.; Ballardini, R.; Scandola, F. J. Phys. Chem. 1984, 88, 2547.
40. Wrighton, M.; Morse, D. L. J. Am. Chem. Soc. 1974, 96, 998.
41. King, K. A.; Spellane, P. J.; Watts, R. J. J. Am. Chem. Soc. 1985, 107, 1431.
42. Hardt, H. D.; Pierre, A. Z. Anorg. Allg. Chem. 1973, 402, 107.
43. Raston, C. L.; White, A. H. J. Chem. Soc., Dalton Trans. 1976, 2153.
44. Ballardini, R.; Indelli, M. T.; Varani, G.; Bignozzi, C. A.; Scandola, F. Inorg. Chim. Acta 1978, 31, L 423.
45. Melby, L. R.; Rose, N. J.; Abramson, E.; Caris, J. C. J. Am. Chem. Soc. 1964, 86, 5117.
46. Bauer, H.; Blanc, J.; Ross, D. L. J. Am. Chem. Soc. 1964, 86, 5125.
47. Knight, H. B.; Swern, D. Org. Synth. 1954, 34, 90.
48. Haimerl, A., unpublished data.
49. Schuster, G. B. Acc. Chem. Res. 1979, 12, 336.
50. Schuster, G. B.; Horn, K. A. In "Chemical and Biological Generation of Excited States"; Adam, W.; Cilento, G., Eds.; Academic Press: New York, 1982, chapter 7.
51. Tirouflet, J.; Dabard, R.; Laviron, E. Bull. Soc. Chim. France 1963, 1655.

52. Willits, C. O.; Ricciuti, C.; Knight, H. B.; Swern, D. Anal. Chem. 1952, 24, 785.
53. Chassot, L.; Müller, E.; von Zelewsky, A. Inorg. Chem. 1984, 23, 4249.
54. Vogler, A.; Kunkely, H., submitted for publication.
55. Mann, C. K.; Barnes, K. K. "Electrochemical Reactions in Non-aqueous Systems"; Marcel Dekker: New York, 1970.
56. Gouterman, M. In "The Porphyrins"; Dolphin, D., Ed.; Academic Press: New York, 1978, Vol. III, chapter 1.
57. Bartocci, C.; Sostero, S.; Traverso, O. J. Chem. Soc., Faraday I 1980, 76, 797.
58. Scandola, F.; Ballardini, R.; Indelli, M. T. Sol. Energy Rand D Eur. Community, Ser. D 1982, 1, 66; Chem. Abstr. 1983, 98, 44028t.
59. Borgarello, E.; Pelizetti, E.; Ballardini, R.; Scandola, F. Nouv. J. Chim. 1984, 8, 567.
60. Hemingway, R. E.; Park, S.-M.; Bard, A. J. J. Am. Chem. Soc. 1975, 97, 200.
61. Crosby, G. A.; Whan, R. E.; Alire, R. M. J. Chem. Phys. 1961, 34, 743.
62. Filipescu, N.; Sager, W. F.; Serafin, F. A. J. Phys. Chem. 1964, 68, 3324.
63. Bhaumik, M. L.; El-Sayed, M. A. J. Chem. Phys. 1965, 42. 787.
64. Dawson, W. R.; Kropp, J. L.; Windsor, M. W. J. Chem. Phys. 1966, 45, 2410.
65. Winston, H.; Marsh, O. J.; Suzuki, C. K.; Telk, C. L. J. Chem. Phys. 1963, 39, 267.
66. Yee, E. L.; Gansow, O. A.; Weaver, M. J. J. Am. Chem. Soc. 1980, 102, 2278.
67. Sabbatini, N.; Ciano, M.; Dellonte, S.; Bonazzi, A.; Bolletta, F.; Balzani, V. J. Phys. Chem. 1984, 88, 1534.
68. Sabbatini, N.; Bonazzi, A.; Ciano, M.; Balzani, V. J. Am. Chem. Soc. 1984, 106, 4055.

RECEIVED November 3, 1986

Chapter 11

Electron Spin Resonance Studies of Primary Processes in the Radiolysis of Transition Metal Carbonyls

Martyn C. R. Symons[1], J. R. Morton[2], and K. F. Preston[2]

[1]Department of Chemistry, The University, Leicester, United Kingdom
[2]Division of Chemistry, National Research Council of Canada, Ottawa, Ontario K1A 0R6, Canada

The effect of ionizing radiation on molecular or ionic solids is to eject electrons, which often subsequently react at sites in the material well removed from the residual electron-loss centre. These electron-loss and electron-gain centres, or breakdown products thereof, are paramagnetic and have been extensively studied by e.s.r. spectroscopy. Results for a wide range of organo metals both as pure compounds and as dilute solid solutions are used to illustrate this action. Aspects of the electronic structures of these centres are derived from the spectra and aspects of redox mechanisms are discussed.
    Particular attention is given to recent work on electron-loss centres formed in $CFCl_3$ and related solvents, to studies on metal carbonyl derivatives, and to work on methyl cobalamine and its derivatives.

This is a large field, and since it is impossible to do justice to all but a small portion of it, we start with a general outline of the types of studies involved, and go on to illustrate and probe these generalisations with a few examples drawn mainly from our own studies. This is a form of laziness and is not in any sense intended as an implication of superiority! We start by asking what are the 'primary' events in radiolyses.

Primary Effects of Ionizing Radiation

From a chemist's viewpoint, the most important act of ionizing radiation (usually X-rays, γ-rays or high energy electrons) is electron ejection. Initially the ejected electrons have sufficient energy to eject further electrons on interaction with other molecules, but the electrons ultimately become thermalised and then are able to interact "chemically". We consider first various reaction pathways for these electrons, and then consider the fate of the "hole" centres created by electron ejection. [We refer to electron-gain and electron-loss centres rather than to radical-anions and -cations since, of course, the substrate may comprise ions rather than neutral molecules.

0097-6156/87/0333–0169$06.50/0

We consider, primarily, events in solids since most e.s.r. studies have been carried out on radicals trapped in solids. Only relatively persistent organometallic radicals have been studied by liquid-phase e.s.r. with *in situ* radiolysis, because of the technical difficulties involved. In most solid systems at low temperature radical centres are physically trapped in the rigid matrix and hence can be studied by e.s.r. without difficulty. However, although radicals as such may be immobile, this does not necessarily apply to electron-gain or -loss centres, particularly if these are charged, since electron-transfer may be facile.

Pure Materials

We envisage two extreme situations, one in which electron-gain and -loss centres are immobile, and one in which one or other is mobile *via* e^--transfer. From the viewpoint of solid-state physics, both are initially mobile after ionization, since the electron can be viewed as being in the conduction band and the loss centre as being part of the valence band. This situation is frequently very short-lived, since the electrons and 'holes' become trapped as discrete entities *via* various distortions or reactions. The two extremes are illustrated by reactions of KNO_3 and CH_3NO_2. The former gives trapped $NO_3\cdot$ and $\cdot NO_3{}^{2-}$ radicals *via* reactions (1) and (2). The key

$$NO_3^- \rightarrow NO_3\cdot + e^- \qquad (1)$$
$$NO_3^- + e^- \rightarrow \cdot NO_3{}^{2-} \qquad (2)$$

factor in such trapping is that the electron or 'hole' should remain close to one particular NO_3^- ion for a time long compared with the time taken for the $NO_3\cdot$ or $\cdot NO_3{}^{2-}$ radicals to relax to their ground-state geometries. Indeed trapping is probably initially induced at specific sites because the required distortion is incipiently present. The centres then remain trapped until they acquire sufficient thermal energy to distort back to the shape and size of a neighbouring NO_3^- ion, when electron-transfer can occur [1].

In contrast, when CH_3NO_2 is exposed to radiation, neither $CH_3NO_2^-$ nor $CH_3NO_2^+$ centres are formed [2]. We can be sure of this since the e.s.r. spectra of both these radical-ions have been thoroughly studied, and give well-defined spectra [3-4]. Instead, the major radicals detected by e.s.r. methods are $\cdot CH_3$ and $\cdot NO_2$. These radicals, which are also formed by photolysis, are presumably formed by electron-return into an outer orbital followed by homolysis (3 and 4). In this case, electron or 'hole' transfer, or

$$CH_3NO_2\cdot^+ + CH_3\dot{N}O_2^- \rightarrow (CH_3NO_2)^* \qquad (3)$$
$$CH_3NO_2{}^* \rightarrow \cdot CH_3 + \cdot NO_2 \qquad (4)$$

both, occur more rapidly than relaxation to give the radical ions.

Changes in shape are not, of course, the only factors that can prevent electron-return. Other factors, such as a change in 'solvation' or chemical reactions such as protonation, deprotonation, unimolecular break-down, rearrangement, etc., are summarised in Schemes 1 and 2. Some consequences of electron return are presented in Scheme 3. Here, AB stands for any species suffering the effects of radiation, including positive or negative ions as well as neutral molecules.

$$e^- \rightarrow e_t^-$$

$$e^- + AB \rightarrow \cdot AB^- \longrightarrow \text{RELAXATION (BOND BENDING, STRETCHING)}$$

$$\cdot AB^- \longrightarrow \text{SOLVATION}$$

$$\cdot AB^- \longrightarrow \text{BOND BREAKING} \quad \begin{cases} A\cdot + B^- \\ A^- + B\cdot \end{cases}$$
$$\text{(D.E.C.)}$$

$$\cdot AB^- \xrightarrow{H^+} \cdot ABH$$

All **prevent** $\cdot AB^- + AB \rightarrow AB + \cdot AB^-$ AND HENCE, MOBILITY.

SCHEME 1
Changes which prevent the effective **mobility** of $\cdot AB^-$ [e_t^- represents an electron trapped or solvated].

$$\cdot AB^+ \rightarrow \text{RELAXATION (BOND BENDING, CONTRACTING)}$$

$$\cdot AB^+ \rightarrow \text{SOLVATION}$$

$$\cdot AB^+ + AB \rightarrow (AB) \dot{-} (AB)^+ \, \sigma^* \text{ species}$$

$$\cdot AB^+ \rightarrow \text{Loss of } H^+$$

All **prevent** $AB + \cdot AB^+ \rightleftharpoons \cdot AB^+ + AB$ AND HENCE, MOBILITY.

SCHEME 2
Changes which prevent the effective **mobility** of $\cdot AB^+$

$$AB \xrightarrow{\gamma} \cdot AB^+ + e^-$$

$$\cdot AB^+ + e^- \rightarrow (AB)^* \rightarrow (A\cdot\text{---}B\cdot)$$

$$(A\cdot\text{---}B\cdot) \rightarrow A\cdot + B\cdot$$

$$\rightarrow AB$$

$$\rightarrow (AB)' \text{ etc.}$$

SCHEME 3
Electron Ejection and Return. $(AB)^*$ represents AB in an excited state and $(A\cdot\text{---}B\cdot)$ represents $A\cdot$ and $B\cdot$ trapped close together, as in the primary 'solvent' cage.

Dilute Solutions

A major advantage of studying pure compounds is that single crystals
can be used, and hence e.s.r. parameters, which are generally aniso-
tropic, can be accurately extracted. Furthermore, if the crystal
structure is known, and if, as is frequently the case, the paramagnetic centres retain the orientation of the parent species, the
directions of the g- and electron-nuclear hyperfine tensor
components can be identified relative to the radical frame.
    Sometimes it is possible to dope crystals with impurities which
act as electron or 'hole' traps, but if 'powder' e.s.r. spectra
suffice, it is convenient to use specific solvents to encourage
either specific electron-capture or hole-capture by dilute solutions
of suitable compounds.

Specific Electron Capture. It is now customary to use solvents such
as methanol (usually CD_3OD) or methyl tetrahydrofuran (MTHF) as
solvents if electron-capture by AB is required. These solvents form
good glasses at 77 K, and for sufficiently dilute solutions of the
substrate, AB, electron ejection occurs overwhelmingly from solvent
molecules, so that $\cdot AB^+$ centres are not formed. Electrons are
fairly mobile, and hence $\cdot AB^-$ radicals are formed provided AB has,
effectively, a positive electron affinity. The hole centres, such
as CD_3OD^+, are not mobile because proton transfer to surrounding
solvent molecules occurs rapidly at all temperatures.
    Some examples of specific electron-capture by various organo-
metallic compounds in such solvents are summarised in reactions
[5]-[10] [refs. (5)-(10) respectively]. In some cases, the parent
anion ($\cdot AB^-$) is detected by e.s.r. spectroscopy, but in others
subsequent reactions have occurred.

$$Mn_2(CO)_8(PR_3)_2 + e^- \rightarrow R_3P(CO)_4Mn\dot{-}Mn(CO)_4PR_3{}^- \qquad [5]$$
$$Rh_2(O_2CMe)_4 + e^- \rightarrow [Rh_2(O_2CMe)_4]\cdot{}^- \qquad [6]$$
$$Ti(Cp)_2Cl_2 + e^- \rightarrow [\cdot Ti(Cp)_2Cl_2] \qquad [7]$$
$$Ph_3AsMe^+ + e^- \rightarrow \cdot As(Ph)_3Me \qquad [8]$$
$$(MeO)_3PO + e^- \rightarrow \cdot PO(OMe)_3{}^- \qquad [9]$$
$$ \rightarrow \cdot PO(OMe)_2 + MeO^- \Big\}$$
$$P(OMe)_3 + e^- + MeOH \rightarrow H\dot{P}(OMe)_3 + MeO^- \qquad [10]$$

In reactions [5]-[8] pure electron addition occurs, but in reaction
[9] addition and dissociative electron capture giving loss of MeO^-
occur concurrently. Furthermore, $\dot{C}H_3$ radicals are also formed,
together, presumably, with $(MeO)_2PO_2{}^-$; this being an alternative
dissociative route. Reaction [10] occurs in methanol, there being
no clear sign of the parent anion, $\cdot P(OMe)_3{}^-$. This protonation step
is also accompanied by dissociative electron capture to give
$\cdot P(OMe)_2$ radicals.

Specific Electron Loss. Certain solvents, such as CCl_4, $CFCl_3$ or
SF_6 trap ejected electrons with high efficiency, and irreversibly,
but the electron-loss centres are mobile *via* electron transfer, and
hence can readily reach solute molecules (S) even in very low
concentration. The sequence of reactions is summarised in
reactions [11]-[14] for the most commonly used matrix, $CFCl_3$.

$$CFCl_3 \rightarrow CFCl_3 \cdot^+ + e^- \qquad [11]$$
$$CFCl_3^- + e^- \rightarrow \cdot CFCl_3^- \rightarrow \cdot CFCl_2 + Cl^- \qquad [12]$$
$$CFCl_3 \cdot^+ + CFCl_3 \rightarrow CFCl_3 + CFCl_3 \cdot^+ \qquad [13]$$
$$CFCl_3 \cdot^+ + S \rightarrow CFCl_3 + S \cdot^+ \qquad [14]$$

Fortunately, for this solvent, the electron-capture centres give very broad e.s.r. features at 77 K, and hence the spectra for $S \cdot^+$ cations are readily distinguished. We know of no instance in which $S \cdot^+$ cations are not formed provided the ionization potential of S is less than that of the solvent. There are two complicating factors, one is unimolecular break-down or rearrangement of the radical cations, and the other is weak complexation with a solvent molecule. The latter is readily detected because specific interaction with one chlorine or one fluorine nucleus occurs, and the resulting hyperfine features are usually well-defined.

The majority of studies using this procedure have involved organic radical-cations (11), but a few organometallic systems have been investigated. For example, the cations $\cdot SnR_4^+$ (R=H, Me) (12,13), $Me_3Sn \cdot SnMe_3^+$ (14), $\cdot SiMe_4^+$ (15), $Me_3Si \cdot SiMe_3^+$ (16), $R_2S \cdot^+$ (17), $R_2S \dot{-} SR_2^+$ (17), $R_2Hg \cdot^+$ (18), and $R_3P \cdot^+ + R_3P \dot{-} PR_3^+$ (19).

The set of cations $\cdot CH_4^+$ (20), $\cdot SnH_4^+$ (19), $\cdot CMe_4^+$ (21), $\cdot SiMe_4^+$ (15), $\cdot GeMe_4^+$ (15) and $\cdot SnMe_4^+$ (12) make some interesting structural contrasts. In all cases, there have to be Jahn-Teller distortions after electron-loss, but the SOMO's, as estimated from the e.s.r. parameters, differ markedly from one group to another. In summary, $\cdot CH_4^+$ distorts to give two strongly and two weakly coupled protons, the SOMO comprising a $2p$ orbital on carbon (Ia). In contrast, $\cdot SnH_4^+$ gives two distinct cations at 77 K, one with two strongly coupled protons (Ib) but the other (II), having only one strongly coupled proton.

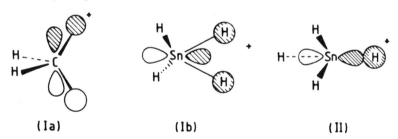

(Ia)   **(Ib)**   **(II)**

The outstanding difference, however, is that for both forms of $\cdot SnH_4^+$, the tin contributes a high s-character to the SOMO, as can be judged by the large hyperfine coupling for ^{117}Sn and ^{119}Sn nuclei (Figure 1). Again in contrast with this, the tin contribution for $\cdot SnMe_4^+$, which also distorts as in (II), is primarily p-orbital in character. The latter change can be understood in terms of the shape of the R_3Sn- residue. For H_3Sn- this remains pyramidal, as in II, so the Sn-H σ^* orbital retains a high s-character. However, for Me_3Sn-, this unit has, we suggest, become essentially planar for the cation, as in III (12).

The "dimer-cations" $Me_3Si \cdot SiMe_3^+$, $R_3P \dot{-} PR_3^+$, etc., which can be classified as σ, and σ^* radicals respectively, are of particular importance and hence are discussed separately.

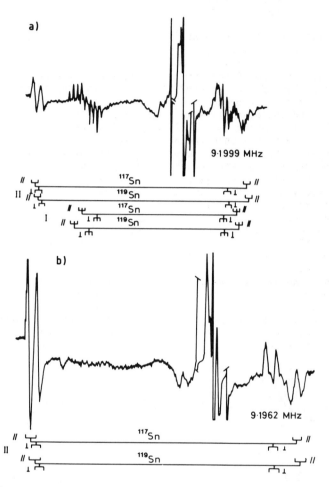

FIGURE 1

First derivative X-band e.s.r. spectra for a solid solution of
4 mol% SnH_4 in $CFCl_3$ recorded (a) immediately after exposure to
^{60}Co γ-rays at 77 K, showing features assigned to C_{2v} (I) and C_{3v}
(II) structures of SnH_4^+ radical cations, and (b) after the
storage of the irradiated sample for one week at 77 K, showing
conversion of I into II.

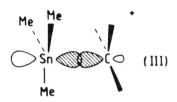

(III)

σ- and σ*-Radicals

We define these radicals as species having SOMO's comprising primarily (or formally) a single $\underline{\sigma}$ bond containing either one ($\underline{\sigma}^1$) or three ($\underline{\sigma}^2\underline{\sigma}^{*1}$) electrons. We symbolise these as $A \cdot B^+$ and $A \dot{-} B^-$ respectively. The latter class have been known for some time, and are typified by the alkali-halide V_K centres such as $Cl \dot{-} Cl^-$. They can be formed by electron addition [15] or by electron loss followed by reaction, as for example, in [16].

$$A-B + e^- \rightarrow A \dot{-} B^- \qquad \ldots [15]$$
$$R_3P \cdot^+ + R_3P \rightarrow R_3P \dot{-} PR_3^+ \qquad \ldots [16]$$

This is frequently, but not necessarily, a "dimerisation" as in [16]. The tendency to react in this way is very marked and constitute chemically important reaction pathways in many cases. A biologically important example is [17], and it is noteworthy that this radical is converted from an electron-acceptor (RS·) into an electron-donor (RS$\dot{-}$SR)$^-$.

$$RS \cdot + RS^- \rightarrow RS \dot{-} SR^- \qquad \ldots [17]$$

The $\underline{\sigma}^1$-structure is less well known, but it seems probable that, after electron loss, a distortion that leads to a $\underline{\sigma}^1$-type radical is a necessary step in dissociation. For example, the $Me_3Sn \cdot CH_3^+$ cation, formed at 77 K, readily gives methyl radicals on annealing [18] ($\underline{12},\underline{13}$). Examples of such species include the

$$Me_3Sn \cdot CH_3^+ \rightarrow Me_3Sn^+ + \cdot CH_3 \qquad \ldots [18]$$

'dimers' $Me_3Si \cdot SiMe_3^+$ and $Me_3Sn \cdot SnMe_3^+$ formed by electron loss. In principle, $\underline{\sigma}^1$ radicals can also be formed by electron capture followed by reaction, but we know of only two examples studied by e.s.r. spectroscopy [19,20] ($\underline{22},\underline{23}$).

$$(MeO)_3B \cdot^- + (MeO)_3B \rightarrow (MeO)_3B \cdot B(OMe)_3^- \qquad \ldots [19]$$
$$Ag \cdot + Ag^+ \rightarrow Ag \cdot Ag^+ \qquad \ldots [20]$$

We now turn attention to some specific examples, ranging from transition-metal carbonyl derivatives to some biological systems. There is a large literature on paramagnetic transition-metal cyanide derivatives generated by ionizing radiation which would make an interesting contrast with the carbonyls, but we have reluctantly omitted these from our discussion. Some pertinent references are given in ($\underline{24}$)-($\underline{27}$).

Transition-Metal Carbonyls

Spectroscopic observations made on irradiated transition metal
carbonyls show that, in addition to the anticipated processes of
electron-loss and electron-gain, metal-carbon bond fission occurs
with a resultant loss of carbon monoxide in many instances. The
coordinatively unsaturated carbonyl radicals so produced are highly
reactive species, which if they escape cage recombination with CO,
may react further with the parent carbonyl to give cluster carbonyls
or with impurities such as oxygen. Electron spin resonance
spectroscopy has been used to great advantage in identifying the
paramagnetic products of the γ-irradiation of pure solid metal
carbonyls and of their solid solution in both single crystals and
powders. Unfortunately, the technique cannot detect diamagnetic
products which may constitute a significant portion of the reaction
products. In many instances, notably $Mo(CO)_6$ and $W(CO)_6$, prolonged
γ-irradiation does not yield products detectable by e.s.r. spectroscopy.

Group VA (V, Nb, Ta). The γ-radiolysis of $V(CO)_6$ has been investigated by e.s.r. spectroscopy in a matrix of krypton (28) and in a
single crystal (29) of $Cr(CO)_6$. $V(CO)_6$ is itself paramagnetic,
although its e.s.r. spectrum is not detectable (29,30) at
temperatures above 10 K. Two e.s.r. spectra were detected (28) in
γ-irradiated $Kr-V(CO)_6$ solid solutions at 20 K. The first of these
was intense and was evidently associated with an electronic doublet
having rhombic symmetry. Consideration of the g-, ^{51}V- and ^{13}C-
tensors for this species indicated that it was $V(CO)_5$ having a
distorted trigonal bipyramidal geometry (Figure 2) and a $^2B_2$ (d_{yz})
ground state in C_{2v} symmetry. A second, weaker e.s.r. spectrum in
irradiated $Kr-V(CO)_6$ could not be unequivocally assigned to a
vanadium radical because of the absence of ^{51}V hyperfine structure.
However, the low, isotropic g-tensor (1.9583) and the presence of
four equivalent carbon nuclei showing 7.6 G ^{13}C hyperfine
interactions suggest an assignment to the high spin molecule $V(CO)_4$
which has a ground state $^6A_1$ in T_d symmetry. This identification
relies heavily on the analogy with the established (31) structure of
$Cr(CO)_4^+$ (*vide infra*).

Group VIA (Cr, Mo, W). γ-Radiolysis studies have been carried out
for the simple hexacarbonyls (31,32) and for certain carbonyl
iodides (33) and cyclopentadienyl carbonyl iodides (34). In the
case of the hexacarbonyls (31,32), two free-radical products have
been detected and characterized for $Cr(CO)_6$, weak unidentified EPR
signals have been observed for $Mo(CO)_6$, but irradiated $W(CO)_6$
apparently contains no detectable paramagnetic centres.

An intense, solution-like EPR spectrum is observable in γ-
irradiated $Kr-Cr(CO)_6$ mixtures at 20 K and below (31). ^{53}Cr- and
^{13}C- hyperfine satellites are detectable in natural abundance
(Figure 3) and their intensities indicate a formulation $Cr(CO)_4^{\pm}$ for
the carrier of unpaired spin. Slight anisotropy in the ^{13}C hyperfine structure of the 95% ^{13}C- enriched species could only be
correctly reproduced in simulations under the assumption of
tetrahedral geometry. The centre is thought to be $Cr(CO)_4^+$ with a
$^6A_1$ ground state in T_d symmetry, a rare example of a high-spin metal
carbonyl.

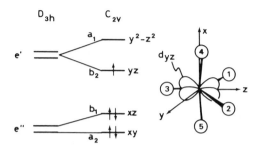

FIGURE 2

Energy level diagram for the distortion of a trigonal-bipyramidal $V(CO)_5$ toward C_{2v} symmetry.

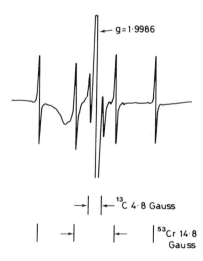

FIGURE 3

EPR spectrum of $Cr(CO)_4^+$ in krypton at 20 K.

A much weaker spectrum in γ-irradiated $Kr-Cr(CO)_6$ mixtures (32) may be attributed to $Cr(CO)_5^-$. This species is isoelectronic with $Mn(CO)_5$ and $Fe(CO)_5^+$, both of which have been characterized by EPR spectroscopy (*vide infra*) (Table I). Unfortunately, only the perpendicular features of this weak spectrum were detectable.

TABLE I   EPR Parameters for $M(CO)_5$ Radicals in Kr at 77 K

	$Cr(CO)_5^-$	$Mn(CO)_5$	$Fe(CO)_5^+$
$g_\parallel$	?	2.0003	2.0022
$g_\perp$	2.0164	2.0340	2.0686
$A_\parallel(M)$[a]	?	65.4	?
$A_\perp(M)$	2.5	(-)34.8	6.8
$A_\parallel(Kr)$[a]	?	6.4	?
$A_\perp(Kr)$	6.1	4.0	3.4
$A_\parallel(C)$[b]	?	3.3	?
$A_\perp(C)$	10.4	<2	12.5

[a] Hyperfine interactions in Gauss for ^{53}Cr, ^{55}Mn, ^{57}Fe or ^{83}Kr.

[b] Hyperfine interaction for unique, axial ^{13}C. Principal values for equatorial carbons not known with certainty, see Ref. 37.

A dominant process in the radiolysis of the carbonyl iodides (33) $M(CO)_5I^-$, (M=Cr, Mo, W) and of the cyclopentadienyl carbonyl iodides (34) $M(cp)(CO)_3I$, (M=Mo, W) is evidently electron capture in the metal-iodine bond. The EPR spectra of the resulting anion radicals are dominated by the large hyperfine coupling to the ^{127}I nucleus. The unpaired electron in these species is more or less localized in a σ^* orbital comprising metal $\underline{d}_{z^2}$ and iodine p_z atomic orbitals. A secondary species obtained by annealing $Mo(cp)(CO)_3I^-$ had considerably reduced ^{127}I hyperfine couplings and larger positive g-shifts. This may be an isomer of the initially formed σ^* complex in which the unpaired electron occupies a π^* metal-iodine orbital.

In the absence of ^{13}C hyperfine interaction measurements, it is not clear to what extent dissociative electron capture with CO loss occurs for such molecules. It is entirely possible, for example, that electron capture by $M(CO)_5I^-$ results in loss of the axial CO ligand, yielding the radical $M(CO)_4I^{2-}$. If so, it is noteworthy that CO is lost in preference to I^-.

Group VIIA (Mn, Tc, Re). A number of mononuclear manganese carbonyl derivatives have been γ- irradiated and examined by e.s.r. spectroscopy. The motivation behind much of this effort was the search for the elusive radical $Mn(CO)_5$. The e.s.r. spectrum of this species is now firmly (35-37) established (Figure 4), although there is still some suggestion that the true "naked" $Mn(CO)_5$ has yet to be observed (37).

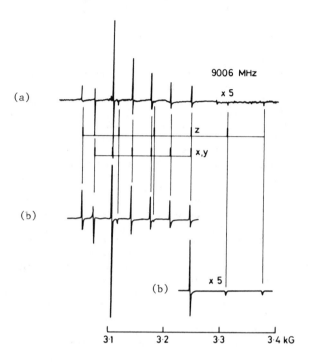

FIGURE 4
(a) ESR spectrum of KrMn(CO)₅ in a Kr matrix at 77 K.
(b) Computer simulation using parameters given in the Table.

γ-Irradiation of $HMn(CO)_5$ is a source ($\underline{37}$) of $Mn(CO)_5$, although as in the u.v.-photolysis ($\underline{36}$), the more important primary act may be CO loss, rather than H-Mn fission. $Mn(CO)_5$ has square pyramidal geometry ($\underline{35}$-$\underline{38}$) (C_{4v} symmetry) and its single unpaired electron spends roughly 60% of the time in the Mn $3\underline{d}_{z^2}$ atomic orbital. In a Kr matrix, a single Kr nucleus contributes significant hyperfine coupling ($\underline{37}$). There is clearly substantial overlap of the Mn $3\underline{d}_{z^2}$ and Kr $4p_z$ orbitals suggesting that this radical is more correctly described as Kr $Mn(CO)_5$.

Interestingly, there is no evidence of paramagnetic hydrido manganese carbonyls in irradiated $HMn(CO)_5$, yet one might have anticipated the formation of $HMn(CO)_4^{\pm}$. By contrast ($\underline{39},\underline{40}$), the γ-irradiation of $Me_3SnMn(CO)_5$ and $Ph_3PbMn(CO)_5$ gave no trace of $Mn(CO)_5$, but yielded the radical anions $XMn(CO)_4^-$ with spectral parameters and structures similar to the parent $Mn(CO)_5$. In both cases, electron capture was followed by exclusive loss of an equatorial CO, leading to a low symmetry (C_S) species. A second radical detected ($\underline{40}$) in the irradiated triphenyl lead derivative is possibly the electron-loss centre. This species exhibits a very large coupling to a ^{207}Pb nucleus, and is probably a σ^* radical.

γ-Irradiation of the pentacarbonyl manganese halides $Mn(CO)_5X$, (X=Cl, Br or I) generates ($\underline{41},\underline{42}$) e.s.r. spectra which have been attributed to electron-loss and electron-gain centres. The latter are electronic doublets in which the unpaired electron occupies a σ_z^* orbital composed of Mn $3\underline{d}_{z^2}$ and halogen p_z orbitals. These radicals have C_{4v} geometry and are either $Mn(CO)_5X^-$ or $Mn(CO)_4X^-$. An electron-gain centre of similar geometry and electronic structure is generated ($\underline{43}$) by radiolysis of the nitrosocarbonyl $Mn(CO)_4NO$. Spectra associated with the electron-loss centres $Mn(CO)_nX^+$ (n=4 or 5) are less well-defined and pose analytical difficulties ($\underline{41}$). However, there is little doubt that these are high-spin radicals, probably electronic sextets.

Spectroscopic evidence ($\underline{44},\underline{45}$) has been adduced for the formation of electron-gain centres upon γ-irradiation of the binuclear carbonyls $Mn_2(CO)_{10}$ and $Re_2(CO)_{10}$. A study ($\underline{45}$) of a single crystal of irradiated $Mn_2(CO)_{10}$ has shown that the radical anion contains two equivalent ^{55}Mn nuclei whose hyperfine tensors lie 118° apart. This has led to the suggestion that the anion radical contains a bridging CO and that its correct formulation is $Mn_2(CO)_9^-$. The observation of a bridged $Mn_2(CO)_9$ species in u.v.-photolyzed material lends some support to this hypothesis ($\underline{46}$).

The cationic species detected ($\underline{44}$) in irradiated $Mn_2(CO)_{10}$ has an intriguing structure. Electron-loss occurs at one manganese leading to a high-spin (S = $^5/_2$) configuration for that atom and disparate hyperfine couplings for the two metal nuclei. The radical is clearly an analogue of the high-spin cations obtained from $Mn(CO)_5X$ (*vide supra*).

A complex EPR spectrum detected ($\underline{47}$) in the γ-irradiated Vahrenkamp molecule $Mn_2(CO)_8(\mu-AsPh_2)_2$ is thought to belong to the cation of the molecule. From an analysis of the ^{55}Mn and ^{75}As hyperfine structure it was concluded that the $\underline{d}^6\underline{d}^5$ dimer radical has its single unpaired electron in a σ^* MO composed of Mn $3\underline{d}_{x^2-y^2}$ orbitals.

Group VIIIA (Fe, Ru, Os, Co, Rh, Ir, Ni, Pd, Pt). Free radical products have been detected in γ-irradiated solutions of $Fe(CO)_5$ in methyl tetrahydrofuran (48), trichlorofluoromethane (48), krypton (37) and in single crystals (49,50) of $Cr(CO)_6$. Two distinct free radicals were observed in $Cr(CO)_6$, one of which had parameters essentially identical to those of an ephemeral species detected (32) in liquid solutions of $Fe(CO)_5$ in H_2SO_4 (Figure 5). This species is evidently the cation radical $Fe(CO)_5^+$. From the single crystal measurements (49) of the principal directions and values of the g-, the ^{57}Fe- and the five ^{13}C- hyperfine tensors it was possible to establish with certainty that the species has square pyramidal (C_{4v}) geometry with the unpaired electron located in an a_1 orbital having considerable Fe $3d_{z^2}$ character. In a krypton matrix (37), the "empty" axial position is occupied by a single Kr nucleus which is thought to be an integral part of a $KrFe(CO)_5^+$ molecule. Parameters for $Fe(CO)_5^+$ and the isoelectronic radicals $Mn(CO)_5$, $Cr(CO)_5^-$ are assembled in Table I. The analogous osmium radical is a product (51) of radiolysis of $Os(CO)_5$ dissolved in $Cr(CO)_6$.

The axially symmetric cation radical observed (48) in γ-irradiated CCl_3F-$Fe(CO)_5$ mixtures has a $g_\perp$ value sufficiently far removed from that of $Fe(CO)_5^+$ that one must consider an alternative identification. A likely possibility in view of Burdett's predictions (52) is a square planar $Fe(CO)_4^+$ species.

A second free radical detected (50) at low temperatures (<20 K) in γ-irradiated $Cr(CO)_6$-$Fe(CO)_5$ crystals is undoubtedly the anion $Fe(CO)_5^-$. This species is bent and is, in essence, an acyl radical $Fe(CO)_4$-C-O^- with the unpaired spin located largely in an sp^2 hybrid of the unique carbon atom. Acceptance of the extra electron into the degenerate π_{CO} orbitals (followed by a Jahn-Teller first-order distortion) rather than the empty, but energetic Fe $3d_{x^2-y^2}$ orbital is surely a consequence of the square pyramidal geometry imposed upon the $Fe(CO)_5$ by the host $Cr(CO)_6$ lattice. By contrast, in MTHF glass, it would appear (48) that γ-irradiation leads to a trigonal bipyramidal (D_{3h}) anion of $Fe(CO)_5$ in which the ninth d-electron enters the Fe $3d_{z^2}$ orbital, as expected (53). This species decomposes on annealing leading, it is thought, to $Fe(CO)_4^-$. However, an equally reasonable alternative was proposed in which both species are identified as $Fe(CO)_4^-$ with different geometries (53).

γ-Irradiation of the trinuclear carbonyl $Fe_3(CO)_{12}$ in MTHF solution generated a number of free radical intermediates arising from electron capture (48). Amongst these were $Fe_3(CO)_{12}^-$ and $Fe_3(CO)_{11}^-$, free radicals which have also been generated and characterized by chemical reduction (54).

Another intermediate in the radiolysis of iron carbonyls is probably $Fe(CO)_4$, a species which has been detected by infra-red spectroscopy (55) and by magnetic circular dichroism (56). Observation by the latter technique implies a triplet electronic ground-state for the radical, yet it has eluded detection by EPR in spite of careful searches from 0 to 25 kG at X-band (32).

Electron-capture centres have been identified amongst the products of γ-irradiation of (57) $Fe(CO)_2(NO)_2$ and of (34) $Fe(cp)(CO)_2X$ (X=Cl or I). The anion radicals $Fe(cp)(CO)_2X^-$ are formally Fe d^7 species and, not surprisingly, have their single unpaired electron essentially localized in a Fe- halogen σ^*

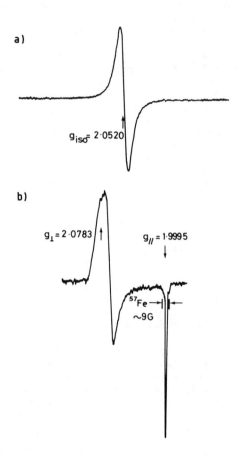

FIGURE 5
First-derivative X-band e.s.r. spectra assigned to $Fe(CO)_5^+$
(a) in H_2SO_4 at 300 K, (b) at 77 K.

orbital. On annealing, these species lose halide ion to give ($\underline{34}$) $Fe(cp)(CO)_2$, a radical in which the unpaired electron is confined to a $\underline{d}_{z^2}$ orbital lying perpendicular to the "plane" of the molecule. A second radical observed in certain polar solvents after annealing is the conjugate acid $HFe(cp)(CO)_2^+$ which has appreciable spin-density on the proton and probably contains an Fe-H bond.

The anion radical $Fe(CO)_2(NO)_2^-$ detected in γ-irradiated single crystals ($\underline{57}$) is isoelectronic with the persistent radicals $Co(CO)_2(NO)_2$ and $Fe(CO)(NO)_3$ ($\underline{58},\underline{59}$). The added electron enters the π^* system of the NO ligands giving an e.s.r. spectrum reminiscent ($\underline{60}$) of that of N_2^-.

A number of cobalt carbonyl derivatives have been subjected to γ-irradiation and subsequent EPR spectroscopic investigation. Perhaps the most fruitful of these studies was the radiolysis of $HCo(CO)_4$ in a Kr matrix ($\underline{61},\underline{62}$). Free radicals detected in the irradiated material corresponded to processes of H-Co fission, electron capture, H-atom additions and clustering. Initial examination at 77 K or lower temperatures revealed the presence of two radicals, $Co(CO)_4$ and $HCo(CO)_4^-$, having similar geometries (IV and V) and electronic structures. Both have practically all of the unpaired spin-density confined to nuclei located on the three-fold axis, in Co $3\underline{d}_{z^2}$, C $2\underline{s}$ or H $1\underline{s}$ orbitals. Under certain conditions, a radical product of hydrogen-atom addition, $H_2Co(CO)_3$, was observed; this species is believed to have a distorted trigonal bipyramidal structure in which the H-atoms occupy apical positions.

Upon annealing, two new binuclear cobalt radicals were formed ($\underline{62}$). One of these is thought to be the anion of $Co_2(CO)_8^-$ with a staggered D_{3d} geometry and a σ^* SOMO consisting of an antibonding combination of Co $\underline{d}_{z^2}$ orbitals. Entirely analogous electron adducts are generated in the radiolysis ($\underline{63}$) of the binuclear compounds $Co_2(CO)_6(ER_3)_2$, (ER_3 = P-$\underline{n}$-Bu$_3$, P(OMe)$_3$, or As-$\underline{i}$-Bu$_3$). The other species is a hydrogen-bridged dimer with a structure like that of the $HFe_2(CO)_8^-$ ion.

Radiolysis of thallium(I) tetracarbonylcobaltate generates several interesting free radicals ($\underline{64}$). In addition to the electron-loss centre Tl^{2+}, three cobalt-containing species have been identified. Two of these are thought to be isomers of the nineteen-electron species $Co(CO)_4^{2-}$. In one of them, the unpaired electron is largely confined to the Co $4\underline{s}$ orbital; the other species is believed to result from electron capture at one of the carbonyl ligands, a process which also occurs (*vide supra*) in the radiolysis ($\underline{50}$) of $Fe(CO)_5$. The third radical observed in irradiated $TlCo(CO)_4$ is probably the peroxyl $Co(CO)_4O_2$ formed from adventitious traces of oxygen.

Nickel carbonyl radicals show an even greater tendency than cobalt carbonyls to cluster in a krypton matrix. Three binuclear nickel carbonyls have been detected by EPR spectroscopy in the products of γ-irradiated $Ni(CO)_4$ in Kr, yet no mononuclear species has been positively identified ($\underline{65}$). ^{13}C hyperfine structure has helped in establishing tentative structures for these dimer radicals. Two of them are believed to be the cations, $Ni_2(CO)_8^+$ and $Ni_2(CO)_6^+$, each containing two bridging CO ligands. $Ni_2(CO)_8^+$ evidently has a low-symmetry structure (C_{2v}) resembling that of $Co_2(CO)_8$. $Ni_2(CO)_6^+$ appears to have D_{2h} geometry with the bridging ligands lying above and below the plane containing the remaining atoms. Unpaired spin-density in this radical is essentially confined (by symmetry) to $Ni(3\underline{d}_{xy})$ and in-plane C 2p orbitals. A third dimer radical in irradiated $Ni(CO)_4$ has low g-values and a unique carbon nucleus. This is thought to be $Ni_2(CO)_7^-$ which possesses a single CO bridge.

Carbon Centred α- and β- Radicals

The structures under consideration are indicated in Inserts VI and VII. Such radicals are usually secondary products of radiolysis, formed, for example, by extraction of hydrogen from a $R_2C(H)$- precursor. However, closely related radicals can be formed, for example, by electron-loss from vinyl- or allyl- derivatives ($\underline{66},\underline{67}$) or from substituted aromatic cations ($\underline{68},\underline{69}$) [see, for example, VIII-X].

(VI)   (VIIa)   (VIIb)

$H_2C \doteq CHSiMe_3$

(VIII)   (IX)   (X)

Perhaps the most important conclusion to be drawn from results for α-substituents (VI) is that delocalisation onto the 'metal' atoms in groups such as $-SiL_3$ or $-PL_3^+$ is undetectably small ($\underline{70},\underline{71}$). Indeed, the $R_2\dot{C}$- moiety displays hyperfine interaction with ^1H and ^{13}C that suggest normal planarity at carbon with essentially unit spin-density thereon, and coupling to the metal atom (specifically, ^{31}P) is small and probably negative. This implies that spin-density is acquired by spin-polarisation of the C-M $\underline{\sigma}$-electrons and not by p_π-$\underline{d}_\pi$ delocalisation, as is so often

implied. For these structures at least, such interaction seems to
be of little importance (see also next section on Cyclic Phosphazene
Radical-Cations and -Anions).

Alkyl radicals with organo-metal substituents in the β-position
(VII) are also important, since they tend to favour the out-of-plane
site indicated, which gives maximum σ-π overlap. This, in turn,
leads to hyperconjugative delocalisation, evidence for which comes
from the large hyperfine coupling constants observed for β-^{31}P
nuclei. Typically, the coupling for α-^{31}P (VI) might be in the
region of (-)20 G, whereas that to β-^{31}P (VII) can be as high as *ca.*
200 G, at low temperatures (<u>71</u>). This preferred conformation is
also indicated by the low coupling to the methylene protons of *ca.*
13 G, indicating that at low temperatures the limiting structure is
achieved. As the temperature is increased, libration sets in,
reducing the ^{31}P coupling and increasing the ^{1}H coupling towards the
average value.

Cyclic Phosphazene Radical-Cations and -Anions

We single these out for discussion because they are very important
organometallic species, and because electron-loss and electron-gain
results in radicals whose e.s.r. spectra should reflect the degree
of delocalisation in these compounds. Indeed, they are often
referred to as being pseudo aromatic (<u>72</u>), the implication being
that p_π-$\underline{d}_\pi$ bonding results in delocalisation of the six π-electrons
that are formally located on nitrogen (XI). If such delocalisation
were important, one might have hoped that electron-addition or -loss
would give structures stabilised by such delocalisation. In fact,
however, e.s.r. results for the anions show clearly that the excess
electron is confined to one phosphorus atom, delocalisation onto the
two ligands (L) being important, but that onto the adjacent nitrogen
atoms being negligibly small (<u>73</u>). In fact, these are typical
phosphorayl radicals, the electron being formally in an <u>s</u>-p hybrid
projecting from phosphorus along a radius in the ring plane, and is
not in any sort of π-orbital.

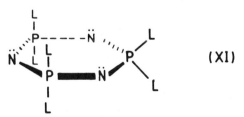

(XI)

Results for the cations are even more surprising (<u>74</u>). So far
as we can judge, the 'hole' is largely confined to one nitrogen,
with no delocalisation onto the other two nitrogen atoms, and
probably none onto the two neighbouring phosphorus atoms other than
by spin-polarisation as discussed above. Since the P_2N units are
essentially coplanar in the parent, it is most surprising that only
one nitrogen p(π) orbital contributes to the SOMO. We suppose that,
as with the radical-anion, there is sufficient local distortion to
fix the electron for a time which is long on the e.s.r. time scale.

Examples of Biological Systems

Mention has already been made to some reactions involving RSH and
RSSR derivatives. These biologically important molecules have been
studied widely by pulse radiolysis methods. Exposure of a range of
proteins containing S-S bonds gave RS∸SR⁻ units as important
components after exposure to ^{60}Co γ-rays (75). These results show
that ejected electrons are relatively free to move within proteins
to sites of high electron affinity. In general, our results suggest
that electrons will normally seek out such centres, especially
transition metal centres, but that electron-loss centres are not
mobile. Thus met-haemoglobin (Fe^{III}) is efficiently converted into
deoxyhaemoglobin (Fe^{II}) on irradiation, but the reverse does not
occur.

Oxyhaemoglobin. An interesting example of specific electron-capture
is the FeO_2 unit of oxyhaemoglobin or myoglobin (76,77). The
resulting ·FeO_2^- species, which is stable at 77 K, has an e.s.r.
spectrum characteristic of low-spin Fe^{III}, but with extensive
delocalisation onto oxygen. Interestingly, the species formed in
the α-subunits differs considerably from that in the β-subunits. On
annealing, it seems that protonation of oxygen occurs to give an
Fe^{III}-O_2H derivative which ultimately loses HO_2^- or H_2O_2 and normal
high-spin met-haemoglobin is formed. Again, only electron-capture
is of any importance on irradiation.

Superoxide Dismutase. Again, only electron-capture is important on
irradiation (78). For the Cu-Zn enzyme, Cu^{II} is converted into Cu^I
form. In the presence of oxygen, ·O_2^- is formed in competition with
Cu^I, and on annealing reacts to re-form Cu^{II}. Thus radiolysis has
proven to be a useful method for checking the mechanism of action of
this dismutase. The conclusion is that the somewhat disputed
mechanism [21,22] is probably correct.

$$Cu^{II} + O_2^- \rightarrow Cu^I + O_2 \qquad \qquad \text{.... [21]}$$
$$Cu^I + O_2^- + (2H^+) \rightarrow Cu^{II} + H_2O_2 \qquad \qquad \text{.... [22]}$$

[However, see refs. (79) and (80) for an alternative viewpoint.]

Methyl-Cobalamine and Related Compounds. There are three major
types of cobalamine in animals and man, the methyl and hydroxo
derivatives and the deoxyadenosyl derivative (coenzyme B_{12}). The
unique metal-carbon bond in the methyl derivative and in B_{12} is
extremely photolabile, and their photolyses have been widely
studied. However, radiolytic processes have not been widely studied.
    Use of CD_3OD or methyl tetrahydrofuran solvents to encourage
electron capture, resulted in a complex set of reactions for methyl
cobalamine. Initial addition occurred into the π* corrin orbital,
but on annealing a cobalt centred radical was obtained, the e.s.r.
spectrum of which was characteristic of an electron in a $\underline{d}_{x^2-y^2}$
orbital (involving the corrin ring) rather than the expected $\underline{d}_{z^2}$
orbital. However, the final product was the normal Co^{II} species
formed by loss of methyl. Formally, this requires loss of :CH_3^-,
but this step seems highly unlikely. Some form of assisted loss,
such as protonation, seems probable.

In contrast, the very similar compound methyl (pyridine) cobaloxime captured electrons into the Co-Me σ^* orbital, probably with concomitant loss of pyridine (82). Well-defined doublet splittings were obtained when the $^{13}CH_3$ derivative was used, and the spectrum had the form expected for a $\underline{d}_{z^2}$ configuration (Figure 6). The contrast in reactivity between these similar compounds is remarkable.

Conclusions

We conclude that for organometallic derivatives, radiolysis can be used as an excellent method for inducing specific electron-loss or electron-addition. Furthermore, this can be done at very low temperatures such that, often, the primary gain and loss species are formed and can be characterised by e.s.r. spectroscopy. Thus this technique is a useful complement to more conventional studies of redox reactions.

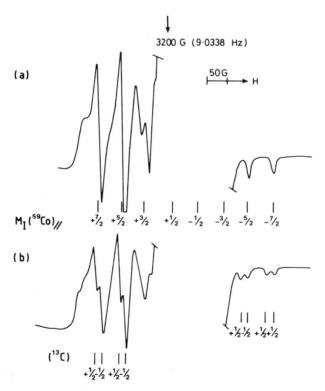

FIGURE 6
First-derivative X-band e.s.r. spectra for dilute solutions of methylcobaloxime after exposure to ^{60}Co γ-rays at 77 K:
(a) in MTHF, showing features assigned to CH_3-Co(dmg)$_2$ complex,
(b) using ($^{13}CH_3$) showing the extra doublet splitting.

Literature Cited

1. Atkins, P.W.; Symons, M.C.R. "The Structure of Inorganic Radicals"; Elsevier: Amsterdam, 1967.
2. Symons, M.C.R., unpublished results.
3. Chachaty, M.C. C.R. Acad. Sci. Paris, 1966, 262C, 686.
4. Rao, D.N.R.; Symons, M.C.R. J. Chem. Soc., Faraday Trans. I, 1985, 81, 565.
5. Symons, M.C.R.; Wyatt, J.L.; Peake, B.M.; Simpson, J.; Robinson, B.H. J. Chem. Soc., Dalton Trans., 1982, 2037.
6. Eastland, G.W.; Symons, M.C.R. J. Chem. Soc., Dalton Trans., 1984, 2193.
7. Symons, M.C.R.; Mishra, S.P. J. Chem. Soc., Dalton Trans., 1981, 2258.
8. Symons, M.C.R.; McConnachie, G.D.G. J. Chem. Soc., Faraday Trans. I, 1984, 80, 211.
9. Nelson, D.J.; Symons, M.C.R. J. Chem. Soc., Perkin Trans. 2, 1977, 286.
10. Symons, M.C.R. Mol. Phys., 1972, 24, 885.
11. Symons, M.C.R. Chem. Soc. Rev., 1984, 393.
12. Symons, M.C.R. J. Chem. Soc., Chem. Comm., 1982, 869.
13. Hasegawa, A.; Kaminaki, S.; Wakabayashi, T.; Hayashi, M.; Symons, M.C.R.; Rideout, J. J. Chem. Soc., Dalton Trans., 1984, 1667.
14. Symons, M.C.R. J. Chem. Soc., Chem. Comm., 1981, 1251.
15. Walther, B.W.; Williams, F. J. Chem. Soc., Chem. Comm., 1982, 270.
16. Wang, J.T.; Williams, F. J. Chem. Soc., Chem. Comm., 1981, 666.
17. Rao, D.N.R.; Symons, M.C.R.; Wren, B.W. J. Chem. Soc., Perkin Trans. 2, 1984, 1681.
18. Rideout, J.; Symons, M.C.R. J. Chem. Soc., Chem. Comm., 1985, 129.
19. Hasegawa, A.; McConnachie, G.D.G.; Symons, M.C.R. J. Chem. Soc., Faraday Trans. I, 1984, 80, 1005.
20. Knight, L.B.; Steadman, G.; Feller, D.; Davidson, E.R. J. Am. Chem. Soc., 1984, 106, 3700.
21. Iwasaki, M.; Toriyama, K.; Nunome, K. J. Am. Chem. Soc., 1981, 103, 3591.
22. Hudson, R.L.; Williams, F. J. Am. Chem. Soc., 1977, 99, 7714.
23. Shields, L.; Symons, M.C.R. Mol. Phys., 1966, 11, 57.
24. Symons, M.C.R.; Wilkinson, J.G. J. Chem. Soc., Dalton Trans., 1972, 1086.
25. Symons, M.C.R.; Wilkinson, J.G. J. Chem. Soc., Dalton Trans., 1973, 965.
26. Danon, J.; Muniz, R.P.A.; Garide, A.O.; Wolfson, I. J. Mol. Structure, 1967, 1, 127.
27. Symons, M.C.R.; Aly, M.M.; West, D.X. J. Chem. Soc., Dalton Trans., 1979, 1744.
28. Morton, J.R.; Preston, K.F. Organometallics, 1984, 3, 1386.
29. Boyer, M.P.; LePage, Y.; Morton, J.R.; Preston, K.F.; Vuolle, M.J. Can. J. Spectry., 1981, 26, 181.
30. Bratt, S.W.; Kassyk, A.; Perutz, R.N.; Symons, M.C.R. J. Am. Chem. Soc., 1982, 104, 490.
31. Fairhurst, S.A.; Morton, J.R.; Preston, K.F. Chem. Phys. Lett., 1984, 104, 112.

32. Morton, J.R.; Preston, K.F., unpublished data.
33. Symons, M.C.R.; Bratt, S.W.; Wyatt, J.L. J. Chem. Soc., Dalton Trans., 1982, 991.
34. Symons, M.C.R.; Bratt, S.W.; Wyatt, J.L. J. Chem. Soc., Dalton Trans., 1983, 1377.
35. Howard, J.A.; Morton, J.R.; Preston, K.F. Chem. Phys. Lett., 1981, 83, 226.
36. Symons, M.C.R.; Sweany, R.L. Organometallics, 1982, 1, 834.
37. Fairhurst, S.A.; Morton, J.R.; Perutz, R.N.; Preston, K.F. Organometallics, 1984, 3, 1389.
38. Church, S.P.; Poliakoff, M.; Timney, J.A.; Turner, J.J. J. Am. Chem. Soc., 1981, 103, 7515.
39. Morton, J.R.; Preston, K.F.; Thibodeau, D.L. J. Magn. Reson., 1982, 48, 55.
40. Anderson, O.P.; Fieldhouse, S.A.; Forbes, C.E.; Symons, M.C.R. J. Organomet. Chem., 1976, 110, 247.
41. Anderson, O.P.; Fieldhouse, S.A.; Forbes, C.E.; Symons, M.C.R. J. Chem. Soc., Dalton Trans., 1976, 1329.
42. Lionel, T.; Morton, J.R.; Preston, K.F. Chem. Phys. Lett., 1981, 81, 17.
43. Lionel, T.; Morton, J.R.; Preston, K.F. J. Phys. Chem., 1982, 86, 367.
44. Bratt, S.W.; Symons, M.C.R. J. Chem. Soc., Dalton Trans., 1977, 1314.
45. Lionel, T.; Morton, J.R.; Preston, K.F. Inorg. Chem., 1983, 22, 145.
46. Hepp, A.F.; Wrighton, M.S. J. Am. Chem. Soc., 1983, 105, 5934.
47. Kawamura, T.; Enoki, S.; Hayashida, S.; Yonezawa, T. Bull. Chem. Soc. Japan, 1982, 55, 3417.
48. Peake, B.M.; Symons, M.C.R.; Wyatt, J.L. J. Chem. Soc., Dalton Trans., 1983, 1171.
49. Lionel, T.; Morton, J.R.; Preston, K.F. J. Chem. Phys., 1982, 76, 234.
50. Fairhurst, S.A.; Morton, J.R.; Preston, K.F. J. Chem. Phys., 1982, 77, 5872.
51. Lionel, T.; Morton, J.R.; Preston, K.F. J. Magn. Reson., 1982, 49, 225.
52. Burdett, J.K. Adv. Inorg. Chem. Radiochem., 1978, 21, 113.
53. Rossi, A.R.; Hoffmann, R. Inorg. Chem., 1975, 14, 365.
54. Krusic, P.J.; San Filippo Jr., J.; Hutchinson, B.; Hance, R.L.; Daniels, L.M. J. Am. Chem. Soc., 1981, 103, 2129.
55. Poliakoff, M.; Turner, J.J. J. Chem. Soc., Dalton Trans., 1973, 1351.
56. Barton, T.J.; Grinter, R.; Thompson, A.J.; Davies, B.; Poliakoff, M. J. Chem. Soc., Chem. Commun., 1977, 841.
57. Conture, C.; Morton, J.R.; Preston, K.F.; Strach, S.J. J. Magn. Reson., 1980, 41, 88.
58. Morton, J.R.; Preston, K.F.; Strach, S.J. J. Phys. Chem., 1980, 84, 2478.
59. Atherton, N.M.; Morton, J.R.; Preston, K.F.; Vuolle, M.J. Chem. Phys. Lett., 1980, 70, 4.
60. Brailsford, J.R.; Morton, J.R.; Vannotti, L.E. J. Chem. Phys., 1969, 50, 1051.
61. Fairhurst, S.A.; Morton, J.R.; Preston, K.F. J. Magn. Reson., 1983, 55, 453.

62. Fairhurst, S.A.; Morton, J.R.; Preston, K.F. Organometallics, 1983, 2, 1869.
63. Hayashida, S.; Kawamura, T.; Yonezawa, T. Inorg. Chem., 1982, 21, 2235.
64. Symons, M.C.R.; Zimmerman, D.N. J. Chem. Soc., Dalton Trans., 1975, 2545.
65. Morton, J.R.; Preston, K.F. Inorg. Chem., 1985, 24, 3317.
66. Eastland, G.W.; Kurita, Y.; Symons, M.C.R. J. Chem. Soc., Perkin Trans. 2, 1984, 1843.
67. Kira, M.; Nakazawa, H.; Sakurai, H. J. Am. Chem. Soc., 1983, 105, 6983.
68. Rao, D.N.R.; Chandra, H.; Symons, M.C.R. J. Chem. Soc., Perkin Trans. 2, 1984, 1201.
69. Rao, D.N.R.; Symons, M.C.R. J. Chem. Soc., Perkin Trans. 2, 1985, 991.
70. Begum, A.; Lyons, A.R.; Symons, M.C.R. J. Chem. Soc. (A), 1971, 2388.
71. Lyons, A.R.; Symons, M.C.R. J. Chem. Soc., Faraday Trans. II, 1972, 68, 622.
72. Allcock, H.R.; Birdsall, W.J. J. Am. Chem. Soc., 1969, 91, 7541; Inorg. Chem., 1971, 10, 2495.
73. Mishra, S.P.; Symons, M.C.R. J. Chem. Soc., Dalton Trans., 1976, 1622; Symons, M.C.R. J. Chem. Res. (S), 1978, 358.
74. Eastland, G.W.; Symons, M.C.R., unpublished results.
75. Rao, D.N.R.; Stephenson, J.M.; Symons, M.C.R. J. Chem. Soc., Perkin Trans. 2, 1983, 727.
76. Dickinson, L.C.; Symons, M.C.R. Chem. Soc. Reviews, 1983, 12, 387.
77. Symons, M.C.R.; Petersen, R.L. Proc. Roy. Soc. B, 1978, 201, 285.
78. Symons, M.C.R.; Stephenson, J.M. J. Chem. Soc., Faraday Trans. I, 1983, 79, 2983.
79. Plonka, A.; Metodiewa, D.; Gasyna, Z. Biochim. Biophys. Acta, 1980, 612, 299.
80. Plonka, A.; Metodiewa, D.; Zgivski, A.; Hilewicz, M.; Leyko, W. Biochem. Biophys. Res. Comm., 1980, 95, 978.
81. Rao, D.N.R.; Symons, M.C.R. J. Chem. Soc., Faraday Trans. I, 1983, 79, 269.
82. Rao, D.N.R.; Symons, M.C.R. J. Chem. Soc., Faraday Trans. I, 1984, 80, 423.

RECEIVED November 26, 1986

Chapter 12

Sonochemistry of Organometallic Compounds

Kenneth S. Suslick

**School of Chemical Sciences, University of Illinois at Urbana-Champaign,
Urbana, IL 61801**

The chemical effects of ultrasound are reviewed with a focus on organometallic systems. Acoustic cavitation is the principal source of sonochemistry, but its nature is quite dependent on the local environment. Cavitational collapse in homogeneous liquids generates hot spot heating of ≈5200K, which can cause multiple ligand dissociation, clusterification, and initiation of homogeneous catalysis. Cavitation near solids has much different effects due to microjet impact on the surface; enhancement of a wide variety of liquid-solid reactions and heterogeneous catalysis occur. These unusual reaction patterns are compared to other high energy processes.

The purpose of this chapter will be to serve as a critical introduction to the nature and origin of the chemical effects of ultrasound. We will focus on organo-transition metal sonochemistry as a case study. There will be no attempt to be comprehensive, since recent, exhaustive reviews on both organometallic sonochemistry (1) and the synthetic applications of ultrasound (2) have been published, and a full monograph on the chemical, physical and biological effects of ultrasound is in press (3).

The chemical and biological effects of ultrasound were first reported by Loomis more than 50 years ago (4). Within fifteen years of the Loomis papers, widespread industrial applications of ultrasound included welding, soldering, dispersion, emulsification, disinfection, refining, cleaning, extraction, flotation of minerals and the degassing of liquids (5),(6). The use of ultrasound within the chemical community, however, was sporadic. With the recent advent of inexpensive and reliable sources of ultrasound, there has been a resurgence of interest in the chemical applications of ultrasound.

A number of terms in this area will be unfamiliar to most chemists. Cavitation is the formation of gas bubbles in a liquid and occurs when the pressure within the liquid drops significantly below the vapor pressure of the liquid. Cavitation can occur from a variety of causes: turbulent flow, laser heating, electrical discharge, boiling, radiolysis, or acoustic irradiation. We will be concerned

0097-6156/87/0333-0191$06.00/0

exclusively with <u>acoustic cavitation</u>. When sound passes through a liquid, it consists of expansion (negative-pressure) waves and compression (positive-pressure) waves. These cause pre-existing bubbles to grow and recompress. Acoustic cavitation can lead, as discussed later, to implosive bubble collapse and associated high-energy chemistry. To indicate ultrasonic irradiation ("sonication"), we will use as our symbol: -)-)-)→. <u>Sonocatalysis</u> will be restricted in its use only to the creation of a catalytically competent intermediate by ultrasonic irradiation; we will not refer to a simple sonochemical rate enhancement of an already ongoing reaction by this term.

Mechanisms of the Chemical Effects of Ultrasound

Ultrasound spans the frequencies of roughly 20 KHz to 10 MHz (human hearing has an upper limit of <18 KHz). Since the velocity of sound in liquids is ≈1500 m/sec, ultrasound has acoustic wavelengths of roughly 7.5 to 0.015 cm. Clearly no direct coupling of the acoustic field with chemical species on a molecular level can account for sonochemistry. Instead, the chemical effects of ultrasound derive from several different physical mechanisms, depending on the nature of the system.

The most important of these is acoustic cavitation, which involves at least three discrete stages: nucleation, bubble growth, and, under proper conditions, implosive collapse. The dynamics of cavity growth and collapse are strikingly dependent on local environment, and cavitation in a homogeneous liquid should be considered separately from cavitation near an interface.

The tensile strength of a <u>pure</u> liquid is determined by the attractive intermolecular forces which maintain its liquid state; the calculated tensile strength of water, for example, is in excess of -1000 atmospheres (<u>7</u>). In practice however, the measured threshold for initiation of cavitation is never more than a small fraction of that. Indeed, if the observed tensile strengths of liquids did approach their theoretical limits, the acoustic intensities required to initiate cavitation would be well beyond that generally available, and no sonochemistry would be observed in homogeneous media! Cavitation is initiated at a nucleation site where the tensile strength is dramatically lowered, such as small gas bubbles and gas filled crevices in particulate matter, which are present in the liquid.

The relevant question for the chemist lies in the actual phenomena responsible for sonochemical reactions. In homogeneous media, the generally accepted sonochemical mechanism involves pyrolysis by a localized "hot-spot" due to the adiabatic heating which is produced by the implosive collapse of a bubble during cavitation. A recent measurement of the temperature generated during this implosive collapse, which we will discuss in detail later, established that the effective temperature in the gas phase reaction zone is ≈5200°K with pressures of hundreds of atmospheres (<u>8</u>)!

When a liquid-solid interface is subjected to ultrasound, transient cavitation still occurs, but with major changes in the nature of the bubble collapse. No longer do cavities implode spherically. Instead, a markedly asymmetric collapse occurs, which generates a jet of liquid directed at the surface, as seen in high speed micro-cinematography by Ellis (<u>9</u>) and Lauterborn (<u>10</u>) (shown

schematically in Figure 1). The jet velocities are greater than 100
m/sec. The origin of this jet formation is essentially a shaped-
charge effect. The impingement of this jet can create a localized
erosion (and even melting) responsible for surface pitting and ultrasonic cleaning (11). A second contribution to erosion created by
cavitation involves the impact of shock waves generated by the implosive collapse. The magnitude of such shock waves can be as high as
10^4 atmospheres, which will easily produce plastic deformation of
malleable metals (12). The relative magnitudes of these two effects
depends heavily on the specific system under consideration.

Acoustic streaming is another non-linear acoustic phenomenon
important to the effect of ultrasound on surfaces (13). This time-
dependent flow of liquid induced by a high intensity sound field is
independent of cavitation. Its origins lie in the conservation of
momentum. As a liquid absorbs energy from a propagating acoustic
wave, it must also acquire a corresponding momentum, thus creating
force gradients and mass transport. Such streaming will occur at
moving solid surfaces or at vibrating bubbles. Thus, when a liquid-
solid interface is exposed to ultrasound, improved mass transport is
expected due to acoustic streaming. This will occur even when the
sound field is a stable standing wave in the absence of cavitation.

Enhanced chemical reactivity of solid surfaces are associated with
these processes. The cavitational erosion generates unpassivated,
highly reactive surfaces; it causes short-lived high temperatures and
pressures at the surface; it produces surface defects and deforma-
tions; it forms fines and increases the surface area of friable solid
supports; and it ejects material in unknown form into solution.
Finally, the local turbulent flow associated with acoustic streaming
improves mass transport between the liquid phase and the surface, thus
increasing observed reaction rates. In general, all of these effects
are likely to be occurring simultaneously.

Experimental Considerations

A variety of devices have been used for ultrasonic irradiation of
solutions. The two most commonly in use are the ultrasonic cleaning
bath and the direct immersion ultrasonic horn. In both cases the
original source of the ultrasound is a piezoelectric material, usually
a lead zirconate titanate ceramic (PZT), which is subjected to a high
voltage, alternating current with an ultrasonic frequency (usually 15
to 50 KHz). In general we find that the typical cleaning bath does
not have sufficient acoustic intensities for most chemical applications, with the exception of reactions involving extremely reactive
metals (e.g., lithium). For this reason, we have adapted a much more
intense source, the direct immersion ultrasonic horn, for inert atmosphere work (shown in Figure 2) and for moderate pressures (<10 atm.).
These configurations may be used for both homogeneous and heterogeneous sonochemistry. Ultrasonic horns are available from several
manufacturers (14) at modest cost and are used primarily by biochem-
ists for cell disruption. A variety of sizes of power supplies and
titanium horns are available, thus allowing flexibility in sample
size. The acoustic intensities are easily and reproducibly varied;
the acoustic frequency is well-controlled, albeit fixed (typically at
20 KHz). Since power levels are quite high, counter-cooling of the

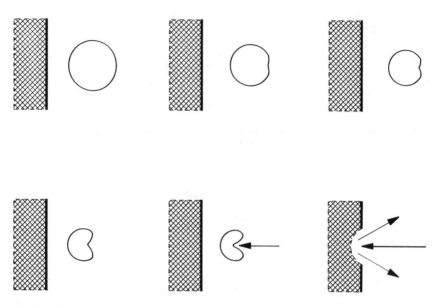

Figure 1. Cavitation Near a Surface. The sequence of a single bubble collapsing follows from left to right, top to bottom.

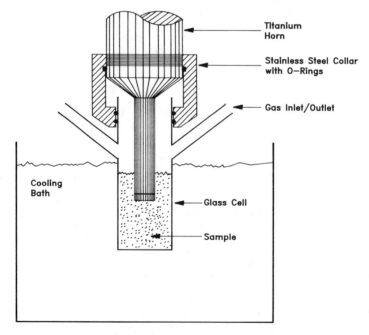

Figure 2.   Direct Immersion Ultrasonic Horn Equipped for Inert Atmosphere Work.

reaction solution is essential to provide temperature control; cooling of the piezoelectric ceramic may also be necessary, depending on the specific apparatus.   One potential disadvantage in corrosive media is the erosion of the titanium tip; this is generally a very slow process without chemical consequences, given the high tensile strength and low reactivity of Ti metal.   Alternatively, a thin teflon membrane can be used to separate the Ti horn from the reaction solution.

A rough, but useful, comparison between typical sonochemical and photochemical efficiencies is shown in Table 1.   Homogeneous sono-chemistry is typically _more_ efficient than photochemistry, and heterogeneous sonochemistry is better still by several orders of magnitude.   Unlike photochemistry, whose energy inefficiency is inherent in the production of photons, ultrasound can be produced with nearly perfect efficiency from electric power.   Still, a primary limitation of sonochemistry remains its energy inefficiency: only a small fraction of the acoustic power is involved in the cavitation events.   This might be significantly improved, however, if a more efficient means of utilizing the sound field to generate cavitation can be found.

Table I.   Comparisons Between Sonochemical
and Photochemical Apparatuses

	Photochemistry	Homogeneous Sonochemistry	Heterogeneous Sonochemistry
Source	250W Quartz-Halogen	200W Cell Disrupter (at 60% power)	200W Cleaning Bath
Approximate Cost (1986)	$1800	$1900	$700
Typical Rates	7 µmol/min	10 µmol/min	500 µmol/min
Electrical Efficiency	2 mmol/KWH	5 mmol/KWH	200 mmol/KWH

Large-scale ultrasonic irradiations are extant technology.   Liquid processing rates of >200 L/min are routinely accessible from a variety of modular, flow reactors with acoustic powers of tens of KW per unit (14).   The industrial uses of these units include 1) degassing of liquids, 2) dispersion of solids into liquids, 3) emulsification of immiscible liquids and 4) large-scale cell disruption.   While these units are of limited use for most laboratory research, they are of potential importance in eventual industrial application of sonochemical reactions.

Sonochemistry is strongly affected by a variety of external parameters, including acoustic frequency, acoustic intensity, bulk temperature, static pressure,   choice of ambient gas, and choice of solvent. These are important considerations in the effective use of ultrasound to influence chemical reactivity, and are also easily understandable in terms of the cavitational hot spot mechanism. A few of the more surprising results are mentioned here, but the reader is referred to one of the recent reviews (1-3) for further details.

The frequency of the sound field is almost irrelevant to most sonochemistry. Unlike photochemistry, there is no direct coupling of the irradiating field with the molecular species in sonochemistry. The effect of changing sonic frequency is simply one of altering the resonant size of the cavitation event. The overall chemistry is therefore little influenced over the range where cavitation can occur (from tens of Hz to a few MHz) (15). The effect of the bulk solution temperature lies primarily in its influence on the bubble content before collapse. With increasing temperature, in general, sonochemical reaction rates are slower. This reflects the dramatic influence which solvent vapor pressure has on the cavitation event: the greater the solvent vapor pressure found within a bubble prior to collapse, the less effective the collapse (16),(17). When secondary reactions are being observed (as in corrosion or other thermal reactions occurring after initial sonochemical events), temperature can play its usual role in thermally activated chemical reactions. The maximum temperature reached during cavitation is strongly dependent on the ambient gas. Both its polytropic ratio ($\gamma = C_p/C_v$, which defines the amount of heat released during the adiabatic compression of that gas) and the thermal conductivity (which controls the degree to which the collapse is adiabatic) have a dramatic impact, so that even the noble gases affect cavitation differently. By the use of these variables, one can fine-tune the energetics of cavitation, and hence exercise a good measure of control over sonochemical reactions.

Homogeneous Sonochemistry

Stoichiometric Reactions. The effects of high-intensity ultrasound on chemical systems is an area of only limited, and in large part recent, investigation. Still, a variety of novel reactivity patterns are beginning to emerge which are distinct from either normal thermal or photochemical activation. Most of the reactions which have been reported are stoichiometric in terms of a consumed reagent, but a few examples of true sonocatalysis have also appeared. We will divide our discussion into homogeneous and heterogeneous systems, in part because of the distinct nature of the cavitation event in each case.

In 1981, we first reported on the sonochemistry of discrete organometallic complexes and demonstrated the effects of ultrasound on metal carbonyls in alkane solutions (18). The transition metal carbonyls were chosen for these initial studies because their thermal and photochemical reactivities have been well-characterized. The comparison among the thermal, photochemical, and sonochemical reactions of $Fe(CO)_5$ provides an excellent example of the unique chemistry which acoustic cavitation can induce, and (because of space limitations in this review) we will focus upon it as an archetype.

Thermolysis of $Fe(CO)_5$, for example, gives pyrophoric, finely divided iron powder (19); ultraviolet photolysis (20) yields $Fe_2(CO)_9$, via the intermediate $Fe(CO)_4$; multiphoton infrared photolysis in the gas-phase (21),(22) yields isolated Fe atoms. Multiple ligand dissociation, generating $Fe(CO)_3$, $Fe(CO)_2$, etc., is not available from ordinary thermal or photochemical processes (but does occur in matrix isolated (23),(24) and gas phase laser (25),(26) photolyses). These observations reflect the dual difficulties inherent in creating controlled multiple ligand dissociation: first, to deliver sufficient energy in a utilizable form and, second, to quench the highly energetic intermediates before complete ligand loss occurs.

Sonolysis of $Fe(CO)_5$ in alkane solvents in the absence of alternate ligands causes the unusual clusterification to $Fe_3(CO)_{12}$, together with the formation of finely divided iron (18),(27). The rate of decomposition is cleanly first order, and the log of the observed first order rate coefficient is linear with the solvent vapor pressure. This is consistent with a simple dissociation process activated by the intense local heating generated by acoustic cavitation. As discussed earlier, the intensity of the cavitational collapse and the maximum temperature reached during such collapse decreases with increasing solvent vapor pressure. Thus, we would also expect to see the ratio of products change as a function of solvent vapor pressure. This proves to be the case: the ratio of products can be varied over a 100-fold range, with the production of $Fe_3(CO)_{12}$ strongly favored by increasing solvent volatility, as expected, since the sonochemical production of metallic iron requires greater activation energy than the production of $Fe_3(CO)_{12}$.

The proposed chemical mechanism by which $Fe_3(CO)_{12}$ is formed during the sonolysis of $Fe(CO)_5$ is shown in Equations 1 - 4.

$$Fe(CO)_5 \; -)-)-) \blacktriangleright Fe(CO)_{5-n} + n \; CO \qquad (n=1-5) \qquad [1]$$

$$Fe(CO)_3 + Fe(CO)_5 \longrightarrow Fe_2(CO)_8 \qquad [2]$$

$$2 \; Fe(CO)_4 \longrightarrow Fe_2(CO)_8 \qquad [3]$$

$$Fe_2(CO)_8 + Fe(CO)_5 \longrightarrow Fe_3(CO)_{12} + CO \qquad [4]$$

$Fe_2(CO)_9$ is not generated during the synthesis of $Fe_3(CO)_{12}$, and sonolysis of $Fe_2(CO)_9$ yields only $Fe(CO)_5$ and finely divided iron. In this mechanism, the production of $Fe_3(CO)_{12}$ arises from initial multiple dissociative loss of CO from $Fe(CO)_5$ during cavitation, followed by secondary reactions with excess $Fe(CO)_5$. The reaction of the putative $Fe_2(CO)_8$ with $Fe(CO)_5$ may proceed through initial dissociation in analogy to the matrix isolation reactivity (28) of $Fe(C_4H_4)_2(CO)_4$.

In the presence of added Lewis bases, sonochemical ligand substitution also occurs for $Fe(CO)_5$, and in fact for most metal carbonyls. Sonication of $Fe(CO)_5$ in the presence of phosphines or phosphites produces $Fe(CO)_{5-n}L_n$, $n=1$, 2, and 3. The ratio of these products is independent of length of sonication; the multiply substituted products increase with increasing initial [L]; $Fe(CO)_4L$ is not sonochemically converted to $Fe(CO)_3L_2$ on the time scale of its production from $Fe(CO)_5$. These observations are consistent with the same primary sonochemical event responsible for clusterification:

$$Fe(CO)_5 \;-)-)-) \rightarrow Fe(CO)_{5-n} + n\ CO \qquad (n=1-5) \qquad [5]$$

$$Fe(CO)_4 + L \longrightarrow Fe(CO)_4L \qquad\qquad\qquad\quad [6]$$

$$Fe(CO)_3 + L \longrightarrow Fe(CO)_3L \qquad\qquad\qquad\quad [7]$$

$$Fe(CO)_3 + CO \longrightarrow Fe(CO)_4 \qquad\qquad\qquad\quad [8]$$

$$Fe(CO)_3L + L \longrightarrow Fe(CO)_3L_2 \qquad\qquad\qquad\;\; [9]$$

We have also observed sonochemical ligand substitution with a variety of other metal carbonyls (27),(29). In all cases, multiple ligand substitution occurs directly from the parent carbonyl. In fact, we have been able to use these reactions as probes of the nature of the reaction conditions created during acoustic cavitation. In order to probe the nature of the sonochemical hot spot, we determined the first order rate coefficients of sonochemical ligand substitution as a function of metal carbonyl vapor pressure. However, the efficacy of cavitational collapse and the temperatures so generated are strongly dependent on the vapor pressure of the solvent system (16),(17). Therefore, sonochemical substitutions at various ambient temperatures were done in solutions of two n-alkanes which had been mixed in the proper proportion to keep the total system vapor pressure constant (at 5.0 torr). In this fashion, the first order rate coefficients were determined as a function of metal carbonyl vapor pressure, for several metal carbonyls. In all cases the observed sonochemical rate coefficient increases linearly with increasing dosimeter vapor pressure and has a non-zero intercept. The linear dependence of the observed rate coefficients on metal carbonyl vapor pressure is expected for reactions occurring in the gas phase: as the dosimeter vapor pressure increases, its concentration within the gas phase cavity increases linearly, thus increasing the observed sonochemical rate coefficients. In addition, the non-zero intercept indicates that there is a vapor pressure independent component of the overall rate. There must be, therefore, an additional reaction site occurring within the liquid phase, presumably in the thin liquid shell surrounding the collapsing cavity (30),(31).

Our data can be used to estimate the effective temperatures reached in each site through comparative rate thermometry, a technique developed for similar use in shock tube chemistry (32). Using the sonochemical kinetic data in combination with the activation parameters recently determined by high temperature gas phase laser pyrolysis (33), the effective temperature of each site can then be calculated (8),(34): the gas phase reaction zone effective temperature is 5200 ± 650°K, and the liquid phase effective temperature is ≈1900°K. Using a simple thermal conduction model, the liquid reaction zone is estimated to be ≈200 nm thick and to have a lifetime of less than 2 μsec, as shown in Figure 3.

Initiation of Homogeneous Catalysis. Having demonstrated that ultrasound can induce ligand dissociation, the initiation of homogeneous catalysis by ultrasound becomes practical. The potential

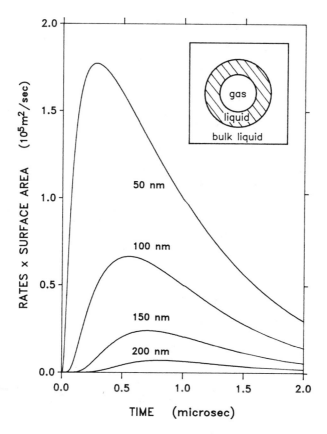

Figure 3.   Temporal and Spatial Evolution of Reaction Rates in the
Liquid Phase Reaction Zone.   Rates were calculated as a function of
time and distance from the bubble surface assuming only conductive
heat transport from a sphere with radius 150μm at 5200K, embedded in
an infinite matrix at 300K.

advantages of such sonocatalysis include 1) the use of low ambient temperatures to preserve thermally sensitive substrates and to enhance selectivity, 2) the ability to generate high-energy species unobtainable from photolysis or simple pyrolysis, 3) the mimicry, on a microscopic scale, of bomb reaction conditions, and 4) the potential ease of scale-up. The transient, coordinatively unsaturated species produced from the sonolysis of metal carbonyls are likely candidates, since similar species produced photochemically are among the most active catalysts known (35).

A variety of metal carbonyls upon sonication will catalyze the isomerization of 1-alkenes to the internal alkenes (18),(27). Initial turnover rates are as high as 100 mol alkene isomerized/mol of precatalyst/h, and represent rate enhancements of ≈10^5 over thermal controls. The relative sonocatalytic and photocatalytic activities of these carbonyls are in general accord. A variety of terminal alkenes can be sonocatalytically isomerized. Increasing steric hindrance, however, significantly diminishes the observed rates. Alkenes without β-hydrogens will not serve as substrates.

The exact nature of the catalytic species generated during sonolysis remains unknown. Results are consistent with the generally accepted mechanism for alkene isomerization in analogous thermal(36) and photochemical systems (37),(38). This involves the formation of a hydrido-π-allyl intermediate and alkene rearrangement via hydride migration to form the thermodynamically more stable 2-alkene complex, as shown in a general sense in Equations 10-14. In keeping with this scheme, ultrasonic irradiation of $Fe(CO)_5$ in the presence of 1-pentene and CO does produce $Fe(CO)_4$(pentene), as determined by FTIR spectral stripping (27).

$$M(CO)_n \ -)-)-) \blacktriangleright \ M(CO)_m + n-m \ CO \qquad [10]$$

$$M(CO)_m + 1-alkene \longrightarrow M(CO)_x(1-alkene) + m-x \ CO \qquad [11]$$

$$M(CO)_x(1-alkene) \longrightarrow M(CO)_x(H)(\pi-allyl) \qquad [12]$$

$$M(CO)_x(H)(\pi-allyl) \longrightarrow M(CO)_x(2-alkene) \qquad [13]$$

$$M(CO)_x(2-ene) + 1-alkene \longrightarrow M(CO)_x(1-ene) + 2-alkene \qquad [14]$$

Recently, we have demonstrated another sort of homogeneous sonocatalysis in the sonochemical oxidation of alkenes by O_2. Upon sonication of alkenes under O_2 in the presence of $Mo(CO)_6$, 1-enols and epoxides are formed in one to one ratios. Radical trapping and kinetic studies suggest a mechanism involving initial allylic C-H bond cleavage (caused by the cavitational collapse), and subsequent well-known autoxidation and epoxidation steps. The following scheme is consistent with our observations. In the case of alkene isomerization, it is the catalyst which is being sonochemical activated. In the case of alkene oxidation, however, it is the substrate which is activated.

$$\diagup\!\!\!\diagdown \quad -)-)-)\blacktriangleright \quad \diagup\!\!\!\diagdown \cdot \quad + \text{ H}^{\cdot} \qquad [15]$$

$$\diagup\!\!\!\diagdown \cdot \quad + \text{ O}_2 \longrightarrow \diagup\!\!\!\diagdown\!\!\!\diagup\text{O}_2^{\cdot} \qquad [16]$$

$$\diagup\!\!\!\diagdown\!\!\!\diagup\text{O}_2^{\cdot} + \diagup\!\!\!\diagdown \longrightarrow \diagup\!\!\!\diagdown\!\!\!\diagup\text{O}_2\text{H} + \diagup\!\!\!\diagdown \cdot \qquad [17]$$

$$\diagup\!\!\!\diagdown\!\!\!\diagup\text{O}_2\text{H} + \diagup\!\!\!\diagdown \xrightarrow{\text{Mo(CO)}_6} \diagup\!\!\!\diagdown\!\!\!\diagup\text{OH} + \triangle\!\!\!\diagdown \qquad [18]$$

Heterogeneous Systems

Liquid-Liquid Reactions. One of the major industrial applications of ultrasound is emulsification. The first reported and most studied liquid-liquid heterogeneous systems have involved ultrasonically dispersed mercury. The use of such emulsions for chemical purposes was delineated by the extensive investigations of Fry and coworkers, (39),(40),(41) who have reported the sonochemical reaction of various nucleophiles with α,α'-dibromoketones and mercury. There are significant synthetic advantages to the use of ultrasound in these systems. For example, such Hg dispersions will react even with quite sterically hindered ketones, yet will introduce only one nucleophilic group even in sterically undemanding systems. Fry believes that the effect of the ultrasound in this system is a kinetic rate enhancement, presumably due to the large surface area of Hg generated in the emulsion.

Liquid-Solid Reactions. The effects of ultrasound on liquid-solid heterogeneous organometallic reactions has been a matter of intense recent investigation. The first use of ultrasound to prepare organo-metallic complexes of the main group metals (e.g. lithium, magnesium, and aluminum) from organic halides, however, originates in the often overlooked work of Renaud (42). The report by Luche in 1980 of the use of an ultrasonic cleaner to accelerate lithiation reactions(43) initiated much of the recent interest (44). Various groups, including Boudjouk's (45) and Ishikawa's (46), have dealt with extremely reactive metals, such as Li, Mg, or Zn, as stoichiometric reagents for a variety of common transformations. The specific origin of the rate and yield improvements has not yet been established in these systems. Ultrasonic cleaning of the reactive metal surface to remove passivating impurities (e.g. water, hydroxide, metal halide, or organolithium) are likely to be important (47).

The activation of <u>less</u> reactive metals continues to attract major efforts in heterogeneous catalysis, metal-vapor chemistry, and synthetic organometallic efforts. Given the extreme conditions generated by acoustic cavitation at surfaces, analogies to autoclave conditions or to metal-vapor reactors may be appropriate. In order to probe the potential generality of ultrasonic activation of hetero-geneous reactions, we examined (48) the sonochemical reactivity of the normally very unreactive early transition metals with carbon monoxide. Even with the use of "activated", highly dispersed transition metal slurries, as investigated by Rieke, (49) the formation of the early transition metal carbonyls still require "bomb" conditions (100-300 atm of CO, 100-300°C) and are prepared in only moderate yields (50). The use of ultrasonic irradiation facilitates the reduction of a variety of transition-metal salts to an active form that will react at low temperatures with low pressures of CO. Reduction of tetrahydrofuran or diglyme solutions of transition metal halides with Na sand in the presence of ultrasound gave fair to good yields of the carbonyl anions for V, Nb, Ta, Cr, Mo, W, Mn, Fe, and Ni, even at 10° and 1 atm. CO. Solubility of the metal halide is necessary for effective reaction. An ultrasonic cleaning bath was found to be of only marginal use when compared to the higher intensity immersion horn. Since these reactions are run at low pressures, they may prove uniquely useful in the production of ^{13}CO labeled carbonyl complexes.

$$MCl_5 + Na + CO \ \text{-)-)-)} \rightarrow M(CO)_6^- \quad (M=V,Nb,Ta) \qquad [19]$$

$$MCl_6 + Na + CO \ \text{-)-)-)} \rightarrow M_2(CO)_{10}^{-2} \quad (M=Cr,Mo,W) \qquad [20]$$

The possible mechanisms which one might invoke for the activation of these transition metal slurries include (1) creation of extremely reactive dispersions, (2) improved mass transport between solution and surface, (3) generation of surface hot-spots due to cavitational micro-jets, and (4) direct trapping with CO of reactive metallic species formed during the reduction of the metal halide. The first three mechanisms can be eliminated, since complete reduction of transition metal halides by Na with ultrasonic irradiation under Ar, followed by exposure to CO in the absence or presence of ultrasound, yielded no metal carbonyl. In the case of the reduction of WCl_6, sonication under CO showed the initial formation of tungsten carbonyl halides, followed by conversion of $W(CO)_6$, and finally its further reduction to $W_2(CO)_{10}^{-2}$. Thus, the reduction process appears to be sequential: reactive species formed upon partial reduction are trapped by CO.

The reduction of transition metal halides with Li has been recently extended by Boudjouk and coworkers for Ullman coupling (benzyl halide to bibenzyl) by Cu or Ni, using a low intensity cleaning bath (51). Ultrasound dramatically decreased the time required for complete reduction of the metal halides (≈12 h without, <40 minutes with ultrasound). The subsequent reactivity of the Cu or Ni powders was also substantially enhanced by ultrasonic irradiation. This allowed significant increases in the yield of bibenzyl (especially for Ni) at lower temperatures, compared to simple stirring.

$$MX_2 + 2\ Li \ \text{-)-)-)} \rightarrow M^* + 2\ LiX \quad (M = Cu, Ni) \qquad [21]$$

$$M^* + C_6H_5CH_2Br \ \text{-)-)-)} \rightarrow MBr_2 + C_6H_5CH_2CH_2C_6H_5 \qquad [22]$$

Another recent application of ultrasound to the activation of transition metals was reported (52) by Bönnemann, Bogdavovic, and coworkers. An extremely reactive Mg species was used to reduce metal salts in the presence of cyclopentadiene, 1,5-cyclooctadiene, and other ligands to form their metal complexes. The reactive Mg species, characterized as $Mg(THF)_3(anthracene)$, was produced from Mg powder in tetrahydrofuran solutions containing a catalytic amount of anthracene by use of an ultrasonic cleaning bath. A plausible scheme for this reaction has been suggested:

$$Mg + C_{14}H_{10} -)-)-) \xrightarrow{THF} Mg(THF)_3(\eta^2-C_{14}H_{10}) \qquad [23]$$

$$2\ Co(acac)_3 + 3\ Mg(THF)_3(\eta^2-C_{14}H_{10}) \longrightarrow 2\ Co^* + 3\ Mg^{2+} \qquad [24]$$

$$2Co^* + 2C_5H_6 + 3\ 1,5-C_8H_{12} \longrightarrow 2Co(Cp)(COD) + C_8H_{14} \qquad [25]$$

The effects of ultrasound on heterogeneous systems are quite general, however, and ultrasonic rate enhancements for many non-metallic insoluble reagents also occurs (2). Ultrasound has been used to enhance the rates of mass transport near electrode surfaces, and thus to enhance rates of electrolysis. This has had some rather useful synthetic applications for the production of both organic and inorganic chalcogenides. The electrochemical reduction of insoluble Se or Te powder by a carbon cloth electrode in the presence of low intensity ultrasonic irradiation produces sequentially E_2^{-2} and E^{-2} (where E is either Se or Te). With generally high current efficiency, these species can be used in a variety of interesting reactions including, for example, the synthesis of $Cp_2Ti(Se_5)$ from Cp_2TiCl_2. Luche as reported the use of ultrasound to accelerate allylations of organic carbonyls by Sn in aqueous media (53). An improved synthesis of $(\eta^6-1,3,5-cyclooctatriene)(\eta^4-1,5-cyclooctadiene)ruthenium(0)$ has been reported which utilizes a cleaning bath to hasten the Zn reduction of $RuCl_3$ in the presence of 1,5-cyclooctadiene (54). The use of ultrasound with Zn is a likely area for routine use in the synthesis and reduction of various organometallic complexes.

The intercalation of organic or inorganic compounds as guest molecules into layered inorganic solid hosts permits the systematic change of optical, electronic, and catalytic properties. The kinetics of intercalation, however, are generally slow, and syntheses usually require high temperatures and very long times. We have recently found that high intensity ultrasound dramatically increases the rates of intercalation (by as much as 200-fold) of a wide range of compounds (including amines, metallocenes, and metal sulfur clusters) into various layered inorganic solids (such as ZrS_2, V_2O_5, TaS_2, MoS_2, and MoO_3) (55). Scanning electron microscopy of the layered solids suggests the origin of our observed rate enhancements. After sonication, in the presence or absence of guest compounds, a significant decrease in the particle size of the host solid occurs. Starting with solids of 60 to 90 μm diameters, after 10 m. sonication the particle size is 5 to 10 μm. Upon further sonication, particle size does not change; the extent of surface damage, however, continues to increase substantially over the next several hours. Both the increase in surface area and the effects of surface damage appear to be important factors in the enhancement of intercalation (56).

Applications to Heterogeneous Catalysis. Ultrasonic irradiation can alter the reactivity observed during the heterogeneous catalysis of a variety of reactions. Sonication has shown such behavior 1) by altering the formation of heterogeneous catalysts, 2) by perturbing the properties of previously formed catalysts, or 3) by affecting the reactivity during catalysis. There is an extensive (but little recognized) literature in this area (57).

However, ultrasonic rate enhancements of heterogeneous catalysis have usually been relatively modest (less than tenfold). The effect of irradiating operating catalysts is often simply due to improved mass transport (58). In addition, increased dispersion during the formation of catalysts under ultrasound (59) will enhance reactivity, as will the fracture of friable solids (e.g., noble metals on C or silica (60),(61),(62) or malleable metals (63)).

We have recently discovered that hydrogenation of alkenes by Ni powder is dramatically enhanced ($>10^5$-fold) by ultrasonic irradiation (64). After sonication of Ni powders under H_2 or Ar, the rapid hydrogenation of alkene solutions occurs even at 25° and 1 atm. of H_2; without sonication, no detectable hydrogenation occurs. This is not a simple surface area effect, since we start with 5 μm sized particles and they do not decrease in size significantly even after lengthy irradiation (as determined by both SEM and BET surface area measurements). There is, however, a very interesting change in surface morphology, as shown in Figure 4, which smooths the initially crystalline surface. In addition, the aggregation of particles increases dramatically upon ultrasonic irradiation, as shown in Figure 5. We have used Auger electron spectroscopy to provide a depth profile of the surface's elemental analysis. The effect of ultrasonic irradiation is to thin substantially the passivating oxide coating to which Ni is quite prone. It appears likely that the changes in surface morphology, in the degree of aggregation, and in the surface's elemental composition originate from inter-particle collisions caused by the turbulent flow and shock waves created by high intensity ultrasound. Previous work on bulk Fe showed similar removal of a passivating layer upon ultrasonic irradiation (65).

Concluding Remarks

The use of ultrasound in both homogeneous and heterogeneous reactions was essentially nil five years ago. Its application in a variety of reactions, especially heterogeneous reactions of highly reactive metals, is now becoming commonplace. The sonochemical generation of organometallic species as synthetic intermediates will continue to find application in nearly any case where interphase mixing is a problem. Much less explored, but potentially quite exciting, is the use of sonochemistry to create high-energy chemistry in condensed phases at room temperature. Unique examples of sonochemical reactivity quite different from thermal or photochemical processes have been noted. The analogies to shock-wave and gas-phase pyrolyses, to "bomb" reactions, and to metal vapor chemistry may prove useful guides in this exploration.

Before ultrasound   15 min ultrasound   120 min ultrasound

Figure 4. The effect of ultrasonic irradiation on surface
morphology of approximately 5 μm as shown by scanning electron
micrographs of Ni powder.

Before ultrasound   15 min ultrasound   120 min ultrasound

Figure 5. The effect of ultrasonic irradiation on particle
aggregation as shown by scanning electron micrographs of Ni
powder (at 10 X less magnification than Figure 4).

Acknowledgments

The graduate students and postdoctoral research associates, all of whom have the author's deepest thanks, are cited in the references. Special thanks are due to Dr. M. L. H. Green and his research group for the work on molecular intercalation. The generosity of funding from the National Science Foundation is gratefully acknowledged. The author is a Sloan Foundation Research Fellow and the recipient of a Research Career Development Award of the National Institute of Health.

References

1. Suslick, K. S. Adv. Organometallic Chem. 1986, 25, 73-119.
2. Suslick, K. S. Modern Synthetic Methods 1986, 4, 1-60.
3. Suslick, K. S. (Ed.) Ultrasound: Its Chemical, Physical and Biological Effects; VCH Publishers, New York 1986.
4. Richards, W. T.; Loomis, A. L. J. Am. Chem. Soc. 1927, 49, 3086-3100; Wood, R. W.; Loomis, A. L. Phil. Mag., Ser. 7, 1927, 4, 417-436.
5. El'piner, I. E. Ultrasound: Physical, Chemical, and Biological Effects; Sinclair, F. A., Trans.; Consultants Bureau: New York, 1964.
6. Rozenberg, L. ed., Physical Principles of Ultrasonic Technology" Vol 1 and 2. Plenum Press: New York, 1973.
7. Crum, L. A. I.E.E.E. Ultrasonics Symposium, 1-11 1982.
8. Suslick, K. S.; Cline, Jr., R. E.; and Hammerton, D. A. J. Amer. Chem. Soc. 1986 108, 5641-2.
9. Felix, M. P.; Ellis, A. T. Appl. Phys. Lett. 19, 484-6 1971.
10. Lauterborn, W.; Bolle, H. J. Fluid Mech. 72, 391-9 1975.
11. Agranat, B. A.; Bashkirov, V. I.; Kitaigorodskii, Y. I. in Physical Principles of Ultrasonic Technology; Rozenberg, L. ed.; Plenum Press: New York, 1973; Vol. 1, pp. 247-330.
12. Hansson, I.; Morch, K. A.; Preece, C. M. Ultrason. Intl. 1977, 267-74.
13. Rooney, J. A. in Methods of Experimental Physics: Ultrasonics; Edmonds, P. D., ed.; Academic Press: New York, 1981; Vol. 19, p. 299-353.
14. Among others: Heat Systems-Ultrasonics, 1938 New Highway, Farmingdale, NY 11735; Branson Sonic Power, Eagle Rd., Danbury, CT 06810; Sonics & Materials, Kenosia Av., Danbury, CT 06810; Lewis Corp., 324 Christian St., Oxford, CT 06483.
15. Margulis, M.; Grundel, L. M.; Zh. Fiz. Khim. 1982, 56, 1445-9, 1941-5, 2592-4.
16. Suslick, K. S.; Gawienowski, J. J.; Schubert, P. F.; Wang, H. H. J. Phys. Chem. 1983, 87, 2299-301.
17. Suslick, K. S.; Gawienowski, J. J.; Schubert, P. F.; Wang, H. H. Ultrason. 1984, 22, 33-6.
18. Suslick, K. S.; Schubert, P. F.; Goodale, J. W. J. Am. Chem. Soc. 1981, 103, 7324-4.
19. Carlton, H. E.; Oxley, J. H. AIChE J. 1965, 11, 79.
20. Geoffroy, G. L.; Wrighton, M. S. Organometallic Photochemistry; Academic Press: New York, 1979.
21. Langsam, Y.; Ronn, A. M. Chem. Phys. 1981, 54, 277-90.
22. Lewis, K. E.; Golden, D. M.; Smith, G. P. J. Am. Chem. Soc. 1984, 106, 3905-12.

23. Poliakoff, M.; Turner, J.J. J. Chem. Soc., Faraday Trans. 2, 1974, 70, 93-9.
24. Poliakoff, M. J. Chem. Soc., Dalton Trans., 1974, 210-2.
25. Nathanson, G.; Gitlin, B.; Rosan, A. M.; Yardley, J. T. J. Chem. Phys. 1981, 74, 361-9, 370-8.
26. Karny, Z.; Naaman, R.; Zare, R. N. Chem. Phys. Lett. 1978, 59, 33-7.
27. Suslick, K. S.; Goodale, J. W.; Schubert, P. F.; Wang, H. H. J. Am. Chem. Soc. 1983, 105, 5781-5.
28. Fischler, I.; Hildenbrand, K.; Koerner von Gustorf, E. Angew. Chem. Int. Ed. Engl. 1975, 14, 54.
29. Suslick, K. S.; Schubert, P. F. J. Am. Chem. Soc. 1983, 105, 6042-4.
30. Suslick, K. S.; Cline, Jr., R. E.; and Hammerton, D. A. I.E.E.E. Ultrason. Symp. Proc. 1985, 4, 1116-21.
31. Suslick, K. S.; Hammerton, D. A. I.E.E.E. Trans. Ultrason. Ferroelec. Freq. Control 1986, 33, 143-6.
32. Tsang, W. in Shock Waves of Chemistry; Lifshitz, A., ed; Marcel Dekker, Inc., New York, 1981; p. 59-130.
33. Lewis N. F.; Golden, D. M.; Smith, G. P.; J. Am. Chem. Soc. 1984, 106, 3905.
34. Suslick, K. S.; and Hammerton, D. A. Ultrasonics Intl. '85, 1985, 231-6.
35. Mitchener, J. C.; Wrighton, M. S. J. Am. Chem. Soc. 1981, 103, 975-7.
36. Casey, C. P.; Cyr, C. R. J. Am. Chem. Soc. 1973, 95, 2248-53.
37. Graff, J. L.; Sanner, R. D.; Wrighton, M. S. Organometallics 1982, 1, 837-42.
38. Chase, D. B.; Weigert, F. J. J. Am. Chem. Soc. 1981, 103, 977-8.
39. Fry, A. J.; Herr, D. Tetrahed. Lett. 1978, 40, 1721-24.
40. Fry, A. J.; Lefor, A. T. J. Org. Chem. 1979, 44, 1270-73.
41. Fry, A. J.; Hong, S. S. J. Org. Chem. 1981, 46, 1962-64.
42. Renaud, P. Bull. Soc. Chim. Fr., Ser. 5, 1950, 17, 1044-5.
43. Luche, J. L.; Damiano, J. C. J. Am. Chem. Soc. 1980, 102, 7926-7.
44. Barboza, J. C. S.; Petrier, C.; Luche, J. L. Tetrahed. Lett. 1985, 26, 829-30, and references therein.
45. Boudjouk, P.; Han, B. H.; Anderson, K. R. J. Am. Chem. Soc. 1982, 104, 4992-3 1982; Boudjouk, P. J. Chem. Ed., 1986, 63, 427-9.
46. Kitazume, T.; Ishikawa, N. Chem. Lett. 1982, 137-40; 1984, 1453-4.
47. Sprich, J. D.; Lewandos, G. S. Inorg. Chim. Acta 1983 76, L241-2.
48. Suslick, K. S.; Johnson, R. E. J. Am. Chem. Soc. 1984, 106, 6856-8.
49. Rieke, R. C. Acc. Chem. Res. 1977, 10, 301-6.
50. Wilkinson, G.; Stone, F. G. A.; Abel, E. W., Eds. Comprehensive Organometallic Chemistry: The Synthesis, Reactions, and Structures of Organometallic Compounds Pergamon Press: Oxford, England, 1982; Vol. 3-6, 8.
51. Boudjouk, P.; Thompson, D. P.; Ohrborn, W. H.; Han, B.-H. Organomet. 1986, 5, 1257-60.
52. Bönnemann, H.; Bogdanovic, B.; Brinkman, R.; He, D. W.; Spliethoff, B. Angew Chem. Int. Ed. Engl. 1983, 22, 728.
53. Petrier, C.; Einhorn, J.; Luche, J. L. Tetrahed. Lett. 1985, 26, 1449-50.
54. Itoh, K.; Nagashima, H.; Ohshima, T.; Oshima, N.; Nishiyama, H. J. Organomet. Chem. 1984, 272, 179-88.

55. Suslick, K. S.; Chatakondu, K.; Green, M. L. H.; and Thompson, M. E.   Manuscript in submission.
56. Suslick, K. S.; Casadonte, D. J.; Green, M. L. H.; and Thompson, M. E.   Ultrasonics, 1987, in press.
57. Mal'tsev, A. N.   Z. Fiz. Khim. 1976, 50, 1641-52.
58. Lintner, W.; Hanesian, D.   Ultrason. 1977, 15, 21-6.
59. Abramov, O. V.; Teumin, I. I.   in Physical Principles of Ultrasonic Technology; Rosenberg, L. D., ed.; Plenum Press: New York, 1973; Vol. 2, pp. 145-273.
60. Boudjouk, P.; Han, B. H.   J. Catal. 1983, 79, 489-92.
61. Townsend, C. A.; Nguyen, L. T.   J. Amer. Chem. Soc., 1981, 103, 4582-3.
62. Han, B. H.; Boudjouk, P.   Organomet. 1983, 2, 769-71.
63. Kuzharov, A. S.; Vlasenko, L. A.; Suchkov, V. V.   Zh. Fiz. Khim. 1984, 58, 894-6.
64. Suslick, K. S.; and Casadonte, D. J.   Manuscript in submission.
65. Alkire, R. C.; Perusich, S.   Corros. Sci. 1983, 23, 1121-32.

RECEIVED November 12, 1986

Chapter 13

Acceleration of Synthetically Useful Heterogeneous Reactions Using Ultrasonic Waves

Philip Boudjouk

Department of Chemistry, North Dakota State University, Fargo, ND 58105

Studies of the effects of low frequency ultrasonic waves
on a broad range of synthetically useful reactions are
summarized. Discussion is centered on the results
obtained in our laboratory where we have concentrated on
the reactions of metals with functionalized organic and
organometallic compounds. Special emphasis is on lithium
and zinc with organic and organosilicon halides.
Catalytic systems (platinum, palladium and nickel) have
been investigated as have been non-metallic reagents. Our
results in these areas are also presented.

The pioneering work on the chemical applications of ultrasound was
conducted in the 1920's by Richards and Loomis in their classic
survey of the effects of high frequency sound waves on a variety of
solutions, solids and pure liquids(1). Ultrasonic waves are usually
defined as those sound waves with a frequency of 20 kHz or higher.
The human ear is most sensitive to frequencies in the 1-5 kHz range
with upper and lower limits of 0.3 and 20 kHz, respectively. A brief
but useful general treatment of the theory and applications of
ultrasound has been given by Cracknell(2).

Early Studies of Chemical Effects of Ultrasonic Waves

Using quartz crystals of 6 to 12 millimeters in thickness and 50 to
80 millimeters in diameter held between two electrodes under oil,
Loomis and his coworkers produced high intensity ultrasonic waves at
frequencies in excess of 100,000 cycles per second. The high
intensities are the result of the ability to generate voltages in
the range of 50,000 with this configuration since the amplitude of
vibration of quartz crystal increases directly with the voltage
applied to it. With power levels of approximately 2 kilowatts
readily available, investigation of the effects of ultrasonic waves
on chemical systems became feasible. Table I summarizes their
observations.
    In spite of the diversity of the chemical effects of sonic
waves discovered by Richards and Loomis, basic research over the
next forty years was sparse and uneven. For the most part, the

emphases have been on inorganic reactions, in particular aqueous solutions, and in nearly all cases the systems were homogeneous.

Table I. First Observations of the Chemical Effects
of Ultrasonic Waves

System	Effects of Ultrasonic Waves
Nitrogen triodide	Rate of explosion accelerated
Superheated liquids	Rate of evaporation accelerated
Conversion of yellow HgI to red HgI	Rate of conversion accelerated
"Atomization" of liquids and solids	Rate of "atomization" of mercury and glass in water accelerated
Gases dissolved in water	Rate of expulsion of gases increased
Hydrolysis of methyl sulfate	Rate of hydrolysis increased
The "iodine clock" reaction	Rate of reduction of iodate by sulfite was increased
Boiling points of liquids	"Apparent" depression of boiling point

Most studies of heterogeneous systems were in the applied areas(3). That the sonicators used by the various research groups were often home-built and quite different in configuration and delivered different frequencies, intensities and wattages may have served to limit study in the field. To complicate the matter further, little was known of the role of these variables in affecting reaction rates. As a result, there are few generalizations upon which a new investigator can rely and sonochemistry, in particular preparative sonochemistry, must be considered an area that is in its earliest stages and in need of much further study.

A detailed review of the literature is outside the scope of this chapter. However, a brief survey of some of the key developments in the application of ultrasonic waves to heterogeneous reactions seems appropriate as an introduction to our work.

There were some studies of heterogeneous reactions using ultrasonic waves in the four decades following the survey of Richards and Loomis. The results were encouraging and clearly held promise for synthetic chemists. In retrospect, it is surprising they did not attract more attention. Table II lists some of the reactions investigated.

Table II. Ultrasonic Acceleration of Some Heterogeneous Reactions

REACTION	YEAR
$Zn + 2HCl \xrightarrow{)))} ZnCl_2 + H_2$	1933([4])
$R-X + Mg \xrightarrow{)))} R-Mg-X$	1950([5])
$Na + \text{(acridine)} \xrightarrow{)))} Na_{1.5}^{+} (C_{13}H_9N)^{-}$	1951([6])
$Na + C_6H_5-Cl \xrightarrow{)))} C_6H_6^{-} Na^{+}$	1957([7])
$CH_3S(O)CH_3 + NaH \xrightarrow{)))} CH_3S(O)CH_2^{-} Na^{+}$	1966([8])
$R_2C\underset{Br}{-}\overset{O}{\overset{\|}{C}}-\underset{Br}{C}R_2 + Hg/HgOAc \xrightarrow{)))} R_2C\underset{H}{-}\overset{O}{\overset{\|}{C}}-\underset{OAc}{C}R_2$	1978([9])

Recent Synthetic Applications of Ultrasound

The report by Luche and coworkers that ultrasonic waves from a
common ultrasonic laboratory cleaner aid the formation of
organolithium and Grignard reagents and also improve the Barbier
reaction spurred much of the current interest in the synthetic
applications of ultrasound([10]):

$$R-Br + Li \xrightarrow[1\ h]{)))} RLi \ (61-95\%)$$

$$R = Pr, \underline{n}-Bu, Ph$$

$$R-Br + R'_2CO + Li \xrightarrow{)))} R'_2RCOH \ (76-100\%)$$

$$R = alkyl, aryl, benzyl, allyl, vinyl$$

$$R'_2CO = ketones \ and \ aldehydes$$

Luche and coworkers extended their studies on the applications
of ultrasound to synthesis to include a variety of systems. Among
these applications are: the syntheses of lithioorganocuprates([11]),
of aldehydes from formamides([12]), of organozinc intermediates([13]),

and of homoallylic alcohols(14). The last development is particularly noteworthy because organozinc intermediates are employed in aqueous media. Ishikawa has found the combination of zinc and ultrasound to be particularly useful in the synthesis of a variety of perfluorinated derivatives(15).

Our entry into sonochemistry was spurred by our need for high yield preparations of symmetrical organics and bimetallics. Our first efforts(16,17) with lithium wire were satisfactory but have since been greatly improved by using of lithium dispersion(18):

$$2 \ R_3MCl \ + \ Li \ \xrightarrow[<1 \ h]{)))} \ R_3M-MR_3 \qquad (85-95\%)$$

$$R = alkyl, \ aryl; \ M = C, \ Si, \ Sn$$

Lithium

Reactions involving lithium appear to be particularly good candidates for ultrasonic acceleration. The reduction of some highly unsaturated cyclic hydrocarbons by lithium was not only accelerated but pushed to completion by ultrasonic irradiation(19):

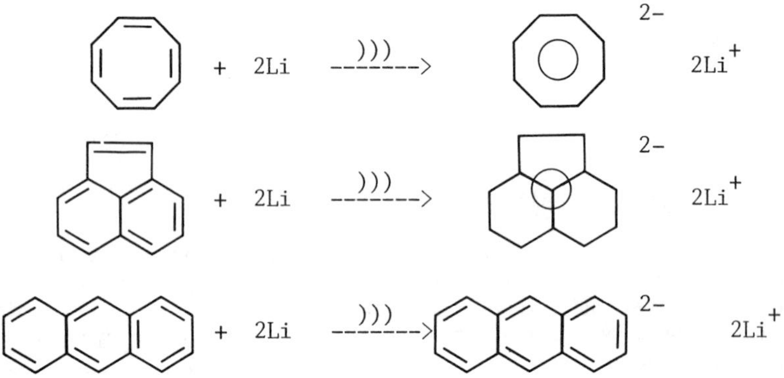

We have also found that ultrasound will promote the liberation of hydrogen from phenylacetylene to give the nucleophile phenylacetylide which can be efficiently quenched with an alkyl halide(19):

$$Ph-C{\equiv}C-H \ + \ Li \ \xrightarrow{)))} \ Ph-C{\equiv}C^- \ Li^+ \ \xrightarrow{MeI} \ Ph-C{\equiv}C-Me$$

as well as the reductive coupling of diphenylacetylene to form the synthetically useful dilithiotetraphenylbutadienedianion(19):

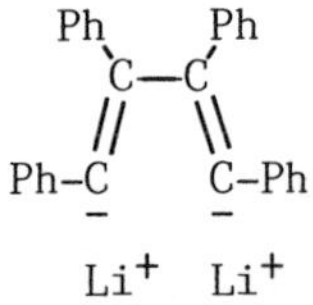

Halides of the less electropositive metals are quickly reduced to highly dispersed and very active metal powders if they are exposed to ultrasonic waves in the presence of lithium and other group I metals(20). Ultrasound not only accelerates the reduction of the halides but also increases the rate of subsequent reactions of these less active metals. These reactions are covered in the chapter by K. Suslick.

Zinc

Zinc promoted reactions are responsive to ultrasonic waves. The classic Simmons-Smith reaction was among our first successes(21). Using zinc powder and diiodomethane we obtained good yields of cyclopropanes from styrene and acenaphthene. However, we could not

$$\text{styrene} + CH_2I_2 + Zn \xrightarrow{\text{)))}} \text{cyclopropane} \qquad 60\%$$

$$\text{acenaphthene} + CH_2I_2 + Zn \xrightarrow{\text{)))}} \text{product} \qquad 75\%$$

consistently control the reaction and it would often get out of hand. Repic and Vogt solved this problem by using mossy zinc with ultrasonic waves to obtain good yields of cyclopropanes from olefins(22). More recently Friedrich found that ultrasound will activate a zinc-copper couple sufficiently to permit the use of dibromomethane as a source of methylene(23). This method is almost as efficient as using freshly prepared zinc powder(24).

We felt the Reformatsky reaction was a worthwhile target because it is the most generally applicable method for converting aldehydes and ketones to β-hydroxyesters(25). The improvements in yield and reaction time exceeded our expectations. Essentially quantitative conversion to the β-hydroxyester was effected in a matter of a few minutes(26). The absence of other products, such as α, β-unsaturated esters, resulting from dehydration, and dimers of the bromo ester and the carbonyl are probably the result of running

$$R_2CO + BrCH_2CO_2Et + Zn \xrightarrow[\text{5-30min}]{\text{)))}} R_2C(OH)CH_2CO_2Et \qquad (94\text{-}100\%)$$

$$R_2CO = \text{alkyl and aryl aldehydes and ketones}$$

the reaction at, effectively, room temperature. An interesting modification of this reaction, in which the ketone was replaced by an imine, leading to a very mild high yield synthesis of β-lactams has recently been published(27).

Reactions of chlorosilanes with carbonyl compounds in the presence of zinc benefit from irradiation with ultrasonic waves. α-Dicarbonyls are converted to bis-siloxyalkenes in very good yields in less than thirty minutes(28). In the absence of

$$R-\overset{\overset{O}{\|}}{C}-\overset{\overset{O}{\|}}{C}-R' + Me_3SiCl + Zn \xrightarrow{\;)))\;} \overset{Me_3SiO}{\underset{R}{}}C=C\overset{OSiMe_3}{\underset{R}{}} \qquad (60\text{-}90\%)$$

ultrasonic waves not even two or three hours of stirring will produce the same yields. The reactions of simple ketones and aldehydes
are even more interesting because, with the appropriate stoichiometry, the carbonyl can not only be silylated, but be persuaded to
couple to another carbonyl group to form bis-siloxyalkanes(29):

$$RCOR' + Me_3SiCl + Zn \xrightarrow{\;)))\;} Me_3SiO-\overset{\overset{R'}{|}}{\underset{R}{C}}-\overset{\overset{R'}{|}}{\underset{R}{C}}-OSiMe_3$$

Some of our results are summarized in Table III.

**Table III.   Reductive Coupling of Carbonyls with Zinc
and Trimethylchlorosilane**

| | | | Yield % | |
R	R'	t(h)	)))	Stir
p-MeO-Ph	Me	8	74	54
p-F-Ph	Me	6	63	48
Ph	Ph	6	84	65
p-Me-Ph	Me	2	76	63
Ph	Me	2	65	55
Ph	H	4	56	42

Carbonyl : Me$_3$SiCl : Zn;   1 : 1 : 5

Benzaldehyde gave modest yields of the coupled product and 15%
of trans-stilbene. By using large excesses of zinc and trimethylchlorosilane, the stilbene yield was increased to 36%. Thus far,
only 1-indanone has produced high yields, 83%, of the unsaturated
dimer(29):

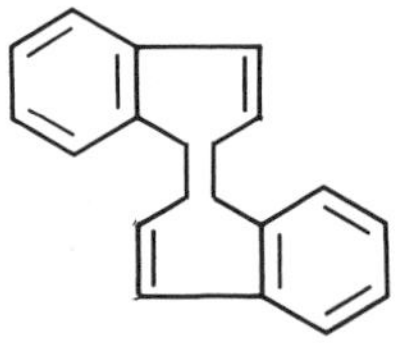

Most of the carbonyls we have tested couple and undergo a pinacol-
type rearrangement via a bis-siloxyalkane intermediate:

$$p\text{-MeO-Ph-}\overset{\overset{Me}{|}}{\underset{Me_3SiO}{C}}\text{-)-}_2 + Me_3SiCl + Zn \xrightarrow[8h]{\;)))\;} (p\text{-MeOPh})_2\overset{\overset{O}{\|}}{\underset{Me}{C}}\text{-}\overset{}{C}\text{-Me} \qquad (100\%)$$

Reactive Intermediates

When zinc and α,α'-dibromo-o-xylene are irradiated with ultrasonic waves at room temperature, synthetically useful quantities of the reactive intermediate, o-xylylene, are generated which can be treated in situ with activated olefins to give good yields of cycloaddition products(30). Chew and Ferrier used this methodolgy to generate o-xylylene for the synthesis of optically pure functionalized hexahydroanthracenes(31). The reaction with lithium takes a different course(19). Rather than generate the o-xylylene intermediate, ionic species are produced. The two fates of α, α'-dibromo-o-xylene are presented in the scheme below:

TMS = Me₃Si

Successes in producing reactive intermediates like o-xylylene and carbene and in preparing bimetallics in high yields using ultrasound led us to attempt to generate West's novel compound, tetramesityldisilene the first example of a stable species with a silicon-silicon double bond(32). We prepared this species in one step and trapped it with methanol(33). The disilene is reactive towards lithium, however, and we have found it very difficult to obtain consistent results. Most often, hexamesitylcyclotrisilane is isolated in very good yield(34).

The cyclotrisilane is one of a few cyclotrisilanes and is, upon photolysis, a useful source of silylenes and disilenes.

Another hindered silane, di-t-butyldichlorodisilane, gives high yields of the reactive divalent species, di-t-butylsilylene, which was characterized by its insertion reactions into Si-H bonds(35):

$$\underline{t}\text{-Bu}_2\text{SiCl}_2 + \text{Li} \xrightarrow{\quad))) \quad} \underline{t}\text{-Bu}_2\text{Si:} \xrightarrow{\quad R_3\text{SiH} \quad} R_3\text{Si-}(\underline{t}\text{-Bu}_2\text{Si})\text{-H}$$

$$R_3 = \text{Et}_3, \text{PhMe}_2$$

When no silane is used to trap the silylene, the mild conditions of the reaction permitted the isolation of the novel ring compound, trans-hexa-t-butylcyclotetrasilane in >15% yield(36).

The successful trapping of di-t-butylsilylene from a dichloro-silane suggested the possibility of a general reaction mechanism. We examined a variety of dihalosilanes using ultrasonic waves and alkali metals(37). The mild conditions permitted led to fewer products than are normally observed in these reactions and the first evidence that silylenes can be important intermediates in metal-dihalosilane reactions in solution. Our results for some dichlorosilanes reacting with lithium metal are summarized below.

$$\text{RR'SiCl}_2 + \text{Li} + R_3\text{Si-H} \xrightarrow{\quad))) \quad} R_3\text{Si-SiRR'-H}$$

R = R' = t-Bu	60%
R = t-Bu, R' = Mes	35–40%
R = R' = Mes	15–20%
R = t-Bu, R' = Ph	<5%
R = R' = Me, Et, Ph	0

It appears that the bulky groups provide enough steric hindrance to slow down intermolecular coupling sufficiently to permit α-elimination to occur. Polysilane compounds are the typical products of these reactions when small organic groups are attached to silicon.

Catalytic Metals

Ultrasound and catalytic processes have a long history. As early as 1940, an unsuccessful attempt to improve the Haber process was reported(38). In 1950 a patent was issued to Richardson for increasing ammonia production by irradiating iron powder suspended in a stream of hydrogen and nitrogen(39). Coating wires fastened to a transducer with iron, platinum or nickel will accelerate not only the synthesis of ammonia but also the hydrogenation of fats as well as the oxidations of sulfides, carbon monoxide, methanol and ethanol(40). Ultrasound has been successfully utilized in several hydrogenation procedures: of carbon monoxide(41); of olefins(42); of olive oil(49); and in the destructive hydrogenation of shale, lignite and petroleum residues(43).

Catalysis and Synthesis in the Laboratory. Research on the practical applications of catalysis was not matched in the laboratory. We began a study of metal and non-metal catalyzed reactions early in our sonochemistry program. Our first project was to develop a convenient method of hydrogenating a wide range of olefins. We chose formic acid as our hydrogen source and found it to be effective. For example, with continuous irradiation, palladium catalyzed hydrogenations of olefins are complete in one hour(44).

$$R_2C=CR_2 \ + \ HCO_2H \ + \ Pd/C \ \xrightarrow{\text{)))}} \ R_2CH-CHR_2 \quad (95-100\%)$$

In the absence of ultrasonic waves, the reactions usually require two or three hours or heating to 80°. Using our procedure, the cyclopropane ring in cyclopropylbenzene was easily opened to give propylbenzene in >95% yield.

Recently we extended this study by replacing formic acid with a stream of hydrogen gas bubbling through the reaction mixture and found that this reaction goes to completion even at 0°(45). A dozen olefins, including vinyl ethers and α, β-unsaturated esters and ketones were quantitatively hydrogenated in one hour or less at 0°. In the absence of ultrasonic waves, no hydrogenation occurs at this temperature.

Our interest in silicon chemistry quite naturally led to a study of the hydrosilation reaction, the addition of the Si-H group across an olefin or an acetylene. This reaction is one of the most useful methods of making silicon-carbon bonds and is an important industrial process. Typically, homogeneous catalysts based on platinum, rhodium or ruthenium are used, and while very efficient, they are not recoverable(46).

The original patent uses platinum as the catalyst and calls for temperatures of 100-300° and pressures of 45-115 psi(47). We found that such rigorous conditions are not required for the hydrosilation reaction with most commercial sources of platinum on carbon. Usually vigorous stirring at slightly elevated temperatures, 40-80°, at 15 psi will give moderate yields of the product. The rates and yields are usually highly dependent on the method of preparation of the catalyst. However, ultrasound permits the reaction to occur at a useful rate at 30° at atmospheric pressure(48):

$$R_3SiH \ + \ {>}C=C{<} \ + \ Pt/C \ \xrightarrow[\text{1-2 h}]{\text{)))}} \ R_3Si-\overset{|}{\underset{|}{C}}-\overset{|}{\underset{|}{C}}-H \quad (30-94\%)$$

$$R = Et, \ Ph, \ EtO, \ Cl$$

Recently we found that freshly prepared nickel powder is an efficient hydrosilation catalyst when continuously irradiated(49).

$$Ni^* \ + \ {>}C=C{<} \ + \ H-SiCl_3 \ \xrightarrow{\text{)))}} \ H-\overset{|}{\underset{|}{C}}-\overset{|}{\underset{|}{C}}-SiCl_3 \quad (60-70\%)$$

In the absence of ultrasonic waves nickel does not promote this reaction at or near room temperature.

Nonmetallic Heterogeneous Reactions

Ultrasound has proven effective in promoting a few heterogeneous nonmetallic reactions. As early as 1933 Moriguchi noted that the reaction of calcium carbonate and sulfuric acid was faster in the presence of ultrasound(4). In the same year Szalay reported that ultrasonic waves depolymerized starch, gum arabic and gelatin(50). Examples of synthetically useful applications are fewer than metallic systems but activity in this arena is increasing.

The ultrasonic preparation of thioamides from amides and phosphorus pentasulfide by Raucher(51) and of dichlorocarbene from chloroform and potassium hydroxide by Regen(52) are some of the more recent examples of nonmetallic applications. We were surprised to find that ultrasound greatly accelerates the reduction of haloaromatics by lithium aluminum hydride, permitting the reaction to be

$$R-Ar-X \; + \; LiAlH_4 \; \xrightarrow{\;)))\;} \; R-Ar-H \quad (70-98\%)$$

conducted at room temperature(53). This reaction requires, in some cases, heating to 100° for several hours to obtain even modest yields of the reduced compounds(54). Lukevics and coworkers extended this methodology to metallic halides finding that reduction to metal hydrides was feasible even in nonpolar solvents(55). Brown has demonstrated that ultrasound accelerated some hydroboration reactions(56). Reactions involving aluminum oxide(57) and potassium permanganate(58) have also been enhanced by ultrasonic waves.

Ultrasound also promotes the reaction of potassium hydride with some silicon hydrides to give silyl anions in excellent yields and

$$R'R_2Si-H \; + \; KH \; \xrightarrow{\;)))\;} \; R'R_2Si:^- \; K^+ \quad (>90\%)$$

$$R = Phenyl \; or \; vinyl$$

with a minimum of byproducts(59). This was not only useful for synthesis but also was an important advantage in obtaining clean spectroscopic data of potentially aromatic silyl anions like the silacyclopentadienyl anion to demonstrate that it is not aromatic(59).

Mechanistic Considerations

Cavitation. To produce a chemical effect in liquids using ultrasonic waves, sufficient energy must be imparted to the liquid to cause cavitation, i.e., the formation and collapse of bubbles in the solvent medium and the consequent release of energy. When ultrasonic waves are passed through a medium, the particles experience oscillations leading to regions of compression and rarefaction. Bubbles form in the rarefaction region which may be filled with a gas, the vapor of the liquid, or be almost empty depending on the pressure and the forces holding the liquid together.

Strictly defined, cavitation refers only to the completely evacuated bubble or cavity, a true void, but since dissolved gases are present unless special steps are taken to remove them, and the vapor of the liquid can also penetrate the cavity, the term cavitation most often encompasses the three kinds of bubbles.

The collapse of these bubbles, caused by the compression region of the ultrasonic wave, produces powerful shock waves. The energy output in the region of the collapsing bubble is considerable, with estimates of $2-3000^\circ C$ and pressures in the 1-10 kilobar range for time periods in the nanosecond regime(60). In summary, the cavitation process generates a transitory high energy environment.

Surface Damage and Reaction Rates. Erosion of surfaces resulting in higher surface area and removal of inhibiting impurities are two effects of cavitation on solids in liquid media, both of which lead to increased reaction rates. The high temperatures and pressures are sufficient to deform and pit metal surfaces (even cause local melting of some metals) and to fracture many nonmetallic solids, in particular, brittle materials.

Mass Transport. Cavitation improves mixing but, on a macroscopic scale, it is probably less effective than a high speed stirrer. On a microscopic scale, however, mass transport is improved at solid surfaces in motion as a result of sound energy absorption. This effect is called acoustic streaming and contributes to increasing reaction rates.

Experimental Considerations

The ultrasonic cleaning bath is the most common source of ultrasound in the laboratory and was the equipment used in most of our investigations. The acoustic intensity is far less than the immersion horn but the low price, less than $200 for a 4" x 9" bath that holds flasks up to 1 liter in size, compared to nearly $2000 for a modest horn setup probably accounts for the difference in popularity.

The Ultrasonic Cleaning Bath. There is little question that the rate enhancements observed from irradiation in the bath would probably be greatly improved if the horn were used. We have observed such differences and are making an effort to quantify them. On the other hand, the cleaning bath provides sufficient intensity to accelerate a wide variety of heterogeneous reactions sufficiently to become very attractive to the synthetic chemist. A healthy range of metals from the main groups and the transition series, including the traditional catalysts, platinum and palladium, as well as nonmetallic solids such as lithium aluminum hydride, potassium hydride and aluminum oxide, are easily activated at room temperature. There is the additional advantage that the bath permits the use of traditional glassware. The lifetimes of the baths in our laboratory are typically >8000 hours.

Location of the Reaction Flask. We found that irradiation from the ultrasonic cleaner is most effective when the flask is positioned in the bath to achieve maximum turbulence of the reagents. This "sweet spot" is the point of maximum cavitation and assures optimum energy transfer to the reaction medium. In practice, this focal point of intensity may move after several hours, possibly because of distortion of the steel bottom caused by local heating of the transducer.

Coupling Medium. Distilled water has proven to be more effective than tap water as the conducting liquid as evidenced by greater cavitation in the reaction flasks (and faster reaction rates). Moreover, distilled water leads to significantly less corrosion of the bath walls. Other low vapor pressure liquids such as ethylene glycol can be used.

Effect of Dissolved Gases. Dissolved gases play a major role in
the optimizing of cavitation by providing nucleation sites for
incipient cavities and by affecting the temperature of the cavitational collapse. Gases with high polytropic ratios, C_p/C_v, increase
the temperature of the cavitational collapse in homogeneous systems
(61). We have found that bubbling argon through the reaction
mixture often improves the effects of ultrasonic waves.

Stirring. Stirring of the reagents will not interfere with cavitation and may prove necessary for acceptable reaction rates. Solids
must not be allowed to collect on the bottom of the flask. If the
solid loading is high and cavitation cannot suspend all of it, a
paddle stirrer should be used.

Temperature. Temperature plays a key role in an unexpected fashion.
Most often, lowering the temperature improves ultrasonically
accelerated reaction rates. This is attributed, in the main, to
a lowering of vapor pressure and consequently an emptier cavity(62).
Implosion of highly evacuated cavities is more energetic than
collapse of vapor filled bubbles. Temperatures can be lowered by
using slush baths as the conducting media in the bath. Temperatures
from 25° to -25° are easy to reach and maintain cavitation in
reaction flasks containing ordinary organic solvents. A cooling fan
mounted on the side of the bath will maintain a temperature of 30 -
35° without additional cooling.

Solvents. Solvents with low vapor pressure will lead to
cavitational implosions of greater energy and potentially faster
reactions. Optimization of polarity and vapor pressure will likely
reap the greatest benefits.

Frequency and Intensity. Most ultrasonic baths operate in the 30 -
80 kHz range. Frequency is rarely an important factor, provided the
frequency is low enough to permit cavitation. The cell disruptors
normally adapted for sonochemical uses operate at 20 kHz. The
intensity must be enough to produce cavitation. Beyond that, optimum
intensities for heterogeneous reactions have not been determined.

Disadvantages of the Ultrasonic Bath. The major disadvantage of the
bath as a source of ultrasonic waves is the low acoustic intensity.
This translates into less than optimum reaction rates and, for some
reluctant systems, no rate enhancements at all. Companion to this
is variability (often on a day-to-day basis) of the intensity and
its focal point, which makes precise rate measurements a formidable
challenge. Ultrasonic baths are not easily adapted to flow
synthesis as are immersion horns.

ACKNOWLEDGMENTS

This chapter is, in large part, a summary of the work of dedicated
graduate students and postdoctoral fellows with whom I have been a
privileged coauthor and to whom I am deeply indebted. The generous
support of this research by the Air Force Office of Scientific
Research is also gratefully acknowledged.

Literature Cited

1. Richards, W.T.; Loomis, A.L. J. Am. Chem. Soc. 1927, 49, 3086–3100.
2. Cracknell, A.P. "Ultrasonics"; Wykeham: London, 1980.
3. Brown, B.; Goodman, J.E. "High Intensity Ultrasonics Industrial Applications"; Van Nostrand: Princeton, 1965.
4. Moriguchi, N.; J. Chem. Soc. (Japan) 1933, 54, 949–957; Chem. Abstr. 1934, 28, 398.
5. Renaud, P. Bull. soc. chim. (France) 1950, 1044–1045.
6. Slough, W.; Ubbelohde, A. R. J. Chem. Soc. 1951, 918.
7. Pratt, M.W.T.; Helsby, R. Nature 1959, 184, 1694–1695
8. Sjöberg, K. Tetrahedron Lett. 1966, 6383.
9. Fry, A.J.; Herr, D. Tetrahedron Lett. 1978, 1721–1724.
10. Luche, J.-L.; Damiano, J.-C. J. Am. Chem. Soc. 1980, 102, 7926–7927.
11. Luche, J.-L; Petrier, C.; Gemal, A.L.; Zikra, N. J. Org. Chem. 47, 1982, 3805–3806.
12. Petrier, C.; Gemal, A.L.; Luche, J.-L. Tetrahedron Lett. 1982, 23, 3361–3364.
13. Luche, J.-L.; Petrier, C.; Lansard, J.P.; Greene, A.E. J. Org. Chem. 1983, 48, 3837–3839.
14. Petrier, C.; Luche, J.-L. J. Org. Chem. 1985, 50, 910–912.
15. Kitazume, T.; Ishikawa, N. Chem. Lett. 1981, 1679–1680.
16. Han, B.-H.; Boudjouk, P. Tetrahedron Lett. 1981, 22, 2757–2758.
17. Boudjouk, P.; Han, B.-H. Tetrahedron Lett. 1981, 22, 3813–3814.
18. Boudjouk, P.; Sooriyakumaran, R.; Han, B.-H. unpublished results.
19. Boudjouk, P.; Sooriyakumaran, R.; Han, B.-H, J. Org. Chem. 1986, 51, 2818–2819.
20. Boudjouk, P.; Thompson, D.E.; Ohrbom, W.H., Han, B.-H. Organometallics, 1986, 5, 1257–1260.
21. Boudjouk, P.; Han, B.-H., Abstracts of Papers, 183rd National Meeting of the American Chemical Society, Las Vegas, NV; American Chemical Society: Washington: DC; ORGN 190.
22. Repic, O.; Vogt, S. Tetrahedron Lett. 1982, 23, 2729–2732.
23. Friedrich, E. C.; Dombek, J. M.; Pong, R. Y. J. Org. Chem. 1985, 50, 4640–4642.
24. Rieke, R. D.; Li, P. T.-J.; Burns, T. P.; Uhm, S. T. J. Org. Chem. 1981, 46, 4323–4324.
25. Rathke, M.W. Org. React. (NY) 1975, 22, 423.
26. Han, B.-H.; Boudjouk, P. J. Org. Chem. 1982, 47, 5030–5032.
27. Bose, A.K.; Gupta, K.; Manhas, M.S. J.C.S. Chem. Commun. 1984, 86–87.
28. Boudjouk, P.; So., J.-H. Synthetic Commun. 1986, 16, 775–778.
29. Boudjouk, P.; Park, M. unpublished results.
30. Han, B.-H.; Boudjouk, P. J. Org. Chem. 1982, 47, 751–752.
31. Chew, S.; Ferrier, R. J. J. C. S. Chem. Commun. 1984, 911–912.
32. West, R.; Fink, M. J.; Michl, J. Science 1981, 214, 1343.
33. Boudjouk, P.; Han, B.-H.; Anderson, K.R. J. Am. Chem. Soc. 1982, 104, 4992.
34. Masamune, S.; Murakami, S.; Lobita, H. Organometallics 1983, 2, 1464–1466.
35. Anderson, K.R. 1986, M.S. Thesis, North Dakota State University, Fargo, ND.

36. Sooriyakumaran, R. 1985, Ph.D. Thesis, North Dakota State University, Fargo, ND.
37. Boudjouk, P.; Samaraweera, U. unpublished results.
38. Pshenitsyn, N.N.; Sidorov, N. V.; Sternina, D. G. J. Appl. Chem. (USSR) 1940, 13, 76-78, Chem. Abstr. 1940, 34, 8189.
39. Richardson, C. N. U.S. Patent 2 500 008, 1972.
40. Aeroprojects, Inc., Brit. Patent 991 759, 1965; Chem. Abstr. 1965, 63, P4994a.
41. Mayer, H.; Marinesco, N. Fr. Patent 893 663, 1944; Chem. Abstr. 1952, 46, 4064i.
42. Mückel, P. Ing.-Tech. 1952, 24, 1534; Chem. Abstr. 1952, 46, 5412f.
43. Saracco, G.; Arzano, F. Chim. Ind., 1968, 50, 314; Chem. Abstr. 1968, 69, 3906k.
44. Boudjouk, P.; Han, B.-H. J. Catal. 1983, 79, 489-492.
45. Boudjouk, P.; Han, B.-H.; So, J.-H. unpublished results.
46. Speier, J. L. Adv. Organomet. Chem. 1979, 17, 407-447.
47. Wagner, G. H. U. S. Patent 2 636 738, 1953.
48. Han, B.-H.; Boudjouk, P. Organometallics 1983, 2, 769-771.
49. Boudjouk, P.; Han, B.-H.; Thompson, D.; Sooriyakumaran, R. Abstracts of Papers, 192nd National Meeting of the American Chemical Society, Anaheim, CA; American Chemical Society: Wash.: DC; INOR 298.
50. Szalay, A. Z. Physik. Chem. 1933, 164A, 231; Chem. Abstr. 1933, 27, 3379.
51. Raucher, S.; Klein, P. J. Org. Chem. 1981, 46, 3558-3559.
52. Regen, S. L.; Singh, A. J. Org. Chem. 1982, 47, 1587-1588.
53. Han, B.-H.; Boudjouk, P. Tetrahedron Lett. 1982, 23, 1643-1646.
54. Brown, H. C.; Krishnamurthy, J. J. Org. Chem. 1969, 34, 3918.
55. Lukevics, E.; Gevorgyan, V. N.; Goldberg, Y. S. Tetrahedron Lett. 1984, 25, 1415-1416.
56. Brown, H. C.; Racherla, U. S. Tetrahedron Lett. 1985, 26, 2187-2190.
57. Varma, R. S.; Kabalka, G. W. Heterocycles 1985, 23, 139-141.
58. Yamawaki, J.; Sumi, S.; Ando, T.; Hanafusa, T. Chem. Lett. 1983, 379-380.
59. Sooriyakumaran, R.; Boudjouk, P. J. Organometal. Chem. 1984, 271, 289-297.
60. Suslick, K. S.; Hammerton, D. A.; Cline, Jr., R. E. J. Am. Chem. Soc. 1986, 108,
61. Mørch, K. A. In "Treatise on Materials Science and Technology", Herbert, H., Ed.; Academic Press: New York, 1979; Vol. 16.
62. Suslick, K. S.; Gawienowski, J. W.; Schubert, P. F.; Wang, H. H. J. Phys. Chem. 1983, 87, 2299-2301.

RECEIVED November 21, 1986

Chapter 14

Preparation of Highly Reactive Metal Powders: Some of Their Uses in Organic and Organometallic Synthesis

Reuben D. Rieke, Timothy P. Burns, Richard M. Wehmeyer, and Bruce E. Kahn

Department of Chemistry, University of Nebraska—Lincoln, Lincoln, NE 68588-0304

In 1972, we reported a general procedure for the preparation of highly reactive metal powders. The basic procedure involved the reduction of a metal salt in a hydrocarbon or ethereal solvent. The reductions are most generally carried out with alkali metals such as potassium, sodium, or lithium. A wide range of methods have been developed to carry out the reductions. The reactivities of these resulting black powders exceed other reports in the literature for metal powders. This high reactivity has resulted in the development of several new synthetic techniques and vast improvements in many older, well established reactions. This review concentrates on the metals Mg, Ni, Zn, Cd, Co, Cu, Fe, and U.

The reaction of organic and inorganic substrates at a metal surface, either in a catalytic fashion or with the consumption of the metal, represents an extremely important area of chemistry. The specific area of generating useful organometallic intermediates by reaction of organic materials with metals had its origin well over one hundred years ago. In the past twenty years, considerable effort has been expended in developing new ways of increasing the reactivity of metals so that new organometallic materials can be produced and studied. These studies have also resulted in some known organometallic materials being generated under very mild conditions. In many cases this has expanded the range of useable substrates and increased yields of known reactions. The new approaches to generate highly reactive metals include the metal vaporization technique, use of ultrasound, and reduction of metal salts. This review will concentrate on the third approach, the reduction of metal salts.

Generation of Activated Metals via the Reduction of Metal Salts

The "freeing" of metals from metal salts by various reducing agents is a process which is as old as civilized man itself. In the past thirty years, several new approaches to reducing metal salts have appeared(1-15). In the past few years, several workers have shown that if care is taken regarding the reducing procedure, finely divided and highly reactive metal powders or slurries can be prepared(16-60).

0097-6156/87/0333-0223$06.75/0

We have concentrated our efforts on reducing metal salts in ethereal or hydrocarbon solvents using alkali metals as reducing agents(16-54,77). Several basic approaches are possible and each has its own particular advantages. For some metals, all approaches lead to metal powders of identical reactivity. However, for some metals one method will lead to far superior reactivity. High reactivity, for the most part, refers to oxidative addition reactions.

Although our initial entry into this area of study was by accident when we happened to mix MgCl$_2$ with potassium biphenylide(61), our early work concentrated on reductions without the use of electron carriers. In this basic approach, reductions are conveniently carried out using an alkali metal and a solvent whose boiling point exceeds the melting point of the alkali metal. The metal salt to be reduced must also be partially

$$MX_n + nK \longrightarrow M^* + nKX$$

soluble in the solvent and the reductions are carried out under an argon atmosphere. The reductions are exothermic and generally are complete within a few hours. In addition to the metal powder, one or more moles of alkali salt are generated. Convenient reducing agents/solvents include potassium/THF, sodium/DME, sodium or potassium/benzene or toluene. The majority of reductions studied to date have been carried out in ethereal solvents such as THF, DME, diglyme, or dioxane. For many metal salts, solubility considerations restrict reductions to ethereal solvents. Also, for some metal salts, reductive cleavage of the ethereal solvents requires reductions in hydrocarbon solvents such as benzene or toluene. This is the case for Al, In, and Cr. When reductions are carried out in hydrocarbon solvents, solubility of the metal salts may become a serious problem. In the case of Cr(21), this was solved by using CrCl$_3$·3THF.

A second general approach is to use an alkali metal in conjunction with an electron carrier such as naphthalene. The electron carrier is normally used in less than stoichiometric proportions, generally 5 to 10 mole percent based on the metal salt being reduced. This procedure allows reductions to be carried out at ambient temperatures or lower in contrast to the previous approach which requires refluxing. A convenient reducing metal is lithium. Not only is the procedure much safer using lithium rather than sodium or potassium, but in many cases the reactivity of the metal powders is greater.

A third approach is to use a stoichiometric amount of preformed lithium naphthalide. This approach allows for very rapid generation of the metal powders as reductions are diffusion controlled. Very low to ambient temperatures can be used for the reduction. In some cases the reductions are slower at low temperatures due to low solubility of the metal salts. This approach frequently leads to the most active metals as the relatively short reduction times at low temperatures leads to reduced sintering of the metal particles and hence higher reactivity. Fujita, et al.(62) have recently shown that lithium naphthalide in toluene can be prepared by sonicating lithium, naphthalene, and N, N, N, N-tetramethylethylene-diamine (TMEDA) in toluene. This allows reductions of metal salts in hydrocarbon solvents. This proved to be especially beneficial with cadmium(49). An extension of this approach is to use the solid dilithium salt of the dianion of naphthalene. Use of this reducing agent in a hydrocarbon solvent is essential in the preparation of highly reactive uranium(54). This will be discussed in detail below.

For many of the metals generated by one of the above three general methods, the finely divided black metals will settle after standing for a few hours leaving a clear, and in most cases colorless solution. This allows the solvent to be removed via cannula. Thus the metal powder can be washed to remove the electron carrier as well as the alkali salt, especially if it is a lithium salt. Moreover, a different solvent may be added at this point providing versatility in solvent choice for subsequent reactions.

The wide range of reducing agents under a variety of conditions can result in dramatic differences in reactivity of the metal. For some metals, essentially the same reactivity is found no matter what reducing agent or reduction conditions are used. In addition to the reducing conditions, the anion of the metal salt can have a profound effect on the resulting reactivity. These effects are discussed separately with each metal.

An important aspect of the highly reactive metal powders is their convenient preparation. The apparatus required is very inexpensive and simple. The reductions are usually carried out in a two-neck flask equipped with a condenser (if necessary), septum, heating mantle (if necessary), magnetic stirrer, and argon atmosphere. A critical aspect of the procedure is that <u>anhydrous</u> metal salts be used. Commerical sources of excellent anhydrous metal salts are available. Alternatively, anhydrous salts can sometimes be easily prepared as, for example, $MgBr_2$ from Mg turnings and 1,2-dibromoethane. In some cases, anhydrous salts can be prepared by drying the hydrated salts at high temperatures in a vacuum. This approach must be used with caution as many hydrated salts are very difficult to dry completely by this method or lead to mixtures of metal oxides and hydroxides. This is the most common problem when metal powders of low reactivity are obtained. The introduction of the metal salt and reducing agent to the reaction vessel is best done in a dry box or glove bag; however, very nonhygroscopic salts can be weighed out in the air and then introduced into the reaction vessel. Solvents, freshly distilled from suitable drying agents under argon, are then added to the flask with a syringe. While it varies from metal to metal, the reactivity will diminish with time and the metals are best reacted within a few days of preparation.

We never have had a fire or explosion caused by the activated metals; however, extreme caution should be exercised when working with these materials. Until a person becomes familiar with the characteristics of the metal powder involved, careful consideration should be taken at every step. To date, no metal powder we have generated will spontaneously ignite if removed from the reaction vessel while wet with solvent. They do, however, react rapidly with oxygen and with moisture in the air. Accordingly, they should be handled under an argon atmosphere. If the metal powders are dried before being exposed to the air, many will begin to smoke and/or ignite. Perhaps the most dangerous step in the preparation of the active metals is the handling of sodium or potassium. This can be avoided for most metals by using lithium as the reducing agent. In rare cases, heat generated during the reduction process can cause the solvent to reflux excessively. For example, reductions of $ZnCl_2$ or $FeCl_3$ in THF with potassium are quite exothermic. This is generally only observed when the metal salts are very soluble and the molten alkali metal (method one) approach is used. Sodium-potassium alloy is very reactive and difficult to use as a reducing agent; it is used only as a last resort in special cases.

Physical Characteristics of Highly Reactive Metal Powders

The reductions generate a finely divided black powder. Particle size analyses indicate a range of sizes from one to two microns to submicron depending on the metal and, more importantly, on the method of preparation.

In cases such as copper, black colloidal suspensions are obtained which do not settle and cannot be filtered. In some cases, even centrifugation is not successful. It should be pointed out that the particle size analyses as well as surface area studies have been done on samples that have been collected, dried, and sent off for analysis and likely have experienced considerable sintering. SEM photographs range from sponge-like material to polycrystalline material (22,23). X-ray powder diffraction studies range from those metals like Al and In which show diffraction lines for both the metal and alkali salt to those of Mg and Co which only show lines for the alkali salt. This would suggest that the metal in this latter case is either amorphous or the particle size is less than 0.1 micron. In the case of Co, a sample heated to 300 °C under argon and then reexamined showed diffraction lines due to Co, suggesting the small crystallites had sintered upon heating(53).

ESCA and AUGER have been carried out on several metals and in all cases the metal has been shown to be in the zerovalent state. Bulk analysis also clearly shows that the metal powders are complex materials containing in many cases significant quantities of carbon, hydrogen, oxygen, halogens, and alkali metal.

A B.E.T. surface area measurement(37) was carried out on the activated Ni powder showing it to have a specific surface area of 32.7 m^2/g. Thus it is clear that the highly reactive metals have very high surface areas which, when initially prepared, are probably relatively free of oxide coatings. The metals contain many dislocations and imperfections which probably add to their reactivity.

Nickel

As was pointed out in the introduction, the properties of the metals produced by these reduction methods are highly dependent on the various parameters of the reduction, i.e., solvent, anion, reducing agent, etc. This turned out to be particularly true in the case of Ni.

Initially we tried the standard approach, reduction of NiI_2, $NiBr_2$, or $NiCl_2$ with 2.0 equivalents of potassium in refluxing THF. Finely divided black nickel powders were obtained; however, they showed rather limited reactivity toward oxidative addition with carbon-halogen bonds. Similar results were found for palladium and platinum.

However, when the reductions were carried out with lithium and a catalytic amount of naphthalene as an electron carrier, far different results were obtained(36-39, 43-48). Using this approach a highly reactive form of finely divided nickel resulted. It should be pointed out that with the electron carrier approach the reductions can be conveniently monitored, for when the reductions are complete the solutions turn green from the buildup of lithium naphthalide. It was determined that 2.2 to 2.3 equivalents of lithium were required to reach complete reduction of Ni(+2) salts. It is also significant to point out that ESCA studies on the nickel powders produced from reductions using 2.0 equivalents of potassium showed considerable amounts of Ni(+2) on the metal surface. In contrast, little Ni(+2) was observed on the surface of the nickel powders generated by reductions using 2.3 equivalents of lithium. While it is only speculation, our interpretation of these results is that the absorption of the Ni(+2) ions on the nickel surface in effect raised the work function of the nickel and rendered it ineffective towards oxidative addition reactions. An alternative explanation is that the Ni(+2) ions were simply adsorbed on the active sites of the nickel surface.

There are few reports of oxidative addition to zerovalent transition metals under mild conditions; three reports involving group 10 elements have appeared. Fischer and Burger reported the preparation of a π-allylpalladium complex by the reaction of palladium sponge with allyl bromide(63). The Grignard-type addition of allyl halides to aldehydes has been carried out by reacting allylic halides with cobalt or nickel metal prepared by reduction of cobalt or nickel halides with manganese/iron alloy-thiourea(64). Klabunde has reported limited reactivity toward oxidative addition reactions of carbon halogen bonds with nickel slurries prepared by the metal vaporization technique(65).

The activated nickel powder is easily prepared by stirring a 1:2.3 mixture of NiL_2 and lithium metal under argon with a catalytic amount of naphthalene (10 mole % based on nickel halide) at room temperature for 12 h in DME. The resulting black slurry slowly settles after stirring is stopped and the solvent can be removed via cannula if desired. Washing with fresh DME will remove the naphthalene as well as most of the lithium salts. For most of the nickel chemistry described below, these substances did not affect the reactions and hence they were not removed. The activated nickel slurries were found to undergo oxidative addition with a wide variety of aryl, vinyl, and many alkyl carbon halogen bonds.

A variety of iodobenzenes and bromobenzenes reacted with the nickel at 85 °C to give the corresponding biphenyls in good to high yields. Substituents in the para position, such as methoxy, chloro, cyano, and acetyl groups, were compatible with the reaction conditions employed. Although the copper mediated Ullmann reaction is a well known method for biaryl synthesis, drastic conditions in the range of 150-280 °C are required. Zerovalent nickel complexes such as bis(1,5-cyclooctadiene)nickel or tetrakis(triphenylphosphine)nickel have been shown to be acceptable coupling reagents under mild conditions; however, the complexes are unstable and not easy to prepare. The method using activated metallic nickel eliminates most of these problems and provides an attractive alternative for carrying out aryl coupling reactions(36,38).

The present reaction may be reasonably explained by the smooth oxidative addition of aryl halides to metallic nickel to give aryl nickel halides, followed by disproportionation to bisarylnickels, which upon reductive elimination afford the dehalogenative coupled products. Providing strong support for this mechanism, the intermediates, arylnickel halide and bisarylnickel ($Ar=C_6F_5$), were isolated as the phosphine complexes.

$$[Ni] + ArX + nL \rightarrow ArNi^{II}XL_n$$

$$2ArNi^{II}XL_n \rightarrow Ar_2Ni^{II}L_n + Ni^{II}X_2 + L_n$$

$$Ar_2Ni^{II}L_n \rightarrow ArAr + [Ni] + L_n$$

The homo-coupling reaction of benzylic halides by metallic nickel proceeded at room temperature. 1,2-Diarylethanes having a variety of functional groups could be easily prepared by this method in good to high yields.

On the other hand, benzylic polyhalides were converted to the corresponding olefins via vicinal dihalide intermediates. Metallic nickel was also shown to be useful for the dehalogenation of vicinal dihalides(36,43).

$$PhRCX_2 \xrightarrow{Ni} \underset{\underset{X\ X}{|\ \ |}}{PhRC-CRPh} \xrightarrow{Ni} PhRC=CRPh \qquad R = Ph,\ H,\ or\ X$$

The homo-coupling of benzylic halides appears to proceed via benzylnickel intermediates similar to that of aryl halides. The intermediate, benzylnickel halide, was successfully trapped with electron deficient olefins.

$$ArCH_2Cl + H_2C=CHY \xrightarrow[DME,\ 85°C]{Ni} ArCH_2CH_2CH_2Y \qquad \begin{array}{l} Ar = Ph,\ Y = CO_2Me;\ 17\% \\ Ar = Ph,\ Y = CN;\ 14\% \end{array}$$

The cross-coupling reactions of benzylic halides and acyl halides produced the expected ketones in good to high yields(39,47). The present method can supplement the corresponding Grignard reaction, which does not work well since benzylmagnesium halides undergo dimerization readily. Moreover, the Grignard approach has the disadvantage of forming byproducts from the addition of Grignard reagents to the formed ketones. The yields of the benzyl ketones are good and the reaction will tolerate a wide variety of functionality.

$$ArCH_2X + RCOX \xrightarrow[DME,\ 85\ °C]{Ni} ArCH_2COR$$

The reaction may be reasonably explained by the smooth oxidative addition of benzylic and acyl halides to nickel to afford benzylnickel halides and acylnickel halides. The metathesis of these complexes could give the acylbenzylnickel complex, which upon reductive elimination would yield the benzyl ketone.

$$ArCH_2X \xrightarrow{Ni} ArCH_2NiX \qquad ArCH_2NiCOR \xrightarrow{-Ni} ArCH_2COR$$
$$RCOX \xrightarrow{Ni} RCONiX \qquad\qquad NiX_2$$

Trapping experiments with electron deficient olefins such as acrylonitrile and 3-buten-2-one gave the expected 1,4-adducts from the proposed acylnickel intermediates. This provides strong support for the proposed mechanism. It was also demonstrated that allylic, vinylic and pentafluorophenyl halides could be cross-coupled with acid chlorides to give the corresponding ketones in good yields.

When alkyl oxalyl chlorides were employed instead of acyl halides, symmetrical dibenzyl ketones were formed in good yields(44). Transition metal carbonyls or metal salts/carbon monoxide have generally been used for

$$ArCH_2X + XCOCO_2R \xrightarrow[DME,\ 85\ °C]{Ni} ArCH_2COCH_2Ar + 1/2\ RO_2CCO_2R$$

dibenzyl ketone syntheses. In contrast, alkyl oxalyl chlorides proved an unusual source of carbon monoxide in this system. Decarbonylation of alkyl oxalyl chlorides with metallic nickel followed by carbonylation of benzylnickel halide intermediates proceeded smoothly. As alkyl oxalyl chlorides are readily available commercially and easy to handle, the present method is superior to that using toxic nickel carbonyl or carbon monoxide/transition metal salt in a laboratory scale preparation. In addition to the observed dibenzylketone, diethyl oxalate (39%) was isolated, which is consistent with the proposed mechanism.

$XCOCO_2R$ Postulated Mechanism

$\downarrow Ni^0$

$$XNiCOCO_2R \longrightarrow XNiCO_2R \quad XNiCO_2R \longrightarrow (RO_2C)_2Ni \longrightarrow RO_2CCO_2R$$

with CO over Ni, and CO added

$$ArCH_2X \xrightarrow{Ni^0} ArCH_2NiX \quad ArCH_2NiX \longrightarrow ArCH_2CONiX \longrightarrow ArCH_2CONiCH_2Ar$$

$$-NiX_2 \quad ArCH_2X \quad -NiX_2 \quad ArCH_2X \quad -NiX_2 \quad -Ni^0$$

$$(ArCH_2)_2Ni \xrightarrow{-Ni^0} ArCH_2CH_2Ar \qquad ArCH_2COCH_2Ar$$

The development of the Grignard-type addition to carbonyl compounds mediated by transition metals would be of interest as the compatibility with a variety of functionality would be expected under the reaction conditions employed. One example has been reported on the addition of allyl halides to aldehydes in the presence of cobalt or nickel metal; however, yields were low (up to 22%). Benzylic nickel halides prepared in situ by the oxidative addition of benzyl halides to metallic nickel were found to add to benzil and give the corresponding β-hydroxyketones in high yields(46). The reaction appears to be quite general and will tolerate a wide range of functionality.

$$ArCH_2X + PhCOCOPh \xrightarrow[DME,\ 85°C]{Ni} ArCH_2-\overset{\overset{\displaystyle Ph}{|}}{\underset{\underset{\displaystyle OH}{|}}{C}}-COPh$$

$$Ar = C_6H_5(78\%); \quad Ar = 4\text{-}CH_3C_6H_4(75\%);$$

$$Ar = 3\text{-}CH_3OC_6H_4(81\%); \quad Ar = 3\text{-}F_3CC_6H_4(73\%)$$

$$Ar = 4\text{-}ClC_6H_4(83\%); \quad Ar = 4\text{-}BrC_6H_4(52\%); \quad Ar = 4\text{-}NCC_6H_4(66\%)$$

Two additional cross-coupling reactions mediated by nickel have recently been developed. 3-Arylpropanenitriles can be prepared in high yields by refluxing a mixture of benzyl halide, bromoacetonitrile, and activated nickel in DME(45). The reaction will tolerate ester, ketone, nitrile, and halogen functionality in the benzylic halides. The second reaction is a Reformatsky-type reaction with α -haloacetonitriles(48). The metallic nickel reacts with α-haloacetonitriles and the resulting organonickel specie adds to aryl and alkyl aldehydes to give β-hydroxynitriles in good yields after hydrolysis. The organonickel appears to add to aldehydes but

$$RCHO + XCH_2CN \xrightarrow[\text{DME, } 85\,°C]{\text{Ni}} \xrightarrow{\text{H+}} R\overset{\overset{\displaystyle OH}{|}}{C}HCH_2CN$$

not ketones which may prove to be quite useful. Also, byproducts seem to be minimal.

Although 1,4-elimination of o-xylene polyhalides with sodium iodide, zinc, copper, or iron metal is a fundamental method for the formation of o-xylylene intermediates, it is difficult to carry out the reaction under mild conditions such as at room temperature. o-(Trimethylsilylmethyl)-benzyltrimethylammonium halides were devised for this purpose and were shown to generate the o-xylylene at room temperature. However, we have successfully generated o-xylylene at room temperature by the reaction of α, α,-dibromo-o-xylene with metallic nickel(51). The Diels-Alder reaction of o-xylylene, thus formed, with electron deficient olefins such as acrylonitrile, methyl acrylate, dialkyl maleate and fumarate, and maleic anhydride afforded tetrahydronaphthalene derivatives in 67-90% yields.

Cadmium

Highly reactive cadmium metal powders and a cadmium-lithium alloy, Cd_3Li, have recently been prepared in our laboratories(49). These cadmium metals readily undergo oxidative addition with a variety of alkyl and aryl halides to yield organocadmium reagents. Importantly, the alkyl and aryl halides can have a variety of functional groups present. Current methods of preparation of organocadmium reagents preclude most types of functional groups(66).

The organocadmium reagents formed undergo the well-known reaction with acid chlorides to form ketones in high yields. In addition, substituted aryl compounds have been produced by cross-coupling reactions of the organocadmium reagents with allyl bromide. Furthermore, this highly reactive metal reacts with α-bromo esters in ethereal solvents forming an organocadmium species that will undergo Reformatsky-type reactions with aldehydes or ketones producing high yields of the β-hydroxyesters.

The highly reactive cadmium can be prepared by two different methods. One approach is a room temperature reduction of $CdCl_2$ with lithium naphthalide in THF or DME. The second approach allows the preparation of the reactive metal in a hydrocarbon solvent. First, lithium naphthalide is prepared in benzene; addition of this solution to $CdCl_2$ produces a highly reactive cadmium powder.

A significant new result was the discovery that a cadmium-lithium alloy, Cd_3Li, could be prepared by refluxing a mixture of $CdCl_2$, Li, and an electron carrier such as naphthalene in THF. By treating the alloy with I_2, the lithium can seemingly be leached out leaving a highly reactive cadmium metal which will react with alkyl and aryl halides to produce organocadmium reagents.

The high reactivity of the cadmium metal powder is clearly demonstrated by the ready oxidative addition of a variety of alkyl and aryl halides. Benzylic halides react in under three hours at room temperature while 1-bromopentane reacted in under 10 h at 85° C. Iodobenzene reacted in refluxing THF in 18 h. These results are significant when compared to the most reactive cadmium metal reported in the literature which only reacted with iodoethane.

Zinc

The direct reaction of zinc metal with organic iodides dates back to the work of Frankland(67). Several modifications have been suggested since that time to increase the reactivity of the metal. The majority of these modifications have employed zinc-copper couples(68-72), sodium-zinc alloys(73), or zinc-silver couples(77). Some recent work has indicated that certain zinc-copper couples will react with alkyl bromides to give modest yields of dialkylzinc compounds(74,75). However, all attempts to react zinc with aryl iodides or bromides have met with failure. The primary use of zinc couples has been in the Simmons-Smith reaction. This reaction has been primarily used with diiodomethane as 1,1-dibromides or longer chain diiodides have proven to be too unreactive even with the most reactive zinc couples.

Highly reactive zinc can be prepared by reduction of anhydrous $ZnCl_2$ with potassium/THF or sodium/DME(17,29). This zinc has been shown to undergo rapid oxidative additions with alkyl bromides to produce near quantitative yields of the corresponding dialkylzinc. It also underwent oxidative addition with phenyl iodide and bromide. Moreover, the zinc was found to be useful in the Reformatsky reaction. Reactions could be carried out in diethyl ether at room temperature to generate near quantitative yields of the β-hydroxyester.

A safer and more convenient method is to reduce $ZnCl_2$ with lithium using naphthalene as an electron carrier(77). In addition to the ease of generation, the resulting zinc powders are, in fact, much more reactive in oxidative addition reactions. Most notable, this metal is exceptionally useful in the Simmons-Smith reaction. Reactions done to date have concentrated on 1,1-dibromides, as the 1,1-diiodides have proven to be too reactive. Refluxing cyclohexene, methylene bromide, and active zinc in diethyl ether generates the corresponding cyclopropane in greater than 90% yields. In fact, the reaction can be extended to longer chain dibromides. Two examples of substituted cyclopropane formation are shown below. This reactive zinc should dramatically increase the utility of the Simmons-Smith reaction.

Cobalt

Cobalt represents an interesting contrast to the many activated metal powders generated by reduction of metal salts. As will be seen, the cobalt powders are highly reactive with regard to several different types of reactions. However, in contrast to the vast majority of metals studied to date, it shows limited reactivity toward oxidative addition with carbon halogen bonds.

Two general approaches have been used to prepare the cobalt powders. The first method(34,37,42) used 2.3 equivalents of lithium along with naphthalene as an electron carrier in DME to reduce anhydrous cobalt chloride to a dark gray powder, l. Use of cobalt bromides or iodides gave a somewhat less reactive form of 1. Slurries of 1 were very reactive toward strongly electrophilic aryl halides such as C_6F_5X (X = Br, I) yielding the solvated species $Co(C_6F_5)_2$ and CoX_2. More reactive cobalt, 2, was prepared by dissolving lithium in DME containing excess naphthalene(53). Excess naphthalene was used to insure rapid and complete dissolution of all the lithium. Addition of cobalt chloride or iodide to cold glyme solutions of lithium naphthalide resulted in rapid formation of black-gray cobalt metal, 2. Addition of dry cobalt halides or their suspensions in DME gave a product with similar properties and reactivity.

In contrast to 1 which rapidly settled to give a clear solution, 2 remained suspended in an opaque black solution. Centrifugation of a suspension of 2 gave a black solution and sediment. Washing the sediment with fresh DME produced a black solution even after several repetitions. Washing the sediment with less polar solvents, such as diethyl ether or light alkanes, gave clear or slightly cloudy colorless solutions. Switching back to DME regenerated the black solutions. This suggests that an intimate mixture of cobalt microparticles and lithium chloride exists. As fresh glyme is added, part of the lithium chloride matrix is dissolved liberating more cobalt particles which color the suspension.

When dry, 2 was pyrophoric and almost completely nonferromagnetic, i.e., very little was attracted to a strong bar magnet. This was in marked contrast to 1 or commercial samples of 325 mesh cobalt powders which were not pyrophoric and were strongly attracted to a magnet held in their vicinity. Qualitatively, the magnetic properties of 2 were suggestive of superparamagnetism. Debye-Scherrer photographs from samples of 2 showed no, or at best weak and diffuse, lines which could not be assigned to any common modification of metallic cobalt. This suggested that the sizes of the cobalt crystallites were less than approximately 30 Å. Surface analyses of the cobalt powders indicate they are composed mainly of metallic cobalt along with the alkali salt as well as considerable carbonaceous matter. Some cobalt oxides or possibly hydroxides were also observed; however, it is highly likely that the majority of the oxide accumulation occurred on sample preparation.

The cobalt powders demonstrated high reactivity with CO. Only a very limited number of transition metals have been demonstrated to react directly with carbon monoxide (under mild conditions) to give reasonable quantities of metal carbonyl complexes. Nickel and iron are two prime examples. It is interesting to note that the first such report for chromium involved finely divided chromium prepared by the reduction of $CrCl_3 \cdot 3THF$ with potassium in benzene(21). Similarly the cobalt powders could be used to prepare $Co_2(CO)_8$ in good yields under mild conditions. A slurry of 2 in hexanes reacted with CO at 1000-1400 psi at 80-110 °C to give cobalt carbonyl in yields up to 79%(42). It is interesting to note that when CO was added to dry 2 in a Parr bomb it catalyzed the disproportionation of CO into CO_2 and carbon along with the evolution of considerable heat(53). Cobalt is considered to bind CO in a nondissociative manner at room temperature; however, at higher temperatures the adsorption becomes dissociative(78). Adsorption of CO on polycrystalline cobalt films is an exothermic process and is dependent on coverage and the type of binding site(79). It would appear that due to the high surface area of the active cobalt powders along with the substantial CO pressures (initial pressure of 1130 psi) enough heat was released to initiate the disproportionation reaction.

Cobalt has been used as a Fischer-Tropsch catalyst in a variety of forms(80). Thus it was not surprising to see that both active forms of cobalt powders were moderate Fischer-Tropsch catalysts. Reacting synthesis gas with **2** in batch reactor conditions at elevated pressure and temperatures generated methane as the primary product.   The lifespans of the catalyst and to a lesser extent the products were affected by whether a support was used or how the cobalt was deposited on the support.   Catalytic activity was not especially high and amounted to 4-7 mol of methane/mol of cobalt.

The highly oxophilic nature of the cobalt powder was readily demonstrated by its reaction with nitrobenzene at room temperature. Reductive coupling was quickly effected by **2** to give azo- and azoxy derivatives.   Nitrobenzene reacted with **2** to give azobenzene in yields up to 37%.   In some cases small amounts of azoxybenzene were also formed.   With 1,4-diiodonitrobenzene, **2** reacted to give low yields of 4,4-diiodoazoxy-benzene and 4,4-diiodoazobenzene.

In marked contrast to the majority of activated metals prepared by the reduction process, cobalt showed limited reactivity toward oxidative addition with carbon halogen bonds.   Iodopentafluorobenzene reacted with **2** to give the solvated oxidative addition products CoI_2 and $Co(C_6F_5)_2$ or $Co(C_6F_5)I$.   The compound $Co(C_6F_5)_2 \cdot 2PEt_3$ was isolated in 54% yield by addition of triethylphosphine to the solvated materials.   This compound was also prepared in comparable yield from **1** by a similar procedure.   This compound had previously been prepared by the reaction of cobalt atom vapor with $C_6F_5I(\underline{81})$.

From the reaction of **2** with iodobenzene at reflux, a low yield of biphenyl was obtained while much of the aryl halide remained unchanged. Similarly, **1** showed little reactivity towards iodobenzene.

Reactions of **2** with alkyl halides were generally more successful for C-C bond formation.   For example, bibenzyl was formed in good yield from the reaction of **2** with benzyl bromide.   Dichlorodiphenylmethane and **1** reacted to give tetraphenylethylene in 63% yield.   Similarly, diiodomethane reacted with **1** to give ethylene.   This area of study is continuing.

Iron

Reduction of anhydrous iron(II) halides with 2.3 equivalents of lithium or reduction of iron(III) halides with 3.3 equivalents of lithium using naphthalene as an electron carrier in THF or DME yields a highly reactive iron powder(34,37,52).   While the iron powders from iron(III) salts are slightly more reactive, the relative ease of handling the much less hygroscopic iron(II) salts make them the preferred choice for preparing highly reactive iron powders.   The iron powders show slow or little settling and the dry powders are pyrophoric.

The iron slurries show exceptional reactivity toward oxidative addition reactions with carbon halogen bonds.   In fact, the reaction with C_6F_5I is so exothermic that the slurry has to be cooled to 0 °C before the addition of C_6F_5I.   The reaction of iron with C_6F_5Br is also quite exothermic, hence, even for this addition, the iron slurry is cooled to about 0 °C.   The organoiron compound formed in the above reactions, solvated $Fe(C_6F_5)_2$, reacts with CO at room temperature and ambient pressure to yield $Fe(C_6F_5)_2(CO)_2(DME)_2$.

Reaction of allyl halides with the iron powders is rapid and exothermic and leads to near quantitative yields of the self-coupled product 1,5-hexadiene.   Similarly, reaction of benzyl chloride with the iron powders at room temperature yields bibenzyl in 60-70% yields along with 20-25% of toluene.   In contrast, reaction of aryl halides with the iron powders leads to reductive cleavage rather than self-coupling.   Similarly, reaction of 1-bromoheptane with the iron powder in THF for three hours at room temperature produced 35% heptane and no self-coupled product. Further reaction of this mixture for one hour at reflux increased the yield to 70% heptane.

The iron slurries react readily with ethyl α-bromoacetate. The resulting organoiron species adds readily to aldehydes and ketones to produce β-hydroxyesters in excellent yields. Addition of a mixture of an aryl aldehyde and an allylic halide to the iron slurry produced good yields of the cross-coupled alcohol.

Finally, it would appear that these highly reactive iron powders will be of value as a general reducing agent. Reaction of nitrobenzene with four equivalents of iron and one equivalent of n-butanol in THF at room temperature was very exothermic. After one hour at room temperature followed by reflux for one hour, the reaction mixture gave 88% aniline upon workup.

Magnesium

Early reports for generating Mg[*] involved reduction of magnesium chloride or bromide by potassium or sodium in refluxing THF or diglyme(16). The reductions required 4-6 h and necessitated handling potassium or sodium. The addition of potassium iodide to the reaction pot prior to reduction resulted in dramatic improvement in reactivity(18,19,22,23).

A current method for producing Mg[*] is based on the use of lithium as the reducing agent. This method is preferred for its inherent safeness. The reductions are carried out at room temperature, using naphthalene as an electron carrier to speed the reaction. 6-12 h are required for complete reduction. If naphthalene will interfere with additional reactions or product isolation, it may be removed by rinsing prior to further reaction. The reactivities are compared in Table I, along with some other methods of activation. It is clear that the magnesium prepared by reduction of $MgCl_2$ is the most reactive.

While the most reactive Mg[*] was produced using the lithium reduction procedure in the presence of alumina, the presently preferred method is entry 11. Though not as reactive as 10, it is more reproducible. The procedures which do not employ alumina also extend equipment life and are thus safer.

In earlier studies, we(19) were able to demonstrate the utility of the potassium recipe with a wide variety of organic halides ranging from p-dihalo substituted benzenes (from which the di-Grignard reagents were also accessible), to alkyl bromides and fluorides, tertiary chlorides, allylic, and vinylic systems. Previous efforts to prepare dimagnesium derivatives of benzene were successful only with dibromo- or bromoiodobenzene, required forcing conditions, and usually resulted in the monomagnesium derivative as the main product(75). However, with the potassium-generated Mg[*], p-dibromobenzene gave good yields of the mono- and di-Grignard reagents. Similiar results were noted for p-bromochlorobenzene and p-dichlorobenzene. 1-Bromooctane reacted readily, as expected, to give the Grignard reagent in 100% yield after 5 min. Carbonation gave nonanoic acid in 81% yield. The tertiary halides, 2-chloro-2-methylpropane and 1-chlorobicyclo 2.2.1 - heptane, gave the corresponding Grignard reagents in 100 and 74% yields, respectively. Carbonation gave pivalic acid (52%) and 1-bicyclo 2.2.1 heptanecarboxylic acid (63%). Use of our method eliminated the need for the lithium sand required by Bixler and Niemann(83). The allyl halide, 3-chloro-2-methylpropene, was converted to 3-methyl-3-butenoic acid after 1 h at room temperature; Wagner(84) obtained an 81% yield of the corresponding Grignard reagent after 10 h at 14-16 °C and upon carbonation isolated a 40% yield of 3-methyl-3-butenoic acid. The vinylic system, 2-bromopropene, was converted to the Grignard reagent in 100% yield in 5 min at room temperature. Carbonation gave methacrylic acid in 40% yield. Normant has recommended a 40-50 °C range for halide addition to magnesium in THF, followed by heating for 0.5-1 h at 70-80 °C($\underline{85}$). Thus, he was able to obtain 95-97% yields of vinylmagnesium bromide. Our method offers the advantage of being able to prepare vinylic Grignard reagents at lower temperatures. 1-Fluorohexane reacted with Mg in THF to give the Grignard reagent in 89% yield after 3 h at 25 °C. $\underline{p}$-Fluorotoluene reacted in refluxing THF to give a 69% yield of the Grignard reagent after 1 h; carbonation gave $\underline{p}$-toluic acid in 61% yield. Previous workers were unable to form Grignard reagents from aryl fluorides($\underline{86}$).

Reactions of Mg* Generated by Lithium/Naphthalene. Although we have not exhaustively duplicated the previous reactions with the newer method of reduction, examination of Table I shows comparable reactivity for the reagents tested. Instead, we have branched out to look at other facets of the chemistry accessible with Mg .

The preparation of Grignard reagents from 3-halo ethers has been a longstanding problem. Although not widely disseminated, the Grignard reagents of 3-halophenoxypropanes are known to cyclize producing the magnesium salt of phenol and cyclopropane. It appears that Hamonet was the first to attempt production of Grignard reagents from such moieties. Although he wasn't successful with the aromatic specie PhO(CH$_2$)$_3$I using ordinary catalysts, he was able to form Grignard reagents from alkyl ethers such as CH$_3$O(CH$_2$)$_3$I($\underline{87}$). Paul($\underline{88}$) reported the formation of Grignard reagents from PhO(CH$_2$)$_3$I and C$_2$H$_5$O(CH$_2$)$_3$Cl, but only after use of AlCl$_3$ as a catalyst. The aromatic specie reacted via two paths to give cyclopropane and 1,6-diphenoxyhexane. Similar results were noted for PhO(CH$_2$)$_3$Br by Erlenmeyer and Marbet($\underline{89}$).

Our results concurred with those above; when 3-bromo-1-phenoxypropane was added to our Mg at room temperature the reaction was quite vigorous($\underline{40}$). Upon carbonation, followed by hydrolysis, the only acidic product isolated was phenol in 70% yield. It was very satisfying to note that at -78 °C the reaction of 3-phenoxypropyl bromide proceeded smoothly. All of the starting material was consumed in a little over an hour. Subsequent carbonation followed by hydrolysis produced 4-phenoxybutanoic acid in over 70% yield.

Furthermore, the 3-phenoxypropyl magnesium bromide reacted with benzoyl chloride at -78 °C, without the benefit of inverse addition, to give 4-phenoxy-1-phenylbutanone in 35% isolated yield. Thus, when the Grignard reaction was carried out at room temperature, the standard cyclopropane formation could be effected. However, by using Mg at low temperatures, cyclization was stopped and the 3-phenoxypropyl Grignard reagent cross-coupled with other acceptors. Previous methods of activation did not allow formation of Grignard reagents at low temperatures.

Another example of the remarkable reactivity of Mg activated by our procedure is its reaction with nitriles. In this respect, the Mg resembles an alkali metal more than an alkaline earth. Benzonitrile reacts with Mg* overnight, in refluxing DME, to give 2,4,6-triphenyl-1,3,5-triazine and 2,4,5-triphenylimidazole in 26 and 27% yield, respectively, based on magnesium. The imidazole was shown to arise, at least in part, from the action of Mg on the triazine. The trimerization of aromatic nitriles to give symmetrical triazines is not unknown, but generally the reactions are catalyzed by strong acids(90-92), less often by strong bases, or least frequently, a weak base and extremely high pressure(93). The action of phenylmagnesium bromide on benzonitrile gives 2,2,4,6-tetraphenyl-1,2-dihydro-1,3,5-triazine along with a small amount of symmetrical triazine(94). Organoalkalis react with benzonitrile to give 2,2,4,6-tetra-phenyl-1,2-dihydro-1,3,5-triazine(95,96). To our knowledge, no one has previously demonstrated a direct reaction of magnesium to give the symmetrical triazine we observed.

Neat butyronitrile reacted under reflux to give 4-amino-5-ethyl-2,6-dipropylpyrimidine isolated in 98% yield based on Mg*. Similarly, acetonitrile gave 4-amino-2,6-dimethylpyrimidine in 152% yield, based on Mg, after reacting in a sealed tube at 130° C for 13 h. The reaction appears catalytic with respect to Mg*. Treatment of alkyl nitriles with metallic sodium(97) or Grignard reagents(98) is known to produce aminopyrimidines, but there have been no reports of such results by the direct action of metallic magnesium on alkyl nitriles.*

We have also found that our Mg* reacts with tosylates to give good yields of the coupled products. For example, benzyl tosylate reacts in refluxing THF to give the expected bibenzyl in 84% yield after 12 h. Neat hexane-1,6-ditosylate reacted with Mg* after 36 h at 90-100 °C to give a polymer with an average carbon chain length of 26.5. Mass spectrometry (EI) revealed masses up to 1135 amu, but much of the sample was unable to be volatilized. Butyl tosylate did not react easily with the Mg*.

Dibenzyl ether reacted with Mg* after five days of reflux in THF to give phenylacetic acid in 42% yield after carbonation. Maercker(99) was able to obtain a 15% yield of 3-butenoic acid from allyl methyl ether after 56 h of reflux.

We have attempted the low temperature formation of Grignard reagents containing nitrile, ketone and ester functional groups. These reactions were largely unsuccessful, except in the case of 8-bromooctanenitrile. Aromatic nitriles were reduced to radical anions. In the case of the alkyl systems, ethyl 3-bromopropionate, 4-bromobutyronitrile, and 5-bromo-valeronitrile, the heteroatoms coordinated strongly to the magnesium surface at the temperatures required to prevent attack of Grignard reagents on the functionality present. Thus, Grignard reagent formation was inhibited by occupation of the active sites on the magnesium surface. However, 8-bromooctanenitrile reacted rapidily with Mg* at -78 °C to give 74% conversion to the Grignard reagent. Carbonation gave 8-cyanooctanoic acid in 35% yield based on Grignard reagent. This appears to be the first example of a Grignard reagent which contains the nitrile functionality. It is surprising to note that the nitrile functionality coordinates so strongly that reaction with organic halides is completely inhibited until 20 °C is reached. 4-Bromobutyronitrile did not react at all with Mg* at -78°C. When the temperature was slowly raised, no reaction was observed until 20° C, at which time polymerization occurred. In confirmatory experiments, 1-bromopentane reacted quantitatively with Mg* in 1 min at -78° C. The resulting Grignard reagent did not react with subsequently added butyronitrile even after 1 h at -78 °C. If butyronitrile was added to Mg* at -78 °C, followed by addition of bromopentane, no Grignard reagent was formed after 2 h.

Copper

The general utility and importance of organocopper reagents in synthesis(100) has instigated considerable research toward developing new and more reactive copper species. The development of organocopper reagents has thus been approached in a variety of ways.

The Ullmann biaryl synthesis(101,102) invokes the reaction of copper powder with aryl halides at relatively high temperatures, typically 100-300 °C, to give biaryl products. The intermediacy of arylcopper species is presumed but not specifically proven due to the instability of the arylcopper at the temperatures required for reaction. The Ullmann reaction has seen appreciable usage as it allows considerable functionality to be incorporated in the products.

The stoichiometric reaction of organolithium or Grignard reagents with copper(I) halides allows direct formation of organocopper(I) species(103). Whereas Ullmann chemistry is limited to the coupling of aryl halides, a variety of species including aryl-, alkynyl-, alkenyl-, and alkylcopper(I) species can be produced which are highly reactive toward other substrates.

Probably the most widely studied and used organocopper reagents are lithium diorganocuprates(103,104). The reaction of an additional equivalent of organolithium reagent with an organocopper(I) species provides a lithium diorganocuprate reagent. The two organolithium species used to prepare the cuprate can be identical giving rise to homocuprates, or different organolithium reagents can be used to give mixed cuprates. Lithium reagents containing assorted heteroatoms have also been used to give heterocuprate reagents which may in some cases have properties superior to typical diorganocuprates(105,106,107). Numerous variations in the nature of the organolithium precursors and in the choice of copper(I) salts(108,109) have been employed in attempts to produce cuprate reagents with high reactivity and selectivity.

The main drawback to both organocopper(I) and lithium diorganocuprates is the limitation on functionality which can be incorporated in the reagents. Since both types of reagents must be derived from organolithium or Grignard precursors, a variety of functional groups cannot be incorporated. In contrast, the Ullmann biaryl synthesis will allow a wide variety of functionalities to be present in the substrates. Thus it is probably not the organocopper species itself which limits the functionality, but the limitations due to forming the organocopper or cuprate using lithium or Grignard reagents. The somewhat severe conditions required for reaction of the aryl halide with the copper powder in traditional Ullmann chemistry, however, do not allow isolation of an intermediate organocopper species. Therefore a more reactive form of copper powder was desirable.

The first general method of active copper preparation involved the reduction of copper(I) iodide with a stoichiometric amount of potassium, using 10% naphthalene as an electron carrier, in DME under an argon atmosphere affording a gray-black copper powder(33). This copper powder was found to react with aryl halides and allyl bromide at temperatures considerably lower than those required for standard Ullmann coupling using copper bronze. In most cases, refluxing DME (85 °C) was sufficient to induce aryl coupling. The reaction of pentafluorophenyl iodide proceeded at room temperature in 30 minutes to yield pentafluorophenylcopper. The same reaction performed in refluxing DME gave decafluorobiphenyl in 83% yield. While this active copper showed reactivity far superior to traditional copper powders, a recently developed activation method has provided an even more reactive copper species.

The reduction of a solution of a trialkylphosphine copper(I) iodide complex (CuIPR$_3$) with preformed lithium naphthalide (LiNp) in THF or DME under argon was found to give a more reactive copper species, which will undergo oxidative addition with a variety of organic substrates at room temperature and below(50). Aryl, alkenyl, and alkynyl halides react readily under mild conditions with the copper solution to give stable organocopper species in good yields. These organocopper species can be either thermally or oxidatively coupled in most cases to give self-coupled products in fair to good yields. Reaction of the preformed organocopper species with a primary alkyl halide provides the alkyl substituted product. Substituted benzophenones and acetophenones have been produced by reaction of arylcopper species with benzoyl chloride or acetyl chloride, respectively. In addition, the presence of the organocopper species has been shown in several examples by reaction with water or deuterium oxide to give the protonated or deuterated products, respectively. Considerable functionality has been incorporated in the arylcopper species which may contain nitro, cyano, ester, alkyl, and phenylketo groups.

Alkyl halides react with the active copper solution under very mild conditions. The reaction proceeds rapidly at room temperature and below to give primarily homo-coupled products as well as some reduction and elimination products. The reaction conditions employed do not give rise to stable alkylcopper products. Reactions of a mixture of two different reactive halides typically gives the expected statistical distribution of homo-coupled and cross-coupled products. Summaries of the homo-coupling and cross-coupling reactions using various organic halides are given in Tables II and III. Table IV lists the yields of stable organocopper species as determined from the protonation and deuteration products. The utility of the copper reagent for effecting 1,4-addition to α,β-unsaturated carbonyl systems is being investigated as well. Preliminary results show that phenylcopper generated using this copper will undergo 1,4-addition with chalcone to give a 40% yield of 3,3-diphenylpropiophenone. The high reactivity toward organic halides and various functionalities which can be tolerated by this form of copper should provide a convenient route to many synthetically important organocopper reagents.

Uranium

The high oxophilicity of the early transition metals, lanthanides, and actinides is well known. This oxophilicity has been exploited in the well known reductive carbonyl coupling reactions pioneered by McMurry(59). The reductive coupling chemistry of uranium, however, is virtually unknown. A report has appeared whereby di-n-butyluranocene has been used to couple aromatic nitro compounds to give azobenzenes(60). In view of this coupling reaction of U(IV), the high oxophilicity of uranium, and the similarities between the chemistry of the actinides and group 4 transition metals, we have investigated the reactions of active uranium with oxygen-containing compounds. In these investigations we have discovered that the oxophilicity of low valent uranium has greatly exceeded our initial expectations. We have also seen that active uranium will form hydridic species which may arise from insertion into C-H bonds which are very weakly acidic.

When active uranium is prepared in DME(33), the low valent uranium species which are produced react rapidly with the solvent yielding gaseous products consisting mainly of ethylene and methyl vinyl ether. When additional lithium naphthalide in glyme is added to the uranium slurry, more products are given off. This can be thought of as uranium mediated reduction of DME by lithium naphthalide. The solvent reactions described above limit the usefulness of active uranium prepared in ethereal solvents, as the uranium is both deactivated by reaction with the solvent, and products generated from reactions of organic substrates with active uranium contain solvent fragments and are complex mixtures.

Table I. Comparison of Various Types of Magnesium

	Source of Mg	$^{\circ}$C	Substrate	% Grignard Reagent 5	10	30	60 min
1	Commercial[a]	25	p-BrC$_6$H$_4$CH$_3$	21	69	93	100
2	Commercial[a,b]	25	p-BrC$_6$H$_4$CH$_3$	0	0	30	100
3	MgCl$_2$ + K	25	p-BrC$_6$H$_4$CH$_3$	100	100	100	100
4	Commercial[a]	25	p-ClC$_6$H$_4$CH$_3$	0	0	0	0
5	Evaporated	25	ClC$_6$H$_5$		85% after 3 hrs.		
6	MgCl$_2$+Li+np[c]	0	ClC$_6$H$_5$	21	52	81	100
7	MgCl$_2$ + K	25	p-ClC$_6$H$_4$CH$_3$	0	0	14	50
8	MgCl$_2$ + K[b]	25	p-ClC$_6$H$_4$CH$_3$	84	94	97	98
9	MgCl$_2$ + K[b]	0	p-ClC$_6$H$_4$CH$_3$	26	46	73	85
10	MgCl$_2$+Li+np[c]	0	p-ClC$_6$H$_4$CH$_3$	30	60	85	96
11	MgCl$_2$+Li+np	0	p-ClC$_6$H$_4$CH$_3$	19	38	74	88

[a]Poly Research 325 mesh. [b]Reduced in presence of KI (1 eq. based on Mg). [c]Reduced in presence of alumina.

Table II. Homocoupling Reactions of Copper

2RX + Cu	% R-R	$^{\circ}$C	Solvent
allyl bromide	99[a]	25	THF
C$_6$F$_5$I	91[a,b]	101	dioxane
C$_6$H$_5$CH$_2$Br	95[a]	25	THF
2-NO$_2$C$_6$H$_4$I	87[c,d]	65	THF
2-NCC$_6$H$_4$Br	76[c,e]	25	THF
1-iodoheptane	83[a]	25	THF/DMF
PhI	66[a]	85	DME
PhC≡CBr	30[c,e]	25	DME
PhCH=CHI	47[a]	25	DME
4-CH$_3$O$_2$CC$_6$H$_4$CH$_2$Br	92[a]	25	THF
4-CH$_3$C$_6$H$_4$I	48[a]	85	DME

[a]GC yield. [b]Best results were obtained by reducing CuCl powder in a solution of lithium naphthalide. [c]Isolated yield. [d]Reducing CuCl·S(CH$_3$)$_2$ gave better results than the phosphine complex. [e]Coupling was induced by injection of O$_2$ into the reaction vessel.

Table III. Cross-Coupling Reactions of Copper

RCu	R'X	R-R'	% yield
C_6F_5Cu	PhCOCl	C_6F_5COPh	25[a]
PhCu	PhCOCl	Ph_2CO	70[b]
$2-NCC_6H_4Cu$	PhCOCl	$2-NCC_6H_4COPh$	95[a]
PhCu	$H_2C=CHCH_2Br$	$PhCH_2CH=CH_2$	36[b]
$2-NCC_6H_4Cu$	H_3CCOCl	$2-NCC_6H_4COCH_3$	46[a]
$2-NCC_6H_4Cu$	$C_6H_5CH_2Br$	$2-NCC_6H_4CH_2C_6H_5$	51[a]
PhCu	$PhCH_2Br$	$PhCH_2Ph$	84[b]
$4-PhCOC_6H_4Cu$	1-iodobutane	$4-CH_3(CH_2)_3C_6H_4COPh$	32[b]
$4-CH_3O_2CC_6H_4Cu$	1-iodobutane	$4-CH_3O_2CC_6H_4(CH_2)_3CH_3$	58[b]
$PhCH_2Br$	$H_2C=CHCH_2Br$	$PhCH_2CH_2CH=CH_2$	50[b]
$CH_3(CH_2)_6I$	$H_2C=CHCH_2Br$	1-decene	45[b]

[a]Isolated yield. [b]GC yield.

Table IV. Stable Organocopper Intermediates

$RX + Cu^o$	°C	solvent	time	% R-H or R-D
PhI	25	THF	10 min	98[a]
$2-NO_2C_6H_4I$	0	THF/DMF	1 min	60[a]
$2-NCC_6H_4Br$	0	THF	1 h	89[b]
$PhC\equiv CBr$	25	DME	18 h	42[a]
C_6F_5I	25	THF	1 min	90[a]
PhCH=CHBr	25	DME	10 min	66[b]
$4-CH_3O_2CC_6H_4Br$	0	THF	3 min	62[a]
$4-H_3CC_6H_4Br$	25	THF	10 min	47[a]
$4-BrC_6H_4COPh$	0	THF	1 min	42[a]
C_6F_5I	25	THF	1 h	90[a,c]
PhCH=CHBr	25	THF	20 h	37[b,c]

[a]GC yield. [b]Isolated yield. [c]Deuterolysis yield.

In order to eliminate competing reaction with the solvent, a method for generating active uranium in hydrocarbon solvents was desired. Thus a hydrocarbon soluble reducing agent [(TMEDA)Li]$_2$ [Nap] (Nap=naphthalene) was prepared. This complex has previously been made from 1,4-dihydronaphthalene(110). We have prepared this complex from lithium, naphthalene and TMEDA in a convenient reaction which is amenable to large scale synthesis.

The lithium and naphthalene are loaded into the reaction flask in a dry box, and the TMEDA and toluene syringed in under argon. The reaction flask is immersed in an ultrasonic cleaner. Sonication is continued until all of the lithium has dissolved. Upon standing, [(TMEDA)Li]$_2$ [Nap] crystallizes as black needles. This complex is indefinitely stable under argon at room temperature but decomposes at $\leq 80°$ C in solution.

The active uranium is prepared by weighing out the reducing agent and UCl$_4$ in a dry box. After removal of the apparatus from the dry box, the solvent (typically xylenes) is added under argon and the mixture stirred for one hour. At this time, the substrates are introduced.

The reaction of active uranium (U*) and benzophenone gives tetraphenylethylene (TPE) and 1,1,2,2-tetraphenylethane (TPA).

$$U^* + Ph_2CO \xrightarrow[\text{10 h Ref.}]{\text{Xylenes}} Ph_2C{=}CPh_2 + Ph_2CHCHPh_2 + Ph_2CH_2$$

Varying amounts of diphenylmethane (DPM) are also seen. While the reductive coupling of benzophenone to give TPE was expected, the production of TPA was not. U* failed to react with TPE, TPA, or DPM. This indicated that the TPA was formed during the coupling step on the metal surface and not from TPE or DPM.

In order to further probe the mechanistic course of the above reaction, the possibility of a pinacolic intermediate was addressed. To test this hypothesis, the reaction of active uranium with benzopinacol was conducted. In this reaction equal amounts of TPE and TPA were seen, whereas when benzophenone was reacted with U* more TPE than TPA was seen (TPE/TPA was about 4:3). The yield of TPE was similar in the two reactions. Benzopinacol contains two acidic hydrogens which could account for the greater yield of TPA. When benzopinacol-d$_2$ was reacted with U*, $\geq$73% of the deuterium was incorporated into the TPA. This clearly shows that the benzylic hydrogens of TPA may originate from a uranium hydride.

However, the source of the uranium hydride in the reaction of benzophenone with U* wasn't clear. When the reaction was worked up with D$_2$O, no deuterium incorporation in TPA or DPM was seen, nor was any seen when using toluene-d$_8$ as a solvent. All of the naphthalene could be recovered, which led us to conclude that TMEDA (presumably coordinated to uranium) was reacting with the low valent uranium generating a uranium hydride which became incorporated into the TPA or DPM.

This reaction with TMEDA is temperature dependent. When the reaction of U* with benzophenone is carried out at 70 °C, TPE is produced exclusively, with no TPA or DPM seen. As the reaction temperature is increased above 70 °C, the amounts of TPA and DPM produced increase. It appears that the hydrogenated products TPA and DPM arise when benzophenone reacts with U* containing uranium hydrides. These hydrides may be formed from either substrates containing acidic hydrogens, or by thermal reaction of the low valent uranium species with coordinated TMEDA.

The inherent chemistry above is not altered by substituting the phenyl rings of the ketone or pinacol with methyl groups. The ketone 4,4'-dimethylbenzophenone, as well as the pinacol 1,1,2,2-tetrakis-(4-methylphenyl)ethane-1,2-diol (TBP) react with U* giving tetrakis-(4-methylphenyl)ethylene and 1,1,2,2-tetrakis(4-methylphenyl)ethane. A 1:1 mixture of benzophenone and 4,4'-dimethylbenzophenone gives the six expected coupling products.

In order to further examine the role of a pinacolic intermediate, a crossover experiment was conducted. In the reaction of a 1:1 mixture of TBP and benzopinacol with U*, a statistical distribution of all 6 coupled products was seen. This surprising result shows that the carbon-carbon bond of the pinacol is broken before the products are formed.

In addition to the reductive coupling reaction of ketones, certain alcohols can also be reductively coupled using active uranium. Benzhydrol is coupled by active uranium to give TPA as the only coupled product. No TPE is seen. Under similar conditions, no coupling of benzyl alcohol is seen. The chemistry of the active uranium is under continued investigation.

Summary

The reduction of metal salts in ethereal or hydrocarbon solvents is a relatively simple procedure which requires very inexpensive equipment. The high reactivity of the resulting metal powders, especially toward oxidative addition with carbon halogen bonds, has allowed the generation of many new organometallic species. In some cases, totally new reactions have been developed. In other instances, existing reactions have been extended to a wider range of substrates or to much milder reaction conditions, resulting in higher yields. In several reactions, the ability to generate the desired organometallic species by direct reaction of the metal and the organic substrate has dramatically extented the range of functionality tolerated.

The origins of the high reactivity of the metal powders are only partially understood and it is anticipated as these factors become established that even more reactive metals will be possible. In spite of the significant progress over the past fifteen years, this area of science is best described as being in its infancy. Significant improvements in existing reactions and scores of new reactions will most probably emerge from this area of study.

Acknowledgments

Grateful acknowledgement is again made to the many coworkers whose efforts are cited in the text. Without their dedicated efforts, this manuscript would not be possible. Funding for work carried out by the author (RDR) has been provided by a grant from NIH: GM 35153.

Literature Cited

1. Brauer, G. "Handbook of Preparative Inorganic Chemistry" 2nd ed.; Academic Press: New York, 1965; Vol. 2.
2. Whaley, T. P. Inorg. Synth. 1957, 5, 195.
3. Smith, T. W.; Smith, S. D.; Badesha, S. S. J. Am. Chem. Soc. 1984, 106, 7247.

4. Whaley, T. P. In "Handling and Uses of Alkali Metals"; ADVANCES IN CHEMISTRY SERIES No. 19, American Chemical Society: Washington, D.C., 1957.
5. Chu, L.; Friel, J. V. J. Am. Chem. Soc. 1955, 77, 5838.
6. Scott, N. D.; Walker, J. F. U.S. Pat. 2177412, 1939.
7. Watt, G. W.; Roper, W. F.; Parker, S. G. J. Am. Chem. Soc. 1951, 73, 5791.
8. Urushibara, Y. Bull. Chem. Soc. Jpn. 1952, 25, 280.
9. Hata, K.; Taira, S.; Motoyama, I. Bull. Chem. Soc. Jpn. 1958, 31, 776.
10. Leprince, J. B.; Collignon, N.; Normant, H. Bull. Soc. Chim. Fr. 1976, 34, 367.
11. Brown, H. C.; Brown, C. A. J. Am. Chem. Soc. 1962, 84, 1493.
12. Mauret, P.; Alphonse, P. J. Org. Chem. 1982, 47, 3322.
13. Gilliland, W. L.; Blanchard, A. A. Inorg. Synth. 1946, 4, 234.
14. Rodier, G.; Moreau, C. Congr. Int. Chim. Appl., 16th 1960, 54, 20448c.
15. Brochet, A. Bull. Soc. Chim. Fr. 1920, 27, 897.
16. Rieke, R. D.; Hudnall, P. M. J. Am. Chem. Soc. 1972, 94, 7178.
17. Rieke, R. D.; Hudnall, P. M.; Uhm, S. J. Chem. Soc. Chem. Commun. 1973, 269.
18. Rieke, R. D.; Bales, S. E. J. Chem. Soc. Chem. Commun. 1973, 739.
19. Rieke, R. D.; Bales, S. E. J. Am. Chem. Soc. 1974, 96, 1775.
20. Rieke, R. D.; Chao, L. Synth. React. Inorg. Met.-Org. Chem. 1974, 4, 101.
21. Rieke, R. D.; Ofele, K; Fischer, E. O. J. Organomet. Chem. 1974, 76, C19.
22. Rieke, R. D. Top. Curr. Chem. 1975, 59, 1.
23. Rieke, R. D. Acc. Chem. Res. 1977, 10, 301.
24. Rieke, R. D.; Wolf, W. J.; Kujundzic, N.; Kavaliunas, A. V. J. Am. Chem. Soc. 1977, 99, 4159.
25. Chao, L.; Rieke, R. D. J. Organomet. Chem. 1974, 67, C64.
26. Chao, L.; Rieke, R. D. Synth. React. Inorg. Met.-Org. Chem. 1974, 4, 373.
27. Chao, L.; Rieke, R. D. Synth. React. Inorg. Met.-Org. Chem. 1975, 5, 165.
28. Chao, L.; Rieke, R. D. J. Org. Chem. 1975, 40, 2253.
29. Rieke, R. D.; Uhm, S. J. Synthesis 1975, 452.
30. Rieke, R. D.; Kavaliunas, A. V.; Rhyne, L. D.; Fraser, D. J. J. Am. Chem. Soc. 1979, 101, 246.
31. Rieke, R. D.; Bales, S. E.; Hudnall, P. M.; Poindexter, G. S. Org. Synth. 1979, 59, 85.
32. Rieke, R. D.; Kavaliunas, A. V. J. Org. Chem. 1979, 44, 3069.
33. Rieke, R. D.; Rhyne, L. D. J. Org. Chem. 1979, 44, 3445.
34. Kavaliunas, A. V.; Rieke, R. D. J. Am. Chem. Soc. 1980, 102, 5944.
35. Rieke, R. D.; Li, P. T.; Burns, T. P.; Uhm, S. T. J. Org. Chem. 1981, 46, 4323.
36. Inaba, S.; Matsumoto, H.; Rieke, R. D. Tetrahedron Lett. 1982, 23, 4215.
37. Kavaliunas, A. V.; Taylor, A.; Rieke, R. D. Organometallics 1983, 2, 377.
38. Matsumoto, H.; Inaba, S.; Rieke, R. D. J. Org. Chem. 1983, 48, 840.
39. Inaba, S.; Rieke, R. D. Tetrahedron Lett. 1983, 2451.
40. Burns, T. P.; Rieke, R. D. J. Org. Chem. 1983, 48, 4141.

41. Rieke, R. D.; Bales, S. E.; Hudnall, P. M.; Burns, T. P.; Poindexter, G. S. Org. Synth. Collective Volume 6, 1986, in press.
42. Rochfort, G. L.; Rieke, R. D. Inorg. Chem. 1984, 23, 787.
43. Inaba, S.; Matsumoto, H.; Rieke, R. D. J. Org. Chem. 1984, 49, 2093.
44. Inaba, S.; Rieke, R. D. Chem. Lett. 1984, 25.
45. Inaba, S.; Rieke, R. D. Synthesis 1984, 842.
46. Inaba, S.; Rieke, R. D. Synthesis 1984, 844.
47. Inaba, S.; Rieke, R. D. J. Org. Chem. 1985, 1373.
48. Inaba, S.; Rieke, R. D. Tetrahedron Lett. 1985, 155.
49. Burkhardt, E.; Rieke, R. D. J. Org. Chem. 1985, 50, 416.
50. Ebert, G. W.; Rieke, R. D. J. Org. Chem. 1984, 49, 5280.
51. Inaba, S.; Forkner, M. W.; Wehmeyer, R. M.; Rieke, R. D. manuscript in preparation.
52. Bryker, W.; Rieke, R. D. manuscript in preparation.
53. Rochfort, G. L.; Rieke, R. D. Inorg. Chem. 1986, 25, 348.
54. Kahn, B. E.; Rieke, R. D. manuscript in preparation.
55. Chao, C. S.; Cheng, C. H.; Chang, C. T. J. Org. Chem. 1983, 48, 4904.
56. Hughes, W. B. J. Org. Chem. 1971, 36, 4073.
57. Nomura, M.; Takebe, K.; Miyake, M. Chem. Lett. 1984, 1581.
58. Boldvini, G. P.; Savoia, D.; Tagliavini, E.; Trombini, C.; Umani-Ronchi, A. J. Organomet. Chem. 1984, 268, 97.
59. McMurry, J. E. Acc. Chem. Res. 1983, 16, 405.
60. Grant, C.; Streitweiser, A. J. Jr. J. Am. Chem. Soc. 1978, 100, 2433.
61. Rieke, R. D.; Hudnall, P. M. J. Am. Chem. Soc. 1973, 95, 2646.
62. Fujita, T.; Watanaba, S.; Suga, K.; Sugahara, K.; Tsuchimoto, K. Chem. Ind.(London) 1983, 4, 167.
63. Fischer, E. O.; Burger, G. Z. Naturforsch. B: Anorg. Chem. Org. Chem. 1961, 16B, 702.
64. Agnes, G.; Chiusoli, G. P.; Marraccini, A. J. Organomet. Chem. 1973, 49, 239.
65. Klabunde, K. J.; Murdock, T. O. J. Org. Chem. 1979, 44, 3901.
66. Cason, J. Chem. Rev. 1947, 40, 15.
67. Frankland, E. Justus Liebigs Ann. Chem. 1849, 71, 171.
68. Gladstone, J. H. J. Chem. Soc. 1891, 59, 290.
69. Job, A.; Reich, R. Bull. Soc. Chim. Fr. 1923, 33, 1414.
70. Kurg, R. C.; Tang, R. J. C. J. Am. Chem. Soc. 1954, 76, 2262.
71. Renshaw, R. R.; Greenlaw, C. E. J. Am. Chem. Soc. 1920, 42, 1472.
72. Noller, C. R. Org. Synth. 1932, 12, 86.
73. Zakharkin, L. I.; Yu, O. Iza. Akad. Nauk, SSSR, Otd. Khim. Nauk. 1963, 193.
74. Galiulina, R. F.; Shabanova, N. N.; Petukhov, G. G. Zh. Obshch. Khim. 1966, 36, 1290.
75. Nesmeyanov, A. N.; Kocheshkov, K. A. "Methods of Elemento-Organic Chemistry"; North Holland Publishing Co.: Amsterdam, 1967; Vol. III, p. 8.
76. Denis, J.; Girard, C.; Conia, J. M. Synthesis 1972, 549.
77. Lenk, B. E.; Kahn, B. E.; Rieke, R. D. manuscript in preparation.
78. Bell, A. Catal. Rev.-Sci. Eng. 1981, 23, 203.
79. Toyoshima, I.; Somorjai, G. A. Catal. Rev.-Sci. Eng. 1979, 19, 105.
80. Young, R. S. "Cobalt: Its Chemistry, Metallurgy and Uses"; ACS Monogr. No. 149, Reinhold: New York, 1960.
81. Anderson, B. B.; Behrens, C. L.; Radonovich, L. I.; Klabunde, K. J. J. Am. Chem. Soc. 1976, 98, 5390.
82. Klabunde, K. J.; Efner, H. F.; Satek, L.; Donley, W. J. Organomet. Chem. 1974, 71, 309.

83. Bixler, N.; Niemann, C. J. Org. Chem. 1958, 23, 742.
84. Wagner, R. B. J. Am. Chem. Soc. 1949, 71, 3214.
85. Normant, H. "Advances in Organic Chemistry"; Interscience: New York 1960, Vol. II, pp. 37-38.
86. Yu, S. H.; Ashby, E. C. J. Org. Chem. 1971, 36, 2123 and references contained therein.
87. Hamonet, I. L. C. R. Hebd. Seances Acad. Sci. 1906, 142, 210 and references contained therein.
88. Paul, R. ibid. 1931, 192, 964.
89. Erlenmeyer, H.; Marbet, R. Helv. Chim. Acta. 1946, 29.
90. Cook, A. H.; Jones, D. G. J. Chem. Soc. 1941, 278.
91. Scholl, R.; Norr, W. Chem. Ber. 1900, 33, 1054.
92. Yanagida, S.; Yokoe, M.; Katagiri, I.; Ohoka, M.; Komori, S. Bull. Chem. Soc. Jpn. 1973, 46, 306.
93. Cairns, T. L.; Larchar, A. W.; McKusick, B. C. J. Am. Chem. Soc. 1952, 74, 5633.
94. Anker, R. M.; Cook, A. H. J. Chem. Soc. 1941, 329.
95. Hofmann, C. Chem. Ber. 1868, 1, 198.
96. Anker, R. M.; Cook, A. H. J. Chem. Soc. 1941, 324.
97. v. Meyer, E. J. Prakt. Chem. 1888, 2 37, 397.
98. Baerts, F. Chem. Zentralbl. 1923, 111, 124.
99. Maercker, A. J. Organomet. Chem. 1969, 18, 249.
100. Posner, G. H. Org. React. 1975, 22, 253.
101. Fanta, P. E. Synthesis 1974, 9.
102. Goshaev, M.; Otroshchenko, O. S.; Sadykov, A. S. Russ. Chem. Rev., Engl. Transl. 1972, 41, 1046.
103. Normant, J. F. Pure & Appl. Chem. 1978, 50, 709.
104. Whitesides, G. M.; House, H. O.; et al. J. Am. Chem. Soc. 1969, 91, 4871.
105. Johnson, C. R.; Dhanoa, D. S. J. Chem. Soc., Chem. Commun. 1982, 358.
106. Bertz, S. H.; Dabbagh, G. J. Org. Chem. 1984, 49, 1119.
107. Bertz, S. H.; Dabbagh, G.; Villacorta, G. M. J. Am. Chem. Soc. 1982, 104, 5824.
108. Lipshutz, B. H.; Kozlowski, J. A.; Wilhelm, R. S. J. Org. Chem. 1983, 48, 546.
109. Lipshutz, B. H.; Wilhelm, R. S. J. Am. Chem. Soc. 1981, 103, 7672.
110. Brooks, J. J.; Rhine, W.; Stucky, G. D. J. Am. Chem. Soc. 1972, 94, 7346.

RECEIVED November 20, 1986

Chapter 15

Living Colloidal Metal Particles from Solvated Metal Atoms: Clustering of Metal Atoms in Organic Media

Matthew T. Franklin and Kenneth J. Klabunde

Department of Chemistry, Kansas State University, Manhattan, KS 66506

A review of preparative methods for metal sols (colloidal metal particles) suspended in solution is given. The problems involved with the preparation and stabilization of non-aqueous metal colloidal particles are noted. A new method is described for preparing non-aqueous metal sols based on the clustering of solvated metal atoms (from metal vaporization) in cold organic solvents. Gold-acetone colloidal solutions are discussed in detail, especially their preparation, control of particle size (2-9 nm), electrophoresis measurements, electron microscopy, GC-MS, resistivity, and related studies. Particle stabilization involves both electrostatic and steric mechanisms and these are discussed in comparison with aqueous systems.

Graham coined the term "colloid" to describe suspensions of small particles in a liquid.(1) Such particles are generally considered to be from 1 to 5000 nm in diameter and are not easily precipitated, filtered, or observed by ordinary optical microscopes. The topic of this paper is metallic colloidal particles, often called metal sols, with special emphasis on non-aqueous media. Some history of gold sols is appropriate.

Gold has held the attention of mankind for thousands of years. Attempts to make the "perfect metal" (gold) from imperfect metals was the realm of alchemists, and over the centuries great medicinal powers were ascribed to gold. And as early as 300 A.D. there are references to the consumption of gold fluid to prolong life.(2) It is doubtful that this was a colloidal solution of gold since aqua regia(3) was unknown to early Chinese alchemists, and more likely amalgams of Hg-Au were actually consumed. Centuries later more recipes for aurum potabile (drinkable gold) appeared which were aqueous gold colloid solutions prepared by dissolving gold in aqua regia followed by treatment (chemical reduction of $AuCl_3$ or $HAuCl_3$ to Au metal particles) with ethereal oils. These solutions were usually then treated with chalk to neutralize the acid before being

NOTE: This chapter is part 15 in a series.

0097-6156/87/0333-0246$06.00/0
© 1987 American Chemical Society

consumed. Fabulous curative powers were attributed to these
solutions, especially toward heart disease. And in 1618 Antoni
published Panacea Aurea: Auro Potabile(4) which centered on the
treatment of venereal diseases, dysentery, epilepsy, tumors and more
with drinkable gold. Additional similar books appeared,(5) and
Helcher pointed out that the addition of boiled starch noticeably
increased the stability of the preparation. In 1802 Richter(6)
mentioned that the shades of color in purple gold solutions and ruby
glass were due to the presence of finely divided gold. He
correlated the colors with particle size in a qualitative way.(4)
And Fulhame(7) noted in 1794 that she could dye silk cloth various
shades of purple with colloidal gold solutions. Cassius(8) and
Glauber(9) were also involved in using colloidal gold as a coloring
agent.
     Faraday published the first scientific investigations of gold
sols.(10) He usually reduced aqueous solutions of $AuCl_3$ with
phosphorous. However, he also experimented with sparking gold
wires. He concluded that gold was present in the solutions as
elemental gold, and that color depended on particle size. He also
discovered flocculation by addition of an electrolyte NaCl. Other
workers also used sparking of metals under water to produce sols of
Pb, Sn, Au, Pt, Bi, Sb, As, Tl, Ag, and Hg.(11-12) As the years
passed aqueous colloidal gold was studied extensively. The classic
work of Svedberg,(13) Zsigmondy,(14) Kohlschutter,(15) and
Turkevich(16) must be noted. Zsigmondy,(14) using a slit
microscope, was able to study the "seeding" phenomenon and found
that gold particles already present preferentially grew as more gold
salt was reduced. This and related work(17-18) improved our
understanding of particle nucleation and growth.

Non-Aqueous Colloidal Metal Solutions. It has been difficult to
prepare colloidal gold in non-aqueous media due to limitations in
preparative methods (low salt solubilities, solvent reactivity,
etc.), and the fact that the low dielectric constant of organic
solvents has hindered stabilization of the particles. In aqueous
solution the gold particles are stabilized by adsorption of innocent
ions, such as chloride, and thus stabilized toward flocculation by
the formation of a charged double layer, which is dependent on a
solvent of high dielectric constant. Thus, it seemed that such
electronic stabilization would be poor in organic media.
     In spite of these difficulties, some limited successes have
been reported. Svedberg(13) struck an electric arc in a glass tube
under organic liquids. Gas flow through the tube carried some of
the metal particles into the solution (liquid methane or isobutanol
were used). Mayer used a similar method using very high voltage
with organic-water mixtures.(19) More recently, Kimura and Bandow
reported a similar method where metals were evaporated and swept
into a cold trap containing ethanol, with some success.(20)
     Additional successes have been reported: Svedberg(21) used an
alternating current discharge to disperse small pieces of metal;
Natanson(22) obtained colloidal copper in an acetone/toluene/1-
pentanol mixture by reducing $CuCl_2$ with zinc powder; Janek and
Schmidt(23) added a gold/citrate aqueous sol to an alcohol/toluene
mixture followed by heating and cooling and found that some gold

colloidal particles remained in the alcohol rich layer; Marinescu(24) reported that sonication of alkali metals at their mp in kerosene yielded pyrophoric colloids; Yamakita(25) used fats, organic acids, alcohols, and other organics as reducing agents for Au_2O_3 and obtained success especially with fats and fatty acids; and Ledwith(26) was able to reduce $AuCl_3$ in water with diazoethane followed by mixing with organics which gave some gold particles in the organic phase.

Protective Colloids. Another approach in preparing and stabilizing metal colloids is by adsorption of macromolecules on their surfaces. A wide variety of materials have been used including gummy gelatinous liquids,(10) albumin,(27) Icelandic moss,(28) latex,(22) polyvinylpyrrolidone,(29) antibodies,(30) carbowax 20M,(31) polyvinylpyridine,(31) and various polymer-water/oil-water mixtures.(32) These studies clearly indicate that "steric stabilization" of metal colloids is also important (along with electronic stabilization).(33)

Results and Discussion

More direct and successful methods for the preparation of non-aqueous metal sols are desirable. Especially valuable would be a method that avoids the metal salt reduction step (and thus avoids contamination by other reagents), avoids electrical discharge methods which decompose organic solvents, and avoids macromolecule stabilization. Such a method would provide pure, non-aqueous metal colloids and should make efficient use of precious metals employed. Such colloids would be valuable technologically in many ways. They would also be valuable to study so that more could be learned about particle stabilization mechanisms in non-aqueous media, of which little is known at the present time.

We have reported numerous studies of the clustering of metal atoms in non-aqueous (organic) media.(34) And very recently our preliminary report of stable non-aqueous gold sols appeared.(35) The approach has been to disperse metal atoms in excess cold organic solvent thus forming solvated metal atoms. Upon warming atoms clustering takes place moderated by solvation. Particle (cluster) growth is eventually inhibited and stopped by strongly bound solvent molecules. As mentioned above, there is some precedent for this approach found in earlier work on the preparation of active metal slurries,(36) dispersed catalysts,(37) and metal atom clustering in polymer oils.(38) However, our initial report(35) was the first describing the preparation of stable metal sols by the solvated metal atom method.

The experimental apparatus used consisted of a stationary metal atom-vapor reactor which has been detailed in the literature earlier.(39) Metal was evaporated (~0.1 to 0.5 g) and codeposited at -196 °C with excess organic solvent vapor (~ 40-150 mL). The frozen matrix was allowed to warm under controlled conditions, and upon melting stirring was commenced. After warming to room temperature stable colloidal solutions were obtained and syphoned out under N_2.

A variety of metals have now been investigated in our laboratory including Fe, Co, Ni, Pd, Pt, Cu, Ag, Au, Zn, Cd, In, Ge, Sn, and Pb. Solvents employed have been acetone, ethanol, THF, diethylether, dimethylsulfoxide, dimethylformamide, pyridine, triethylamine, ispropanol, isopropanol-acetone, toluene, pentane, and water. Acetone-Au will be discussed in detail herein, along with Ag.

$$Au + CH_3\overset{\displaystyle O}{\overset{\|}{C}}CH_3 \xrightarrow{-196\,°C} (CH_3\overset{\displaystyle O}{\overset{\|}{C}}CH_3)_m Au$$

$$\downarrow \text{ warm}$$

$$(CH_3\overset{\displaystyle O}{\overset{\|}{C}}CH_3)_y Au_z \xleftarrow[\text{R.T.}]{\text{warm to}} (CH_3\overset{\displaystyle O}{\overset{\|}{C}}CH_3)_n Au_x$$

stable purple solution

Formation and Stabilization of Colloidal Solutions

Concentration Effects. In order to determine if particle size and distribution could be controlled, we prepared a series of colloidal Au-acetone solutions of varying concentrations. A drop of each solution was placed on a carbon coated copper grid, and after acetone evaporation analyzed by transmission electron microscopy (TEM). Lower concentrations of Au in acetone (0.002 M) yielded smaller particles, and at very low concentrations quite good selectivity to a certain size was possible (1-3 nm). Higher concentrations of gold yielded larger particles with a broader distribution (4-7 nm), and still higher concentration (0.04 M) gave particles 6-9 nm with some as large as 20 nm (see Figures 1-3). Figure 4, a TEM micrograph of Au particles from acetone, shows that the particles appear spherical or oblong in shape and do not show crystalline faces or certain geometrical structures. The non-crystalline nature of these particles is not surprising since, as shown later, they do contain substantial portions of organic material. Their tendency to chain together might also be explained by the presence of organic residues.

The control of particle size by concentration indicates that particle growth is a kinetic phenomenon. It is unlikely that particle growth is reversible; once a Au-Au bond is formed it would not break under these experimental conditions. In a dilute solution of atoms, the frequency of encounters would be lower. As the gold atom-solvent matrix warms, the atoms and subsequent metal particles become mobile. It is the number of encounters that occur before particle stabilization that is important. If metal concentration is high the frequency of encounters is higher and the particles become bigger.

Interestingly, Tunekevich and coworkers([16],[40]) report the reverse effect for gold particle growth in aqueous media. Low concentrations of $HAuCl_3$ were reduced in solution and comparatively

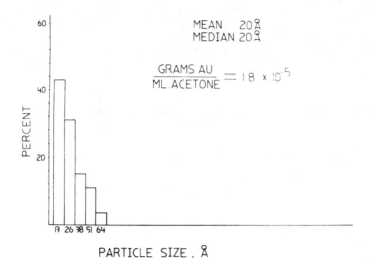

Figure 1.   Particle Size Distribution (Low Au:Acetone Ratio)

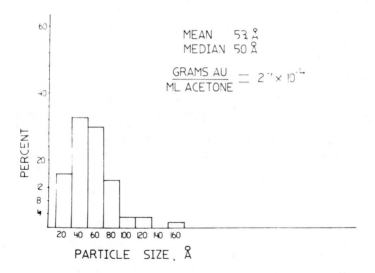

Figure 2.   Particle Size Distribution (Medium Au:Acetone Ratio)

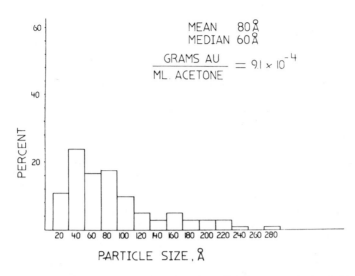

Figure 3. Particle Size Distribution (High Au:Acetone Ratio)

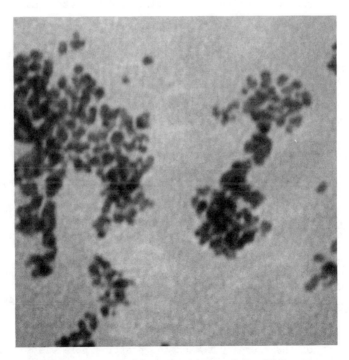

Figure 4. TEM Micrograph of Au Particles From Acetone
(Individual Particles about 8 nm that chain together)

large gold particles with a broad distribution resulted. At higher concentrations of HAuCl$_3$ the mean particle size decreased and the size distribution narrowed. Evidently a certain critical concentration of reduced auric ions was necessary in order that they would agglomerate into a stable particle. Due to the statistical nature of physical events (encounters) this required local concentration would be achieved more readily in a concentrated solution, leading therefore predominantly to rapid nucleation rather than particle growth, which is slower.

Concentration effects for Ag-acetone were also studied in our laboratory. These colloidal solutions were black as compared with purple for Au-acetone. They were also sensitive to light (see later). According to TEM the Ag particles from acetone were much larger (~ 30 nm) compared with Au (2-9 nm). The Ag particles appeared to be denser and perhaps more crystalline. They contained much less organic residue than the Au particles. Particle size for Ag was also dependent on Ag-acetone concentration in the same way as for Au, and the Ag particles were more polydisperse ranging from 20-40 nm.

Matrix Warmup:

Earlier we reported that Ni-pentane matrices upon warming from -196 °C yielded tiny Ni particles that incorporated substantial organic material from fragmented pentane (34). The amount of organic material (mainly C$_1$ fragments) could be increased and the Ni particle sizes decreased by allowing the Ni-pentane matrix to warm slowly.

Somewhat analogous behavior has been observed with these Au-acetone matrices. The most striking finding is that slow warmup (2h from -196 °C to 25 °C) was necessary to achieve a stable purple colloidal solution. Rapid warmup invariably caused excessive particle growth and flocculation/ precipitation of the Au particles. However, incorporation of organic residues does not appear to be the reason for this behavior since the resultant dry Au films (after acetone stripping) did not show significant or systematic changes with initial Au solvent ratio. In the case of Au-acetone and Pd-acetone(41) we suspect two reasons for this behavior: (1) solvent reordering and displacement during particle growth may be a slow process and (2) Au and Pd particles acquire negative charge, which helps stabilize their colloidal nature, and this is probably a slow process (discussed next under the electrophoresis section).

Electrophoresis. Electrophoresis, the movement of charged particles in response to an electric potential, has become very important in biochemistry and colloid chemistry. In the present study an apparatus similar to that described by Burton(42-45) was used. A U-tube with an inlet at the bottom and removable electrodes at the two upper ends was half filled with acetone. The a Au-acetone colloidal solution was carefully introduced from the bottom so that a sharp boundary was maintained between the clear acetone and the dark purple colloid solution. Next, platinum electrodes were placed in the top ends of the U-tube, and a DC potential applied. The movement of the boundary toward the positive pole was measured with time. Several Au-acetone colloids were studied, and electrophoretic velocities determined as 0.76-1.40 cm/h averaging 1.08 cm/h.

Calculation of electrophoretic mobility μ takes the potential into account:

$$\text{electrophoretic velocity} = \frac{1.08}{3600 \text{ s}} = 30 \times 10^{-5} \text{ cm/s}$$

$$\text{electrophoretic mobility} = \mu = 30 \times 10^{-5} \text{ cm s}^{-1} / 12.67 \text{ V}$$

$$\mu = 2.36 \times 10^{-5} \text{ cm/V}\cdot\text{s}$$

This value can be compared with those reported for a variety of aqueous colloidal particles, eg colloidal gold = $30\text{-}40 \times 10^{-5}$ (< 100 nm particle diameter), colloidal platinum = 20×10^{-5} (< 100 nm), colloidal lead = 12×10^{-5} (< 100 nm), and oil droplets = 32×10^{-5} (2000 nm).(43,44). The similarities of these numbers regardless of particle size suggests that the larger the particle is, the more negative charge it acquires.

The relation between electrophoretic mobility μ and the surface properties of the particle (usually modeled as an ionic double layer for aqueous systems) is a classical problem in colloid science.

The Helmholtz-Smolochowski equation is probably the oldest solution to $\mu = \frac{\varepsilon \xi}{4\pi\eta}$ μ = electrophoretic mobility
    ξ = electrokinetic potential (zeta potential)
    ε = dielectric constant of the medium
    η = viscosity of the medium (Stokes)(46)
the problem, but is known to be rather restricted. Huckel(47-48) considered electrophoretic retardation forces as well and proposed:

$$\mu = \frac{\varepsilon\xi}{6\pi\eta}$$

Henry (49) took into account the deformation of the applied D.C. field and proposed:

$$\mu = \frac{\varepsilon\xi}{6\pi\eta} \, f,(\kappa a)$$

The dimentionaless κa is a measure of the ratio between the particle radius and the thickness of the ionic double layer. In the limit κa = ∞ (the double layer is very thin compared with particle radius $f,(\kappa a)$ = 3/2 and the result is the Helmholtz-Smoluchowski equation. In the limit $\kappa a \to 0$, $f,(\kappa a)$ = 1 and Huckel's result is obtained.

Application of any of these expressions to our Au-acetone colloid system poses some problems. First, according to these theories the charge on the particle is supposed to result from the tight adsorption of negative ions and an outer more loosely attracted layer of counterions stabilizes the similarly charged outer layers of other particles, preventing coagulation.(50) Although our Au-acetone system must contain some kinds of positively charge species to preserve electrical neutrality, the nature of this species is unknown. Also, all the equations derived have been worked out for aqueous systems.

The Huckel equation where $\kappa a \to 0$ is the one most likely to be applicable to electrophoresis in non-aqueous media:(44)

$$\mu = \frac{\varepsilon\xi}{6\pi\eta}$$

Solving for ξ: $\xi = \frac{\mu6\pi\eta}{\varepsilon}$

Substituting the appropriate values:([41,51])

$$\xi = 768 \text{ mV}$$

This value is considerably larger than those reported earlier for aqueous metal sols:([43]) colloidal gold = 58 mv, 32 mv; platinum = 44 mv, 30 mv; lead = 18 mv. However, such large values might be expected for a low dielectric medium such as acetone.

Although comparison of these values with those determined in aqueous systems is tenuous, it is clear that the Au particles do possess considerable negative charge. Where this charge comes from is puzzling. It is possible that organic free radicals formed by pyrolysis or by homolylic bond breaking on Au atoms could transfer electrons to Au particles. A number of radiolysis studies of metal colloids in water-acetone solutions indicate that organic radicals do transfer electrons, and the particles act as electron reservoirs (and can behave as catalysts for water reduction).([52,53])

$$(CH_3)_2\overset{\bullet}{C}OH + (Ag)m \rightarrow (CH_3)_2C^+\text{-OH} + (Ag)_m^-$$
$$\downarrow$$
$$CH_3\overset{\overset{O}{\|}}{C}CH_3 + H^+ + (Ag)_m^-$$

If such a process was involved in our system, the generation of H^+ in solution would be expected. We have found no evidence for this (pH measurements, conductivities).

A second possibility is that the Au particles scavenge electrons from the reaction electrodes, walls and solvent. This is the explanation we favor at the present time since we have been able to effect changes in electrophoretic mobilities by supplying electrical potential to the colloid solution as the particles form,([41]) and the fact that such charging has been reported before, for example with oil droplets in water.([43])

Spectroscopic Studies:

UV-visible spectra of the Au-acetone sols showed absorptions at 706 and 572 nm, the latter being attributed to plasmon absorption.([54]) We have found that the plasmon absorption was not a good indicator of true particle size (obtained by TEM) since the 572 band did not shift significantly with particle size.([35]) However, in aqueous solution this band has been used successfully to roughly determine particle size.([55])

NMR studies of acetone stripped from colloidal solutions showed no reaction products.

Conductance Studies:

The conductance of several Au-acetone colloids was measured and compared to pure acetone, and NaI-acetone solutions. As expected the NaI-acetone solutions (0.00075 M up to 1.5 $\underline{M}$) showed greatly increased conductivities (130 to >20,000 μohm s^{-1} cm^{-1}). However, the Au-acetone colloid solutions showed approximately the same conductivities (2.5 to 7.4 μohm s^{-1} cm^{-1}) as acetone itself (4.5 μohm s^{-1} cm^{-1}). We conclude that very little "electrolyte" (ion pairs) was present in the purple Au-acetone colloidal solutions.

<u>Temperature Sensitivity.</u> Samples of Au-acetone colloid were subjected to boiling and freezing. Upon returning to room temperature the colloids remained stable and no flocculation had occurred. These results indicate that steric stabilization($\underline{33},\underline{43}$) (solvation) is a very important mechanism. Charge-stabilized colloids generally flocculate when subjected to such extremes of temperature.($\underline{56}$)

<u>Light Sensitivity.</u> The initially purple-black Ag-acetone colloid solutions turned grey and the Ag precipitated as a spongy grey mass upon exposure to room light for 3-4 days. However, in the dark the colloid solutions remained stable indefinitely.
<u>Particle Formation and Stabilization-Preliminary Conclusions</u>:
     Gold atoms dispersed in excess acetone (or other solvents) begin to cluster upon warming form -196°C. Particle (cluster) growth is moderated by two stabilizing mechanisms: (1) strong solvation which is a form of steric stabilization, and (2) electron scavenging to form negatively charged particles, which is a form of charge stabilization. Ultimate particle size is also affected by initial metal concentration and by warmup time, which suggests that the rates for these processes (stabilization <u>vs</u> growth) are competitive and can be controlled somewhat by time and temperature. Thus, purely kinetic phenomena appear to be involved.

<u>Film Formation.</u> A novel feature of these Au-organic solvent colloids is their film forming properties that can be induced simply by solvent stripping. In this sense they are "living" colloidal particles. Films formed in this way are conductive, but less so than pure metals.($\underline{41}$) The higher resistance of the films is due to the incorporation of substantial portions of the organic solvent, which can partially be removed by heating, and resistivity then decreases.($\underline{41}$)

<u>Elemental Analyses.</u> After solvent (acetone) stripping at room temperature (10^{-3} Torr for 1 h) Au films were scraped out and elemental analyses obtained. Variable results were obtained where Au ranged from 63-83% by weight, C from 6-17%, H form 0.8-1.1, and oxygen (by difference) from 6-17%. A heated sample yielded much lower C and H (0.62 C and < 0.01 H). Silver films showed much lower values for C and H.

<u>Pyrolyses.</u> A film from Au-acetone was pyrolyzed in stages up to 300°C. At intervals mass spectra were recorded, which showed the evolution of acetone (mainly) as well as other products. A similar experiment, where GC-MS was employed allowed identification of several of these minor products as CO_2, H_2O, C_2H_4, C_3H_6, C_3H_8, C_4H_8, C_4H_8O, and C_4H_2 (probably butadiyne). Pyrolysis of adsorbed acetone may be the source of these materials.
     Similar treatment of a Ag film from acetone only evolved some CO_2.

<u>Infrared Studies.</u> A gold residue from acetone was mixed with KBr and compressed to a pellet. Numerous IR scans were accumulated, and the spectrum indicated the presence of adsorbed acetone (mainly,

2960 and 1750 cm^{-1}). However, weaker bands at 2580, 1635, and 570 were also observed which perhaps indicate strongly chemisorbed acetone as Weinberg([57]) has recently reported on a Ru surface. The 570 cm^{-1} band is probably due to ν_{Au-C}.([58])

Resistivity. A film was prepared by dripping a colloidal Au-acetone solution on a glass plate edged with silicon rubber adhesive resin. The acetone was allowed to evaporate, and resistivitiy was measured by trimming the film to a rectangular shape. It was connected to electrodes on each end by vapor depositing an opaque film of copper, and resistance measured with a Digital Multimeter KEITHLEY 178 Model, with the following results: Thickness of film = 4.5 μm, resistance = 46 Ω/cm^2, resistivity $\rho(\Omega \cdot cm) = 1.8 \times 10^{-2}$. This can be compared with bulk gold where $\rho = 2.4 \times 10^{-6}$.([51]) Thus, the gold film from Au-acetone is more than 7000 times less conductive due to the incorporation of organic material.

Film Formation-Preliminary Conclusions:

Removal of solvent allows the colloidal particles to grow to a film. However, strongly adsorbed acetone, and perhaps small amounts of acetone fragments or telomers as well, remain in the film affecting its electrical properties.

Experimental Section

Preparation of a Typical Au-Acetone Colloid. The metal atom reactor has been described previously.([39,59,60]) As a typical example, a W-Al$_2$O$_3$ crucible was charged with 0.50g Au metal (one piece). Acetone (300 mL, dried over K$_2$CO$_3$) was placed in a ligand inlet tube and freeze-pump-thaw degassed with several cycles. The reactor was pumped down to 1 x 10^{-4} Torr while the crucible was warmed to red heat. A liquid N$_2$ filled Dewar was placed around the vessel and Au (0.2g) and acetone (80g) were codeposited over a 1.0 hr period. The matrix was a dark purple color at the end of the deposition. The matrix was allowed to warm slowly under vacuum by removal of the liquid N$_2$ from the Dewar and placing the cold Dewar around the reactor.

Upon meltdown a purple solution was obtained. After addition of nitrogen the solution was allowed to warm for another 0.5 hr to room temperature. The solution was syphoned out under N$_2$ into Schlenk ware. Based on Au evaporated and acetone inlet the solution molarity could be calculated.

Electrophoresis Experiments. The elctrophoresis experiments were carried out by using a glass U-tube of 11.0 cm each with a stopcock on the base to connect a perpendicular glass tube 13 cm long and 35 cm high.([43,45]) Platinum electrodes were attached to the top of the U-tube and through a ground glass joint to the pole of a 12V battery. The acetone was placed in the U-tube and then the colloid solution added slowly through the lower tube. The migration rate was determined based upon the average of the displacement in each side of the U-tube. A typical experiment was carried out for a period of 3 hr. at 25°C.

GC-MS Experiments. GC-MS pyrolyses were carried out using a Porapak Q column 6-ft (flow rate 35 mL/min) attached to a Finnigan 4000 quadrupole GC-MS. The sample was placed in a stainless steel tube 10 cm long connected to a 4 way valve. One of the outlets was attached to a Porapak Q column interfaced with the M.S. The stainless steel tube containing a portion of Au colloid film was placed in a furnace connected to a Variac provided with a digitial quartz pyrometer to measure the temperature. Three pyrolyses were performed at 100, 200 and 350°C with the Au-acetone film.

TEM Studies. Electron micrographs were obtained on JEOL, TEMSCAN -- 100 CX11 combined electron microscope and a HITACHI HV-11B (TEM) operated at 2×10^5 magnification. The specimens for TEM were obtained by placing a drop of the colloid solution on a copper grid coated by a carbon film.

Infrared Red Studies. Infrared spectra were recorded in a Perkin Elmer PE-1330 infrared spectrometer. IR studies of the metal films using either KBr pellets or Fluorolube yielded bands at 2960(s), 2580(m), 1750(s), 1635(m), and 570(w) cm^{-1}.

Acknowledgments

The support of the Office of Naval Research is acknowledged with gratitude. We thank Dr. Galo Cardenas-Trevino for helpful discussions, and Larry L. Seib for assistance with the TEM experiments. Also we want to thank Dr. Ileana Nieves for her assistance in obtaining spectra and Thomas J. Groshens for assistance with the mass spectrometer.

References

1. Graham, T.; *J. Chem. Soc.* (London), 1864, 17, 325.
2. Leicester, H. M., "The Historical Background of Chemistry," Dover Pub., New York, 1956, p 57. Davis, T. L., *Isis*, 1936, 25, 334.
3. Higby, G. J.; *Gold Bull.*, 1982, 15(4), 130.
4. Hauser, E. A.; *J. Chem. Ed.*, 1952, 29, 456; Antoni, F.; "Panacea Aurea: Auro Potabile," Bibliopolio Frobeniano, Hamburg, 1618
5. Helcher, H. H.; "Aurum Potabile oder Gold Tincture," J. Herbord Klossen, Breslau and Leipzig, 1718; Juncker, J.; "Conspectus Chemiae Theoreticopracticae," Magdeburg, 1730, 1, 882; Macquer, P. J.; "Dictionnaire de Chymie, Paris, 1778.
6. Ostwald, W.; *Z. Chem. Ind.* Kolloide, 1908, 4, 5-14 reviews Richter's work.
7. Fulhame, M.; "An Essay on Combustion with a View to the Art of Dyeing and Painting," J. Cooper, London, 1794.
8. Cassius, A.; "De Auro," Leiden, 1685.
9. Glauber, J. R.; "Teutschlands Wohlfahrt," Amsterdam, 1659.
10. Faraday, M.; *Phil. Trans.*, 1857, 147, 145.
11. Mindel, J.; King, C. V.; *J. Am. Chem. Soc.*, 1943, 65, 2112 and references therein.
12. Franklin, M. T.; M. S. Thesis, Kansas State University, 1986 gives a complete historical review.

13. Svedberg, T.; Ann. Phys., 1906, 18, 590; Svedberg, T.; Ber., 1905, 38, 3616.
14. Zsigmondy, Z.; Z. Physikal. Chem., 1908, 56, 65.
15. Kohlschutter, H. W.; "New Synthetic Methods, Vol. 3," Verlag Chemie, Weinheim, 1975, pg. 1.
16. Turkevich, J.; Stevenson, P. C.; Hillier, J.; Faraday Diss., 1951, 11, 55-75; Turkevich, J.; Hillier, J.; J. Phys. Chem., 1953, 57, 670; Chiang, Y. S.; Turkevich, J.; J. Colloid. Sci., 1963, 18, 772.
17. Reitstotter, J.; Kolloidchem. Beichefte, 1917, 9, 221.
18. Zakowski, J.; Kolloidchem. Beihefte, 1926, 23, 117.
19. Mayer, W. M.; Ger. Pat. 1,133,345, July 1962.
20. Kimura, K.; Bandow, S.; J. Chem. Soc. Japan, 1983, 56, 3578.
21. Svedberg, T.; Koll. Zeit., 1907, 1, 229, 257; Svedberg, T.; Koll. Zeit., 1908, 2, 29.
22. Natanson, E. M.; Kolloid. Zhur., 1949, 11, 84.
23. Janek, A.; Schmidt, A.; Kolloid. Zhur., 1930, 52, 280.
24. Marinescu, N.; Bull. Soc. Roumaine Phys., 1934, 36, 181.
25. Yamakita, I.; Repts. Inst. Chem. Res., Kyoto Univ., 1947, 14, 12.
26. Ledwith, A.; Chem. Ind., 1956, 1310.
27. Mayer, E.; Lottermoser, A.; "Uber Anorganische Colloid," Stuttgart, 1901.
28. Gutbier, A.; Huber, J.; Kuhn, E.; Kolloid Zhur., 1916, 18, 57.
29. Jirgensons, B.; Makromol. Chem., 1951, 6, 30.
30. Falk, W. P., Taylor, G. M.; Immunochemistry, 1971, 8, 1081.
31. Lee, P. C.; Meisel, D.; Chem. Phys. Lett., 1983, 99, 262.
32. Kurihara, K.; Kizling, J.; Stenilus, P.; Fendler, J. H.; J. Am. Chem. Soc., 1983, 105, 2574; Ledwith, A.; Chem. Ind. (London), 1956, 1310.
33. Hirtzel, C. s.; Rajagopalam, R.; "Colloidal Phenomena: Advanced Topics," Noyes Pubs., New Jersey, 1985, pp 88-97.
34. Davis, S. C.; Severson, S.; Klabunde, K. J.; J. Am. Chem. Soc., 1981, 103, 3024 and references therein.
35. Lin, S. T.; Franklin, M. T.; Klabunde, K. J.; Langmuir, 1986, 2, 259.
36. Klabunde, K. J.; Murdock, T. O.; J. Org. Chem., 1979, 44, 3901.
37. Klabunde, K. J.; Tanaka, Y.; J. Mol. Catal., 1983, 21, 57; Imizu, Y.; Klabunde, K. J.; J. Am. Chem. Soc., 1984, 106, 2721; Matsuo, K.; Klabunde, K. J.; J. Org. Chem., 1982, 47, 843.
38. Andrews, M. P.; Ozin, G. A.; J. Phys. Chem., 1986, 90, 2922, 2929, 2938.
39. Klabunde, K. J.; Timms, P. L.; Skell, P. S.; Ittel, S., Inorg. Syn., Shriver, D., editor, 1979, 19, 59-86.
40. Turkevich, J.; Hillier, J.; Anal. Chem., 1949, 21, 475; Turkevich, J.; Garton, G.; Stevenson, P. C.; J. Colloid Sci., 1954, Suppl. 1, 26.
41. Cardenas, G.; Klabunde, K. J.; Dale, B.; Langmuir, submitted.
42. Burton, E. F.; Phil. Mag., 1909, 17, 587.
43. Jirgensons, B.; Straumanis, M. E.; "A Short Textbook of Colloid Chemistry," Macmillan, New York, 1962, pp. 132-133, 343.
44. Shaw, D. J., "Introduction to Colloid and Surface Chemistry," Butterworth, 2nd Ed., London, 1970, pp 157-159.
45. Shaw, D. J.; "Electrophoresis," Academic Press, New York, 1969.

46. Overbeek, J. T.; Wiersema, P. H.; "Electrophoresis," Vol. II, M. Bier, Academic Press, New York, 1967, p. 2; Smoluchowski, M.; Bull. Acad. Sci. Cracovic, 1903, 182.

47. Huckel, E.; Physik. Z., 1924, 25, 204.

48. Hiemenz, P. C., "Principles of Colloid and Surface Chemistry," Lagowski, J. J., editor, Marcel Dekker, New York, 1977, 453-466.

49. Overbeek, J. T. G.; Kolloidchem Beihefte, 1943, 54, 287 discusses Henry's work.

50. Verwey, E. J. W.; Overbeek, J. T. G.; "Theory of the Stability of Lyophobic Colloids," Elsevier, Amsterdam, 1948.

51. CRC Handbook of Chemistry and Physics, 65th Ed., CRC Press, Boca Raton, Florida, 1984-85, pp F38 and E51.

52. Hengelein, A.; J. Am. Chem. Soc., 1979, 83, 2209.

53. Henglein, A.; Lillie, J.; J. Am. Chem. Soc., 1981, 103, 1059.

54. Mie, G.; Ann. Phys., 1908, 25, 377; Doyle, W. T.; Phys. Rev., 1958, 111, 1067; Ferrell, T. L.; Calcott, T. A.; Warmack, R. J.; Am. Scientist, 1985, 75, 345.

55. Sadler, P. J.; "Structure and Bonding 29," Dunitz, J. D., et al., editors, Springer Verlag, Berlin, 1976, 178; Caro, R. A.; Nicolini, J. O.; Radicella, R.; Int. J. Appl. Radiol-Isotopes, 1967, 18, 327; Jeppeson, M. A.; Barlow, R. B.; J. Opt. Soc. Am., 1962, 52, 99.

56. Vickery, J. R.; Dept. Sci. Ind. Res. Report Food Investigation Board, 1929 (1930), 24; Also see Chem. Abs., 1931, 25, 5820.

57. Templeton, M. K.; Weinberg, W. H.; J. Am. Chem. Soc., 1985, 107, 774.

58. Puddephatt, R. J.; "The Chemistry of Gold," Elsevier, Amsterdam, 1978, p. 220; Nakamoto, K.; "Infrared and Raman Spectra of Inorganic and Coordination Compounds," 3rd Ed., Wiley, New York, 1978; Scovell, W. M.; Tobias, R. S.; Inorg. Chem., 1970, 9, 945; Scovell, W. M.,; Stocco, G. C.; Tobias, R. S.; Inorg. Chem., 1970, 9, 2682.

59. Klabunde, K. J.; "Chemistry of Free Atoms and Particles," Academic Press, New York, 1980.

60. Timms, P. L. in "Cryochemistry," Moskovits, M.; Ozin, G. A., editors, Wiley, New York, 1976.

RECEIVED December 11, 1986

Chapter 16

Activation of Carbon-Hydrogen Bonds by Metal Atoms

M. L. H. Green and Dermot O'Hare[1]

Inorganic Chemistry Laboratory, South Parks Road, Oxford, OX1 3QR, United Kingdom

Aspects of the apparatus for the synthesis using metal atoms are described. The reactions of the atoms of rhenium, tungsten, and osmium with hydrocarbons including alkanes are described. It is shown that metal atom reactions with alkanes can give isolable organometallic compounds including μ–alkylidene compounds.

The ability of the surfaces of many transition metals to react with alkanes under mild conditions has been long known (1-3). In the last decade there has been increasing interest in the reaction of discrete transition metal compounds with alkanes, where the transition metal inserts into the carbon-hydrogen bonds of the alkane giving alkyl hydride derivatives (4-6). Very recently it has been [1] shown to be possible for transition metal compounds to insert in to the carbon-hydrogen bonds of methane, the most inert of saturated hydrocarbons (7-9). During a study designed to understand the fundamental requirements for a transition metal center to interact with saturated carbon-hydrogen bonds we have investigated the reactions between transition metal atoms and hydrocarbons.

[1]Current address: E. I. du Pont de Nemours & Co., Experimental Station, Wilmington, DE 19898

The general technique of the metal vapor experiments described below was to co-condense the vapors of the transition metal with those of the chosen hydrocarbon or hydrocarbon mixtures. In this paper we briefly outline the technique of metal atom synthesis and then show how it can be applied to alkane activation reactions.

METAL VAPOR SYNTHESIS

The general principles and practice of metal vapor synthesis have been well reviewed (10-11). The experiments described below were carried out using a bell-jar design of apparatus shown in Figure 1 incorporating a positive-hearth electron-gun furnace (Figure 2). This design was first constructed in Oxford. The electron gun furnace can achieve temperatures in excess of 4000°C and can vaporize all the elements at a substantial rate. The positive-hearth mode (12-13) reduces the damage to both the co-condensate and products by reducing the number of reflected electrons emitted from the furnace. Typical rates of evaporation of a selection of the transition metals from a 0.5ml capacity electron gun furnace are given in Table I.

Using the the larger size hearth of capacity 10ml, typically, up to 15g of tungsten or 30g of titanium may be evaporated in a 4h experiment producing multi-gram (5-35g) quantities of products, in favorable reactions. The success of this reactor design owes much to the close proximity of the electron-gun to the large 2000 litres.sec^{-1} cryopump (Figure 1), to maintain the high vacuum required for its operation. The development of this reactor has enabled metal atoms to be considered as 'off the shelf' reagents in synthesis.

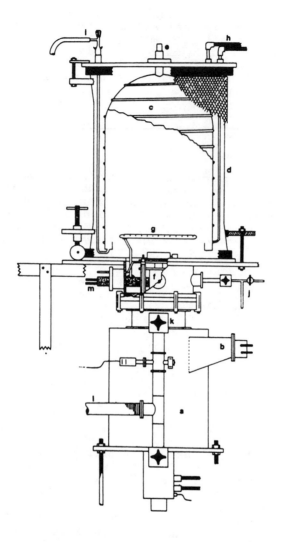

Figure 1 10KW metal atom synthesis apparatus; (a) Varian VK10 cryopump; (b) Gatevalve; (c) Copper matrix support; (d) Glass vacuum vessel; (e) Inlet for washing solvent; (f) 10KW furnace; (g) Ligand inlet ring; (h) Pipes for liguid nitogen coolant; (i) Product exit pipe; (j) Argon inlet; (k) Roughing valve; (l) Flexible pipe to rotary pump; (m) Water cooling pipes.

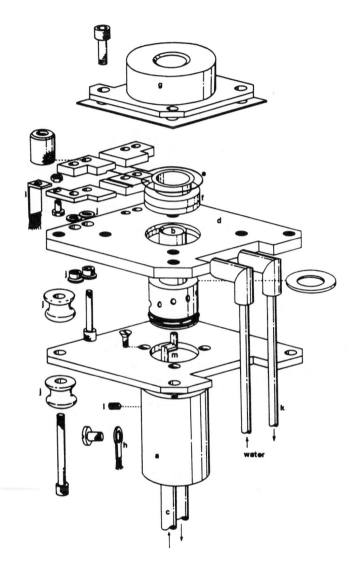

Figure 2   3.5 KW electron beam furnace   components:
(a) Water-cooled furnace block;   (b) Copper   hearth
(c) Cooling pipes;   (d) Top plate   of furnace (earth
potential);   (e)   Filament;   (f) Focus lid;   (g) Lid:
(h) High-tension supply: (i) Low-tension   supply;
(j)   Ceramic insulators;   (k) Cooling pipes for   top
plate: (l) locating stud for hearth adjustment; and
(m) Water conduit.

Table I. Power input and rates of evaporation of metals from a 0.5ml hearth of the positive hearth electron gun furnace

Metal	Power input (KW)[a]	Evaporation Rate (g.h^{-1})
Titanium	1.0	1.3
Zirconium	2.2	0.5
Hafnium	1.8	1.5
Niobium	2.0	0.5
Tantalum	1.8	1.5
Molybdenum	1.2	1.0
Tungsten	1.8	1.5
Rhenium	1.8	1.5
Osmium	1.4	1.5
Uranium[b]	1.4	0.5

[a] Power input depends not only on the temperature required for vaporization, but also on the fluidity of the molten sample and consequent heat loss by very efficient thermal contact with the water cooled hearth. For example, tungsten metal requires a vaporization temperature of ca. 3600°C corresponding to a 1.8KW power input, whilst niobium which vaporizes at ca. 2500°C requires ca. 2.0KW power input because at 2500°C niobium is a more fluid metal than tungsten at 3600°C and so can dissipate the power more efficently by thermal conduction through the water cooled hearth.

[b] Uranium evaporated from a previously prepared 40% U/Re alloy.

Note: Data are from Refs. 14 and 15.

ALKANE ACTIVATION BY METAL ATOMS

Introduction. In a typical experiment, the metal vapor is co-condensed at a liquid nitrogen-cooled surface with the ligand substrate vapor. Upon warming, the reaction mixture melts and is washed from the reactor walls with a suitable solvent. The products are filtered to remove unreacted bulk metal, and finally the products are separated and purified by conventional Schlenk vessel or vacuum line techniques.

Klabunde (16) was the first to describe alkane activation by metal atoms. He showed that co-condensation of nickel atoms and pentane yielded a solid material which contained Ni,C and H. Hydrogenolysis of this material produced C_1-C_5 alkanes. Klabunde suggested that the reactions was analogous to the well-documented heterogeneous cracking processes over nickel catalysts, and that nickel crystallites were responsible for the chemistry observed.

Skell (17) found that co-condensation of zirconium atoms with neopentane gave a residue which upon deuteriolysis gave Bu^tCH_2D and polydeuterated C_2-C_4 hydrocarbons. He suggested that the C_5-product arose from a neopentylzirconium moiety.

The photoexcitation of metal atoms in alkane matrices produces reactive excited-state species capable of activating alkanes. In 1929 Taylor and Hill, and subsequently Steacie (18), showed that atomic Hg in the $^3P_1$ state can abstract an H atom from an alkane.

Billups and Margrave (19) showed in 1980 that photoexcitation of Fe atoms in a methane matrix led to an unstable species which they suggested to be $HFeCH_3$. Photoexcited $Cu(^2P)$ has been found also to react with methane, in a matrix at 12K. The initial photoproduct was proposed to be $HCuCH_3$, which subsequently decomposed photolytically to CuH and CuMe (20).

Activation of carbon-hydrogen bonds by rhenium atoms

Co-condensation of transition metal atoms with arenes such as benzene and toluene is well known to yield bis-arene-metal compounds. However, in many cases the yields based on the metal atoms are less than 40%. Evidence that competing reactions such as carbon-hydrogen activation can occur is provided by the isolation of non-metal-containing products such as biaryl derivatives (21).

Reaction of rhenium atoms with alkyl-substituted arenes forms dirhenium-μ-arylidene compounds (<u>22</u>) (Figure 3). The products require insertion, presumably sequential, into two carbon-hydrogen bonds of the alkyl substituent. These reactions seem highly specific and require only the presence of an alkyl-substituted benzene that possesses a CH_2 or CH_3 substituent. Thus, co-condensation of rhenium atoms with ethylbenzene gives two isomers (see Figure 3) in which the products arise from insertion into the carbon-hydrogen bonds of the methylene or the methyl group. The product distribution in this reaction is in accord with statistical attack at all available sp^3 C-H bonds.

Co-condensation of rhenium atoms with benzene alone gave no isolable products at ambient temperature. It was not expected, however, when rhenium atoms were co-condensed with a benzene:alkane mixture that μ-alkylidene complexes, analogous to the μ-arylidene complexes were formed. Thus, co-condensation of rhenium atoms with a <u>ca</u> 1:1 mixture of benzene and neopentane gave the dirhenium-μ-neopentylidene complex $[\{Re(\eta-C_6H_6)\}_2(\mu-CHBu^t)(\mu-H)_2)]$ in moderate yield (Figure 4) (<u>23</u>). The single crystal X-ray structure determination is shown in Figure 5. The compound $[\{Re(\eta-C_6H_6)\}_2(\mu-CHBu^t)(\mu-H)_2)]$ represents the first isolated crystalline product arising from a metal atom-alkane experiment.

The ability of the rhenium-benzene co-condensate to activate linear- and cyclic- alkanes is quite general. We have co-condensed rhenium atoms with alkane:benzene mixtures using the alkanes; ethane, propane, butane, 2-methylpropane, neopentane, tetramethylsilane, cyclopentane, and cyclohexane, giving the μ-alklyidene compounds shown in Figures 4a and 4b (<u>23</u>). Co-condensation of rhenium atoms with 1:1 mixture of perdeuterobenzene and neopentane gave the compound $[\{Re(\eta-C_6D_6)\}_2(\mu-CHBu^t)(\mu-H)_2)]$ and no ^2H incorporation was detected in either the μ-alkylidene or μ-hydrido positions (^2H n.m.r).

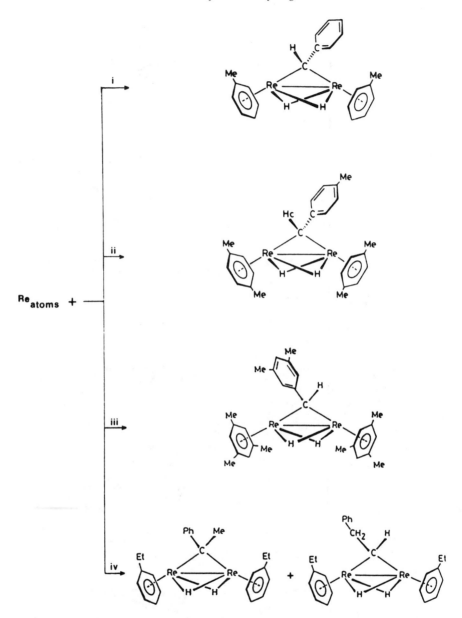

Figure 3 Summary of the reactions of rhenium atoms with alkylbenzenes. Co-condensation of rhenium atoms at −196°C with; (i) Toluene; (ii) p–Xylene; (iii) mesitylene; (iv) ethylbenzene.

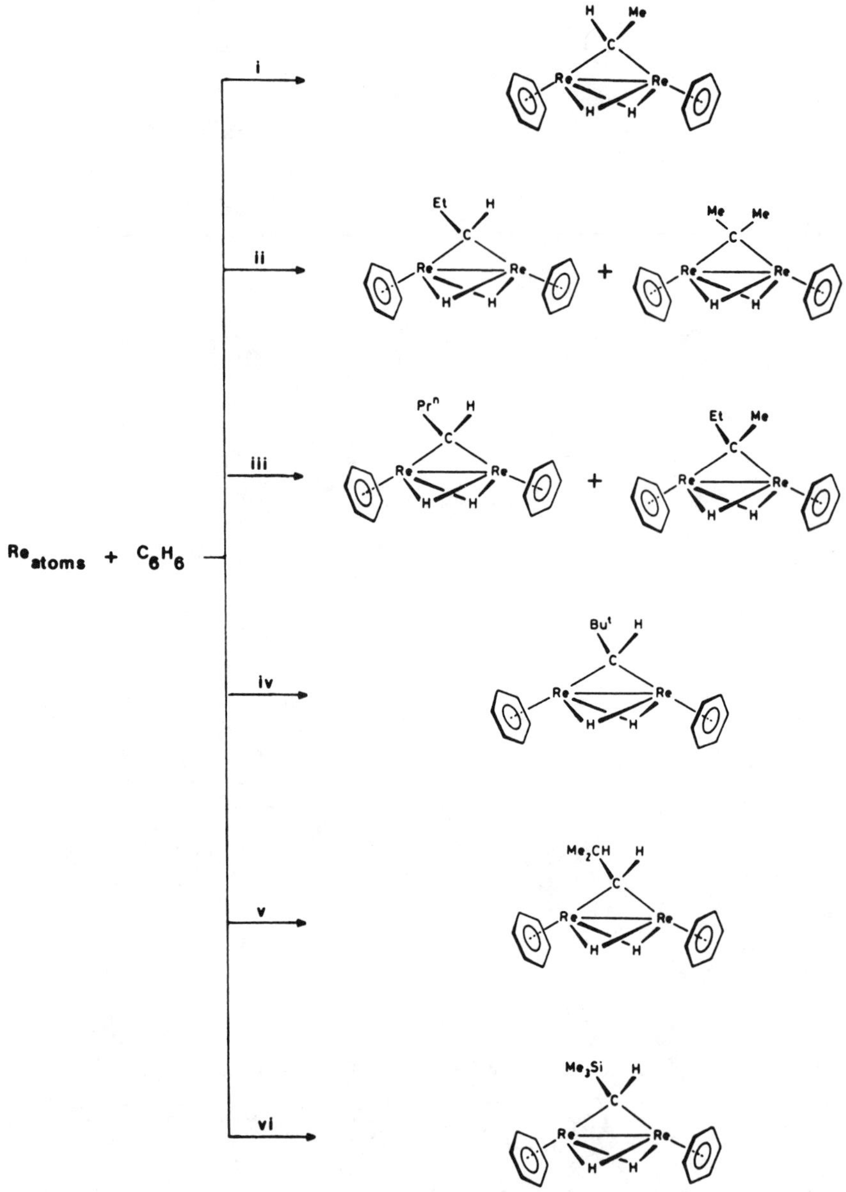

Figure 4a Summary of the reactions of rhenium atoms with acyclic saturated hydrocarbons. Rhenium atoms were co-condensed with the indicated substrates at -196 °C. (i) Ethane; (ii) Propane; (iii) n-Butane; (iv) Neopentane; (v) 2-Methylpropane; and (vi) Tetramethylsilane.

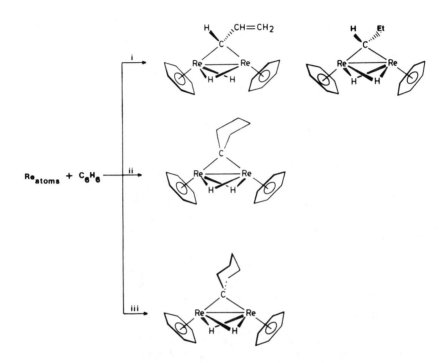

Figure 4b   Summary   of   the   reactions of rhenium atoms   with   cyclic saturated   hydrocarbons.   Rhenium atoms   were   co-condensed   with   the   indicated substrates   at   -195°C;   (i) Cyclopropane;   (ii) cyclopentane; (iii) cyclohexane.

Multi-step alkane dehydrogenation reactions have been observed for complexes of rhenium and iridium. For example, the dehydrogenation of cyclohexane by $[IrH_2(PPh_3)_2(Me_2CO)_2]SbF_6$ giving $[Ir(\eta-C_6H_6)(PPh_3)_2]SbF_6$ (24). We have observed that co-condensation of rhenium atoms with cyclohexane and PMe_3 (20:1) gives the previously described compound $[Re(\eta-C_6H_6)(PMe_3)_2H]$ (25). Similarly, co-condensation of rhenium atoms with cyclopentane and PMe_3 (10:1) gives $[Re(\eta-C_5H_5)(PMe_3)_2H_2]$ (26).

Activation of carbon-hydrogen bonds by tungsten atoms

Co-condensation of tungsten atoms with a mixture of cyclohexane and PMe_3 gives a mixture a products shown in Figure 6 (26). The identity of the individual species has been determined by subsequent indepentant conventional syntheses. Co-condensation of tungsten atoms with cyclohexene and PMe_3 (10:1) gave (2) and (3) only (Figure 6) and there was no evidence for (1) and (4).

When tungsten atoms were co-condensed with a mixture of cyclohexane and perdeuteriocyclohexane (3:1) and PMe_3 the spectroscopic data indicated that there was no evidence for intermolecular hydrogen-deuterium scrambling and this strongly suggests that (1)-(4) are formed by a process involving intramolecular carbon-hydrogen bond activation of the cyclohexane.

In a more remarkable experiment in which tungsten atoms are co-condensed with a mixture of cyclopentane and PMe_3 the major component was identified as $[W(\eta-C_5H_5)(PMe_3)H_5]$ (Figure 7) (27). The other species present were identified as $[W(\eta-C_5H_5)(PMe_3)_2H_3]$, $[W(PMe_3)_4H_4]$ and $[W(PMe_3)_5H_2]$ by comparison of the spectroscopic data with authentic samples. It seems reasonable to propose that the formation of $[W(\eta-C_5H_5)(PMe_3)H_5]$ from cyclopentane proceeds *via* the migration of all five hydrogens from one face of the C_5-ring onto the tungsten in a stepwise intramolecular process. The proposal is supported by the almost complete absence of $[W(\eta-C_5H_5)(PMe_3)H_5]$ when cyclopentane is replaced by cyclopentene in the above experiment.

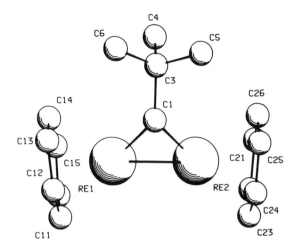

Figure 5   Molecular structure of $[\{Re(\eta-C_6H_6)\}_2(\mu-CHBu^t)(\mu-H)_2)]$.

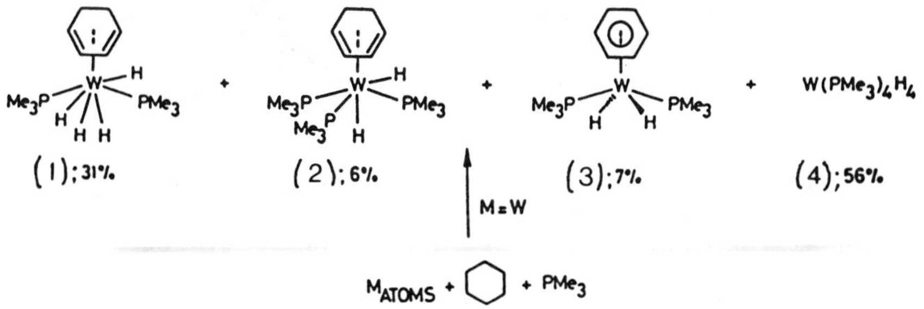

Figure 6   Activation of cyclohexane by tungsten atoms.

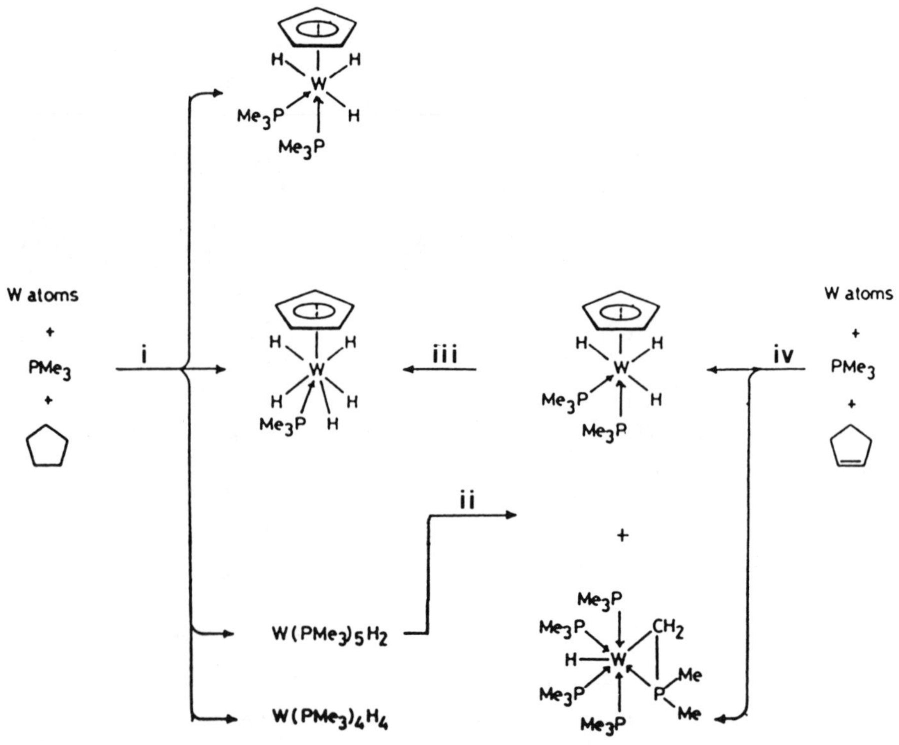

Figure 7   Reaction of tungsten atoms with cyclopentane.
(i and iv) Co-condensation at -196 °C; (ii) Cyclopentadiene in benzene at 60 °C; and (iii)   Dihydrogen at 50
atm.

Activation of carbon-hydrogen bonds by osmium atoms

Co-condensation of osmium atoms with benzene gives moderate yields of $[Os(\eta^4-C_6H_6)(\eta^6-C_6H_6)]$ (28). The molecule is fluxional as demonstrated by variable temperature n.m.r experiments.[1] H magnetization transfer n.m.r experiments show that $[Os(\eta^4-C_6H_6)(\eta^6-C_6H_6)]$ undergoes degenerate haptotropic ring exchange for which the rate constant (k) at 37^oC is $0.2\pm0.03s^{-1}$. Surprisingly, when osmium atoms are co-condensed with mesitylene two products can be isolated. The major yellow crystalline product was identified as $[\{Os(\eta^6-C_6H_3Me_3)\}_2(\mu-CHC_6H_3Me_2-3,5)(\eta-H)_2]$. Fractional crystallization of the reaction mixture gave a second minor product as red crystals which a crystal structure determination showed to be $[\{Os(\eta^6-C_6H_3Me_3)\}_2(\eta-CHC_6H_3Me_2-3,5)]$ as shown in Figure 8 (28).

However, when osmium atoms are co-condensed with a mixture of benzene and 2-methylpropane only trace amounts of $[Os(\eta^4-C_6H_6)(\eta^6-C_6H_6)]$ can be detected in the reaction mixture. More importantly, the tri-nuclear compound $[\{Os(\eta^6-C_6H_6)\}_3(\mu^2-H)_3(\mu^3-(CH_2)_3CH)]$ can be isolated (5) (Figure 8) (29). Co-condensation of osmium atoms with a mixture of perdeuteriobenzene and 2-methylpropane gives the compound $[\{Os(\eta^6-C_6D_6)\}_3(\mu^2-H)_3(\mu^3-(CH_2)_3CH)]$, showing that the three μ^2-H groups arise from the 2-methylpropane. The tri-osmium compound provides a striking model for a surface chemisorbed alkane (30).

CONCLUSION

The co-condensation reactions described above have led to the formation of interesting new compounds and sometimes very unexpected products. The nature of the products formed for example in the osmium atom experiments indicate high degrees of specificity can be achieved. However, the detailed mechanisms of the co-condensation reactions are not known. It seems most likely that in all cases the initial products formed at the co-condensation temperature are simple ligand-addition products and that the insertion of the metal into the carbon-hydrogen bond occurs at some point during the warming up process. In support of this hypothesis we note the virtual absence of any

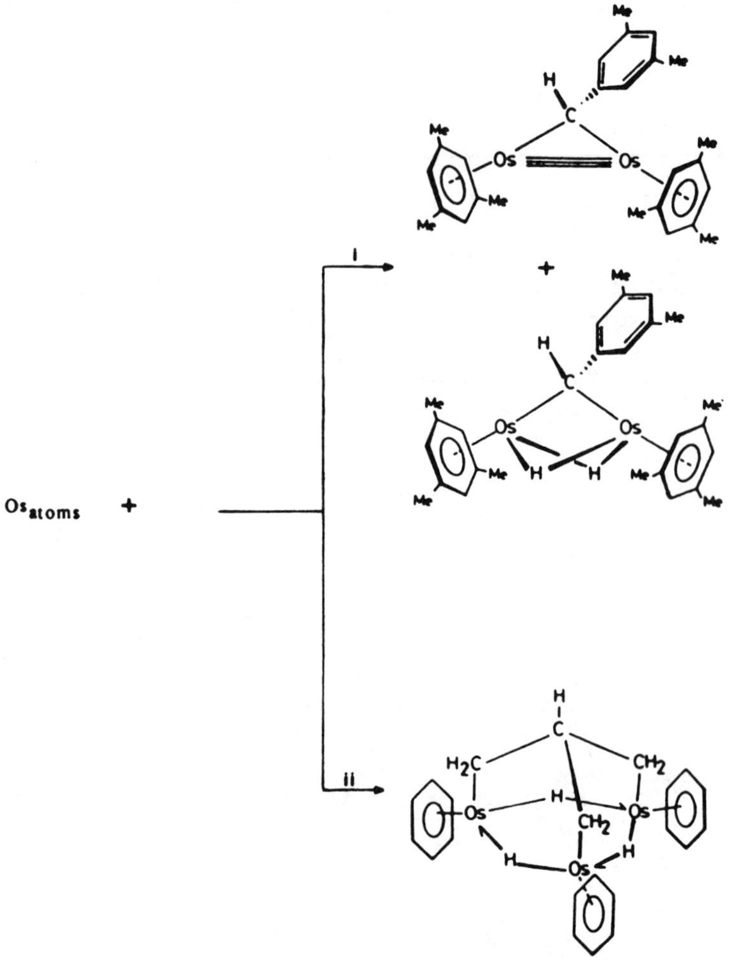

Figure 8 Summary of the reactions of osmium atoms with alkanes and alkylbenzenes. Osmium atoms were co-condensed with the indicated substrates at −196°C. (i) Mesitylene; (ii) Benzene/2-Methylpropane mixture (1:1, w/w).

bis(arene)osmium compound, itself unreactive towards alkanes in the co-condensation reaction of osmium atoms and a benzene:2-methylpropane mixture. Although many of the products are di- or tri- nuclear it seems unlikely that these arise from reactions involving metal dimers or trimers but that they are a consequence of intermolecular reactions of initially formed mononuclear species. However, detailed studies of the product distribution as a function of temperature are needed before we can have any evidence of the nature of the intermediates. Given these assumptions it seems that the reactions of these metal atom co-condensates are similar to that of thermal and photochemical reactions of discrete molecular systems in so far that we believe that they proceed by initial formation of some unsaturated organometallic species which then subsequently inserts into the carbon-hydrogen bond. We note that it is not possible under the conditions of the metal-atom experiments to distinguish between thepossibilities that they are simply thermal reactions or both thermal and/or photochemically-induced reactions.

The conditions of the reactions are such that during the co-condensation the co-condensate receives radiation from the molten metal sample which is comparable to a 1KW tungsten lamp.

It is clear that alkanes should not be treated as 'inert' solvents in reactions involving metal atoms, even when there is an excess of a more established ligand.

ACKNOWLEDGMENTS

We thank the Donors of the Petroleum Research Fund, administered by the American Chemical Society for partial support, and the Royal Commission for the Exhibition of 1851 for a Research Fellowship to D. O'H.

LITERATURE CITED

1. Kemball, C. In *Catalysis Reviews*; Heinemann, H., Ed.; Dekker, M.:New York, 1972; p33.

2. Muetterties, F.L.; Rhodin, T.N.; Band, E.;Brucker, C.F.; and Pretzer, W.R. *Chem. Rev.*, 1979, *79*, 91.

3. Somorjai, G. In *Chemistry in Two-Dimensions*; Cornell University Press: Ithaca, New York, 1981.

4. Shilov, A.E., *Activation of Saturated Hydrocarbons by Transition Metal Complexes*; D. Riedel Publishing Co: Dordrecht; 1984.

5. Green, M.L.H.; and O'Hare, D. *Pure Appl. Chem.*, 1986, 57, 1, 897.

6. Crabtree, R.H.; *Chem. Rev.*, 1985, 85, 245 and refs therein.

7. Watson, P.L. *J. Am. Chem. Soc.*, 1983, 105, 6491.

8. Hoyano, J.K,; McMaster, A.D.; and Graham, W.A.G. *J. Am. Chem.Soc.*, 1983, 105, 7190.

9. Fendrick, C.M.; and Marks, T.J. *J.Am.Chem.Soc.*, 1984, 106, 2214.

10. Blackborrow, J.R., and Young, D. In *Metal Vapor Synthesis in Organometallic Chemistry*; Springer Verlag: Berlin, 1979.

11. Klabunde, K.J. In *Chemistry of Free Atoms and Particles*;Academic Press: New York, 1980.

12. Green, M.L.H. U.K. Patents 5953/74 and 53675/74, U.S. Patent 7,182,742,1980;

13. Green,M.L.H.; and Hammond, V.J. U.S. Patent 4,182,749, 1980.

14. Green, M.L.H. In *Frontiers in chemistry*; Laidler, K.J., Ed.;Pergamon Press: Oxford, 1982; p229.

15. Cloke, F.G.N.; and Green, M.L.H. *J. Chem. Soc., Dalton Trans.*, 1981, 1938;

16. Davis,S.C.; and Klabunde, K.J. *J. Am. Chem. Soc.*, 1978, 100, 5975.

17. Remick, R.J.; Asunta, T.A.; and Skell, P.J. *J. Am. Chem. Soc.*, 1979, 101, 1320.

18. Bymater, S.; and Steacie, E.W.R. *J.Chem.Phys.*, 1951, 19, 172.

19. Billups, W.E.; Konarski, M.M.; Hauge, R.H.; and Margrave, J.L. *J. Am. Chem. Soc.*, 1980, 102, 7393.

20. Ozin, G.A.; McIntosh, D.F.; and Mitchell, S.A. *J. Am. Chem. Soc.*, 1981, 103, 1574.

21. Green, M.L.H. *J. Organomet. Chem.*, 1980, 200, 119.

22. Cloke, F.G.N; Derome, A.E.; Green, M.L.H.; and O'Hare, D. <u>J. Chem. Soc., Chem Commun</u>., 1983, 1312.

23. Bandy, J.A.; Cloke, F.G.N.; Green, M.L.H.; and O'Hare, D. <u>J. Chem. Soc., Chem Commun</u>., 1984, 240.

24. Burk, M.J.; Crabtree, R.H.; Parnell, C.P.; and Uriate, R.J. <u>Organometallics</u>, 1984, <u>3</u>, 816.

25. Green, M.L.H.; O'Hare, D.; and Wallis, J.M. <u>J. Chem. Soc.</u>, <u>Chem Commum</u>., 1984, 233.

26. Green, M.L.H.; O'Hare, D.; and Parkin, G. <u>J. Chem. Soc., Chem Commun.</u>, 1985, 356.

27. Green, M.L.H.; and Parkin, G. <u>J. Chem. Soc., Chem Commun</u>., 1984, 1467.

28. Bandy, J.A.; Green, M.L.H.; O'Hare, D. <u>J. Chem. Soc., Chem Commun.</u>, 1984, 1402.

29. Green, M.L.H.; and O'Hare, D. <u>J. Chem. Soc., Chem Commun</u>., 1985, 355.

30. Avery, N.R.; and Sheppard, N. <u>Proc. R. Soc. Lond.A</u> 1986, <u>405</u>, 1.

RECEIVED November 3, 1986

Chapter 17

Highly Reactive Organosamarium Chemistry via Metal Vapor and Sm(II) Syntheses

William J. Evans

Department of Chemistry, University of California—Irvine, Irvine, CA 92717

The reactivity of the lanthanide metals with unsaturated hydrocarbons has been studied via the metal vapor method in which lanthanide vapor at 400–1500°C is cocondensed with an organic substrate at -196°C. By using this high energy process on a synthetic scale, unusual types of complexes were produced which demonstrated new avenues of reactivity for the lanthanide metals and for lanthanide complexes in general. One of these processes, oxidation addition of C–H to a lanthanide metal center, was further examined using samarium vapor and C_5Me_5H and generated two new highly reactive molecules, $(C_5Me_5)_2Sm$ and $(C_5Me_5)_2Sm(THF)_2$. These species, the first soluble organometallic complexes of Sm(II), are reactive due to the large reduction potential of Sm(II) and have an extensive chemistry with CO and unsaturated organic substrates.

One of the important high energy processes which has served to advance organometallic transition metal chemistry is the metal vapor reaction (1–5). This technique involves heating a metal in an evacuated reaction vessel until it starts to vaporize and then condensing the vapor at low temperature (-100 → -196°C) with an organic reagent. The low temperature matrix which is formed is sometimes studied directly by spectroscopic means. In other cases, the matrix is allowed to warm and the metal vapor reaction products can be isolated on a preparative scale. Metal vapor reactions with organic substrates often have provided new organometallic complexes, intermediates, and reaction pathways which were previously inaccessible via traditional solution reactions.

The metal vapor method also has played an important role in the development of organometallic lanthanide chemistry (6–10). This high energy technique demonstrated that the lanthanide metals had a much greater range of organometallic chemistry than had been assumed previously. The metal vapor technique applied to lanthanides identified reasonable new research goals which could subsequently be pursued by solution techniques. Not only the metal vapor reactions,

0097-6156/87/0333-0278$06.00/0
© 1987 American Chemical Society

but also some of their high energy products have provided access to new areas of organolanthanide chemistry. In this chapter, the utility of metal vapor chemistry in the organolanthanide field is discussed as well as the high reactivity achievable from soluble Sm(II) complexes, compounds which were first isolated by metal vapor techniques.

Metal Vapor Reactions

One major area of organolanthanide chemistry which was advanced by the metal vapor technique involved unsaturated hydrocarbons. Prior to 1975, unsaturated hydrocarbons were not seriously considered as viable substrates/ligands for the lanthanides (11). These soft base ligands were not expected to interact strongly with the lanthanides, which traditionally were viewed as highly ionic, hard acid species. The basis for that view was the limited radial extension of the 4f valence orbitals (12). The 4f orbitals are not nearly as spatially available for covalent metal ligand orbital overlap as the d valence orbitals are in transition metals. Since unsaturated hydrocarbons are very important substrates in organometallic chemistry (13), their absence in organolanthanide chemistry severely limited the scope of the field.

Metal vapor chemistry showed that the lanthanides had quite an extensive chemistry with unsaturated hydrocarbons. Some of the early surveys of metal vapor reactions with unsaturated hydrocarbons included some lanthanide metals and showed that reactivity was present for these metals (14–18). Subsequent synthetic studies in which the products were isolated and characterized led to some of the most unusual organolanthanide complexes currently known (19–28).

Reactions (1)–(4) illustrate some of the unusual features of these complexes (9, 19–21, 25):

$$Ln + H_2C=CH-CH=CH_2 \rightarrow Ln(C_4H_6)_3 \qquad Nd,Sm,Er \qquad (1)$$

$$Ln + H_2C=C(CH_3)-C(CH_3)=CH_2 \rightarrow Ln(C_6H_{10})_2 \qquad La,Nd,Sm,Er \qquad (2)$$

$$Ln + CH_3CH_2C\equiv CCH_2CH_3 \rightarrow Ln_2(C_6H_{10})_3 \qquad Nd,Er \qquad (3)$$

$$Ln + CH_3CH_2C\equiv CCH_2CH_3 \rightarrow Ln(C_6H_{10}) \qquad Sm,Yb \qquad (4)$$

In each case, the metal was vaporized at temperatures ranging from 500–1600°C depending on the specific lanthanide involved and the vapor was cocondensed with the unsaturated hydrocarbon at -196°C. The product was isolated on a preparative scale in an inert atmosphere glove box and the product formula was determined by complete elemental analysis. Yields varied depending on the specific metal/ligand combination, but as much as 2-3 g of isolated product were obtained in some of the systems.

The metal vapor reaction products differed from traditional organolanthanide complexes in many ways. First, the observed stoichiometries had low ligand to metal ratios. For example, the ytterbium and samarium 3-hexyne products (Reaction 4) had formal ligand to metal ratios of one, whereas most organolanthanides are commonly nine or ten coordinate (6–10). Second, the stoichiometries varied in an unusual manner depending on ligand and metal. For example, 2,3-dimethyl substitution of the butadiene ligand changed the ligand to metal ratio from three (Reaction 1) to two (Reaction 2). In the 3-hexyne system, changing from neodymium and erbium to ytterbium and samarium changed the ratio from 1.5 (Reaction 3) to one (Reaction 4). Traditionally, the stoichiometries of organolanthanide complexes are invariant to minor substitutional changes on the ligand and are similar for metals of similar size (10). Third, in contrast to traditional organolanthanides, which have pale colors (11), the metal vapor products were intensely colored materials which displayed strong charge transfer absorptions in the near infrared and visible regions. Fourth, the room temperature magnetic moments for these complexes were often outside the range of "free ion" values previously reported for organolanthanide compounds (11,19). Consistent with this, the NMR spectra of the La, Sm, and Lu products were broad and uninformative whereas these metals, when trivalent, provide sharp interpretable spectra. The lanthanum butadiene product displayed an EPR absorption (La^{3+} is $4f^0$). Finally, the solution behavior of these complexes was unusual. For example, the 3-hexyne product, $[ErC_9H_{15}]_n$, was dimeric in arenes, i.e. n=2 (e.g. $Er_2(C_6H_{10})_3$), but in concentrated solution or in THF it was highly associated with n>10. This is just opposite the trend found for traditional organolanthanides which are more highly associated in arenes than in THF (29,30). Unfortunately, because these metal vaporization products oligomerized rather than crystallized in concentrated solution, these species were not structurally characterized by X-ray diffraction.

Although the structures of these species were not determined, this metal vapor chemistry clearly showed that unsaturated hydrocarbons were viable reagents for lanthanides. Furthermore, this high energy technique showed that new regimes of organolanthanide complexes were accessible under the appropriate conditions. In addition, attempts to understand the synthesis of the products in reactions 1-4 led to new metal vapor syntheses and new organolanthanide chemistry.

Analysis of possible structures and reaction pathways in reactions 1-4 led to various model structures for these complexes (9,25). Some of these involved C-H activation of the substituents attached to the unsaturated carbon atoms. To test the validity of these models, two additional types of metal vapor reactions were examined. In one case, reactions with simpler unsubstituted hydrocarbons were examined. In another case, substrates ideally set up for oxidative addition of C-H to the metal center were examined. As described in the following paragraphs, both of these approaches expanded the horizons of organolanthanide chemistry.

The simple hydrocarbon substrates included ethene, 1,2-propadiene, propene and cyclopropane (22). Their reactivity with Sm, Yb and Er was surveyed. In contrast to the reactions discussed above, lanthanide metal vapor reactions with these smaller hydrocarbons did not provide soluble products (with the exception of the erbium propene product, $Er(C_3H_6)_3$). Information on reaction pathways had to be obtained primarily by analyzing the products of hydrolysis of the metal vapor reaction product.

The analytical data on these small hydrocarbon lanthanide systems indicated that a variety of reactions were occurring. Evidence was observed for alkene insertion into Ln-C bonds, polymerization, C-C bond breaking, dehydrogenation, and oxidative addition of C-H. One study (22) on this topic concluded by saying that it had "defined a set of conditions under which a variety of hydrocarbon activation reactions take place in the presence of the lanthanide metals. Obviously, the challenge in this area is to control this reactivity so that it can be used selectively." These reactions defined new goals for organolanthanide solution chemistry, which was then in the process of being developed. High reactivity of the type found in these metal vapor reactions has been observable in solution and, by choosing the appropriate system, some selectivity can be obtained. The most prominent example of this is the chemistry of the $[(C_5Me_5)_2LuZ]_x$ systems ($Z=CH_3$, H; x=1,2) (31) which are very active polymerization initiators (32,33) and which are such powerful metalation reagents that they can activate C-H bonds in methane (34,35).

Two unsaturated hydrocarbon substrates containing acidic hydrogen were examined by metal vapor methods to deliberately probe C-H oxidative addition reactivity. Terminal alkynes react with Sm, Er and Yb to form alkynide hydride species, the type of product expected from oxidative addition, but a single crystalline product was not isolated (23). The reaction of pentamethylcyclopentadiene, C_5Me_5H, with samarium provided a clearer example of oxidative addition of C-H. In this case, evidence for the direct oxidative addition product, the divalent complex $[(C_5Me_5)SmH(THF)_2]$, was obtained (24). This species was too unstable for full characterization but did provide the desired information on C-H activation. In addition, this species led, via its decomposition, to a new complex, $(C_5Me_5)_2$-$Sm(THF)_2$, which has proven to be extremely important in organolanthanide chemistry. $(C_5Me_5)_2Sm(THF)_2$ was the first soluble organosamarium(II) complex. Due to the large reduction potential of Sm(II), this compound is a high energy species in its own right and its chemistry is discussed in a subsequent section. Also isolated from the Sm/C_5Me_5H reaction was the unsolvated $(C_5Me_5)_2Sm$ (27). As discussed in the Sm(II) section, this molecule has a rather unusual structure. Both of these samarium(II) complexes have considerably extended the scope of organolanthanide chemistry and both were a direct result of metal vapor studies.

In addition to broadening organolanthanide chemistry by demonstrating new reactivity patterns and providing new complexes, the lanthanide metal vapor studies have demonstrated new catalytic activity for these metals. In the course of characterizing the lanthanide metal vapor 3-hexyne products, it was discovered that these complexes had the capacity to initiate catalytic hydrogenation of alkynes to alkenes and alkanes (20,21,24). A variety of lanthanide metal vapor products were found to generate catalytic hydrogenation systems including the products of lanthanide metal cocondensations with 1-hexyne, 2,3-dimethylbutadiene, $(CH_3)_3SiC{\equiv}CSi(CH_3)_3$ and $P(C_6H_5)_3$ (26). Using 3-hexyne as a substrate, these catalytic systems generally gave high yields of cis-3-hexene; many had >95% stereospecificity.

Although catalytic hydrogenation of alkynes can be accomplished in many other ways, this catalytic system was significant because it was the first time an f element complex had been observed to catalytically activate hydrogen in homogeneous solution. The result had additional importance because the metal hydride complexes presumably involved in the catalyses provided the first evidence for the existence of discrete molecular lanthanide hydrides. The existence of lanthanide hydride complexes subsequently was demonstrated crystallographically (36) and fully characterized lanthanide hydrides were shown to have catalytic activity in hydrogenation reactions (37). Subsequent studies of other organolanthanide hydride systems have identified complexes with very high levels of catalytic hydrogenation activity (38).

Organosamarium(II) Chemistry

Of the three readily accessible divalent lanthanide ions, Eu(II), Yb(II), and Sm(II), samarium is the most strongly reducing. A reduction potential of about -1.5V vs NHE is generally assigned to the Sm(III)/ Sm(II) couple (39). Despite this high reactivity, organosamarium(II) chemistry had not been examined because the known systems $[(C_5H_5)_2Sm(THF)]_n$ (40) and $[(CH_3C_5H_4)_2Sm]_n$ (41) were insoluble. The Sm/C_5Me_5H metal vapor reaction changed this situation and provided $(C_5Me_5)_2Sm(THF)_2$ (24) and $(C_5Me_5)_2Sm$ (27,42) as well as their C_5Me_4Et analogs (27). These complexes are soluble species which have an extensive, high energy reducing chemistry.

Structurally, $(C_5Me_5)_2Sm(THF)_2$ had the bent metallocene arrangement expected (43) for a dicyclopentadienyl complex with two additional ligands (44) (Figure 1a). In contrast, $(C_5Me_5)_2Sm$, the first structurally-characterized direct f element analog of ferrocene, had an unusual structure (42) (Figure 1b). Steric and simple electrostatic arguments and even a molecular orbital analysis (45) predicted that $(C_5Me_5)_2Sm$ would be structurally similar to $(C_5Me_5)_2Fe$ (46) (Figure 1c), i.e., it would have parallel cyclopentadienyl rings. It did not. The ring centroid-Sm-ring centroid angle of 140.1° was very close to the 136.5° value observed in $(C_5Me_5)_2$-Sm(THF)$_2$. $(C_5Me_5)_2Eu$ (47) and $(C_5Me_5)_2Yb$ (Haaland, A.; Andersen, R. A., personal communication, 1986) are isomorphous with $(C_5Me_5)_2Sm$ in the solid state and the ytterbium complex is also bent in the gas phase (48). The origin and chemical consequences of this bent structure presently are under investigation.

The $(C_5Me_5)_2Sm(THF)_2$ metal vapor product provided the first opportunity to see if Sm(II) complexes ($\mu = 3.5$-3.8 μ_B) could be characterized by ^1H NMR spectroscopy (24). Fortunately, the paramagnetism doesn't cause large shifting and broadening of the resonances and hence samarium provides the only Ln(III)/Ln(II) couple in which both partners are NMR accessible. Once the existence and identity of $(C_5Me_5)_2Sm(THF)_2$ was known, a solution synthesis was developed from KC_5Me_5 and $SmI_2(THF)_x$ (44). This system is the preferred preparative route and also provides another soluble organosamarium(II) complex, $[(C_5Me_5)Sm(THF)_2(\mu-I)]_2$, under appropriate conditions. This is another example of how solution studies subsequently catch up to the research targets often identified first in metal vapor reactions.

One of the best examples of the high reactivity achievable via organosamarium(II) chemistry is the reaction of $(C_5Me_5)_2Sm(THF)_2$ with CO (49). This compound reacts under a variety of conditions with CO to give complex mixtures of products. Under 90 psi of CO in THF, a crystallographically characterizable product separates (Reaction 5).

The remarkable dimeric product contains four Sm(III) centers and the equivalent of six CO molecules, i.e., six CO molecules have been

$$4(C_5Me_5)_2Sm(THF)_2 + 6CO \rightarrow \qquad\qquad (5)$$

reduced by four electrons as they are homologated into two ketene-carboxylate units.

The most interesting feature of this CO reduction is that complete cleavage of one CO triple bond must have occurred to give the central, oxygen-free carbon of the C=C-C skeleton. Although CO cleavage is thought to occur in heterogeneous Fischer-Tropsch systems (50-55), this rarely occurs in homogeneous systems. Complete CO cleavage has not been observed in the extensive studies of CO reduction by molecular early transition metal (56,57) actinide (58), and lanthanide (59) hydride systems.

The literature system most closely related to equation 5 is the two electron reduction of two CO molecules by alkali metals which is reported to form an insoluble alkyne diolate (Equation 6) (60). The relationship between reactions 5 and 6 can be seen by comparing

$$2K + 2CO \rightarrow 2[KOC\equiv COK] \qquad\qquad (6)$$

KOCCOK with the atom connectivity on the right hand (or left hand) side of the ketenecarboxylate dimer, SmO(CO)CCOSm. Both the K and Sm systems contain MOCCOM units involving two electron reduction of two CO molecules. The samarium system, a two electron/three CO system, has an extra CO formally "inserted" between a C and O bond. The insolubility of the KOC≡COK product may be responsible for the lack of further homologation in this system compared to the samarium system. Hence, $(C_5Me_5)_2Sm(THF)_2$, although strongly reducing like the alkali metals, is differentiated from those reducing agents by its solubility. The C_5Me_5 ligands provide solubility to the reducing agent as well as to intermediate reduction products and allow $(C_5Me_5)_2Sm(THF)_2$ to accomplish more than is possible with the alkali metals.

The reactivity of $(C_5Me_5)_2Sm(THF)_2$ with CO can also be compared with that of $(C_5Me_5)_2Ti$. Both organometallic reagents are soluble, strongly reducing complexes of oxophilic metals. As shown in reaction 7, decamethyltitanocene forms a carbonyl complex rather than reducing the CO (61,62). Since the 4f valence orbitals of Sm(II) are not as suitable for carbonyl complex formation as those of Ti(II), this is less likely for samarium and reduction occurs.

$$(C_5Me_5)_2Ti + 2CO \quad (C_5Me_5)_2Ti(CO)_2 \tag{7}$$

Considering the comparisons above and the fact that neither $(C_5Me_5)_2Yb(R_2O)$ (63,64) nor $(C_5Me_5)_2Eu(THF)$ react with CO, $(C_5Me_5)_2Sm(THF)_2$ appears to be a unique reducing agent in the periodic table. The combination of strong reducing power, oxophilicity, solubility, and lack of d valence orbitals is not duplicated by any other reducing agent presently available.

$(C_5Me_5)_2Sm(THF)_2$ also shows high reactivity in its reactions with $C_6H_5C\equiv CC_6H_5$ and $C_6H_5N=NC_6H_5$ and the subsequent reactions of these products with CO. The $C_6H_5C\equiv CC_6H_5$ reaction (65), which forms $[(C_5Me_5)_2Sm]_2C_2(C_6H_5)_2$ (Equation 8) may have a parallel in alkali metal chemistry, except that in the heterogeneous alkali metal systems, polymerization or dimerization of the intermediate radical

$$(C_5Me_5)_2Sm(THF)_2 + C_6H_5C\equiv CC_6H_5 \quad [(C_5Me_5)_2Sm](C_6H_5)C=\overset{\bullet}{C}(C_6H_5)$$

$$\xrightarrow{(C_5Me_5)_2Sm(THF)_2} \quad \begin{matrix} (C_5Me_5)_2Sm \\ \diagdown \\ C_6H_5 \end{matrix} C=C \begin{matrix} C_6H_5 \\ \diagup \\ Sm(C_5Me_5)_2 \end{matrix} \tag{8}$$

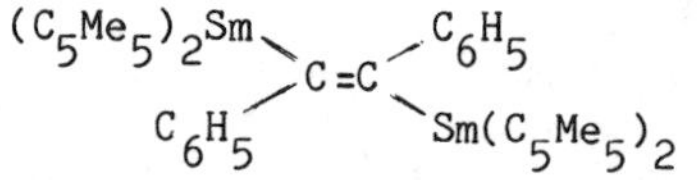

generally occurs (66). The more soluble $(C_5Me_5)_2Sm(THF)_2$ may more readily trap the radical to give the bimetallic product. $[(C_5Me_5)_2Sm]_2C_2(C_6H_5)_2$ is unusual in that the room temperature magnetic susceptibility indicates that it contains Sm(III) centers whereas its black color differs from virtually all other Sm(III) complexes which are pale orange, yellow, or red.

$[(C_5Me_5)_2Sm]_2C_2(C_6H_5)_2$ reacts with CO in a remarkably facile stereospecific synthesis of a tetracyclic hydrocarbon (67) (Equation 9).

$$2(C_5Me_5)_2Sm(THF)_2 + C_6H_5C\equiv CC_6H_5 \xrightarrow{-4 \ THF}$$

$$[(C_5Me_5)_2Sm]_2C_2(C_6H_5)_2 \xrightarrow{2 \ CO}$$

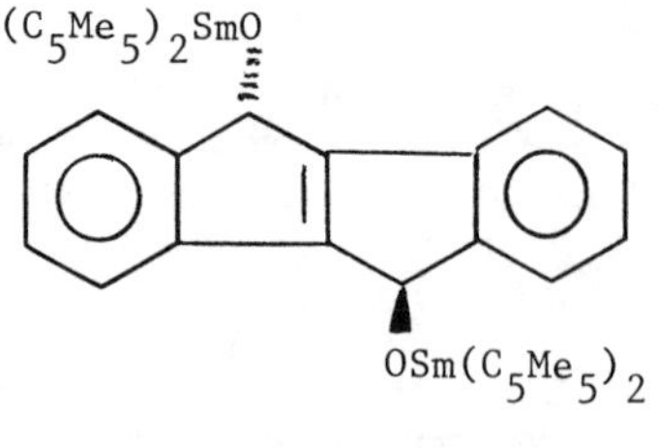

$(C_5Me_5)_2SmO$

$OSm(C_5Me_5)_2$

$$\tag{9}$$

Figure 2 shows one possible explanation of the CO and C—H activation found in this synthesis. Activation of CO by insertion into the Sm—C

The remarkable dimeric product contains four Sm(III) centers and the equivalent of six CO molecules, i.e., six CO molecules have been

$$4(C_5Me_5)_2Sm(THF)_2 + 6CO \rightarrow \qquad\qquad (5)$$

reduced by four electrons as they are homologated into two ketene-carboxylate units.

The most interesting feature of this CO reduction is that complete cleavage of one CO triple bond must have occurred to give the central, oxygen-free carbon of the C=C–C skeleton. Although CO cleavage is thought to occur in heterogeneous Fischer–Tropsch systems (50–55), this rarely occurs in homogeneous systems. Complete CO cleavage has not been observed in the extensive studies of CO reduction by molecular early transition metal (56,57) actinide (58), and lanthanide (59) hydride systems.

The literature system most closely related to equation 5 is the two electron reduction of two CO molecules by alkali metals which is reported to form an insoluble alkyne diolate (Equation 6) (60). The relationship between reactions 5 and 6 can be seen by comparing

$$2K + 2CO \rightarrow 2[KOC\equiv COK] \qquad\qquad (6)$$

KOCCOK with the atom connectivity on the right hand (or left hand) side of the ketenecarboxylate dimer, SmO(CO)CCOSm. Both the K and Sm systems contain MOCCOM units involving two electron reduction of two CO molecules. The samarium system, a two electron/three CO system, has an extra CO formally "inserted" between a C and O bond. The insolubility of the KOC≡COK product may be responsible for the lack of further homologation in this system compared to the samarium system. Hence, $(C_5Me_5)_2Sm(THF)_2$, although strongly reducing like the alkali metals, is differentiated from those reducing agents by its solubility. The C_5Me_5 ligands provide solubility to the reducing agent as well as to intermediate reduction products and allow $(C_5Me_5)_2Sm(THF)_2$ to accomplish more than is possible with the alkali metals.

The reactivity of $(C_5Me_5)_2Sm(THF)_2$ with CO can also be compared with that of $(C_5Me_5)_2Ti$. Both organometallic reagents are soluble, strongly reducing complexes of oxophilic metals. As shown in reaction 7, decamethyltitanocene forms a carbonyl complex rather than reducing the CO (61,62). Since the 4f valence orbitals of Sm(II) are not as suitable for carbonyl complex formation as those of Ti(II), this is less likely for samarium and reduction occurs.

$$(C_5Me_5)_2Ti + 2CO \quad (C_5Me_5)_2Ti(CO)_2 \qquad (7)$$

Considering the comparisons above and the fact that neither $(C_5Me_5)_2Yb(R_2O)$ (63,64) nor $(C_5Me_5)_2Eu(THF)$ react with CO, $(C_5Me_5)_2Sm(THF)_2$ appears to be a unique reducing agent in the periodic table. The combination of strong reducing power, oxophilicity, solubility, and lack of d valence orbitals is not duplicated by any other reducing agent presently available.

$(C_5Me_5)_2Sm(THF)_2$ also shows high reactivity in its reactions with $C_6H_5C\equiv CC_6H_5$ and $C_6H_5N=NC_6H_5$ and the subsequent reactions of these products with CO. The $C_6H_5C\equiv CC_6H_5$ reaction (65), which forms $[(C_5Me_5)_2Sm]_2C_2(C_6H_5)_2$ (Equation 8) may have a parallel in alkali metal chemistry, except that in the heterogeneous alkali metal systems, polymerization or dimerization of the intermediate radical

$$(C_5Me_5)_2Sm(THF)_2 + C_6H_5C\equiv CC_6H_5 \quad [(C_5Me_5)_2Sm](C_6H_5)C=\overset{\bullet}{C}(C_6H_5)$$

$$\xrightarrow{\quad (C_5Me_5)_2Sm(THF)_2 \quad} \quad \begin{array}{c} (C_5Me_5)_2Sm \\ \diagdown \\ C_6H_5 \end{array} C=C \begin{array}{c} C_6H_5 \\ \diagup \\ Sm(C_5Me_5)_2 \end{array} \qquad (8)$$

generally occurs (66). The more soluble $(C_5Me_5)_2Sm(THF)_2$ may more readily trap the radical to give the bimetallic product. $[(C_5Me_5)_2Sm]_2C_2(C_6H_5)_2$ is unusual in that the room temperature magnetic susceptibility indicates that it contains Sm(III) centers whereas its black color differs from virtually all other Sm(III) complexes which are pale orange, yellow, or red.

$[(C_5Me_5)_2Sm]_2C_2(C_6H_5)_2$ reacts with CO in a remarkably facile stereospecific synthesis of a tetracyclic hydrocarbon (67) (Equation 9).

$$2(C_5Me_5)_2Sm(THF)_2 + C_6H_5C\equiv CC_6H_5 \xrightarrow{\quad -4 \text{ THF} \quad}$$

$$[(C_5Me_5)_2Sm]_2C_2(C_6H_5)_2 \xrightarrow{\quad 2 \text{ CO} \quad}$$

$$\qquad (9)$$

Figure 2 shows one possible explanation of the CO and C–H activation found in this synthesis. Activation of CO by insertion into the Sm–C

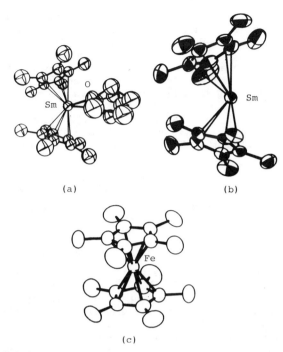

Figure 1. Side view of the molecular structures of (a) $(C_5Me_5)_2$-$Sm(THF)_2$. (b) $(C_5Me_5)_2Sm$, and (c) $(C_5Me_5)_2Fe$. All unlabelled atoms are carbon atoms. Only one oxygen is visible in this view of (a). (Adapted from references 42, 44 and 46).

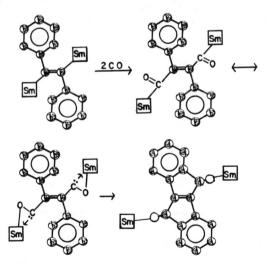

Figure 2. A possible sequence for the reaction of $C_6H_5C\equiv CC_6H_5$, $(C_5Me_5)_2Sm(THF)_2$ and CO. $(C_5Me_5)_2Sm$ is shown as Sm in a box.

bonds could give two dihaptoacyl units with considerable Sm-O interaction and carbene character on the acyl carbon atoms (68). Insertion of these carbene-like centers into ortho C-H bonds would give
the two five membered rings. If this synthesis is generally applicable, it would be a valuable way to make polycyclic hydrocarbons in
a stereospecific manner from simple starting materials, namely
alkynes and CO, using the high energy activation potential of Sm(II).

$(C_5Me_5)_2Sm(THF)_2$ reacts with $C_6H_5N=NC_6H_5$ to form another trivalent complex with an unusual color, $[(C_5Me_5)_2Sm]_2N_2(C_6H_5)_2$ (Equation 10). This complex also has an unusual structure (Evans, W. J.;
Drummond, D. K.; Bott, S. G.; Atwood, J. L. Organometallics, in
press).

$$2(C_5Me_5)_2Sm(THF)_2 + C_6H_5N=NC_6H_5 \qquad \begin{array}{c} (C_5Me_5)_2Sm \diagdown \quad \diagup C_6H_5 \\ N=N \\ \diagup \quad \diagdown \\ C_6H_5 \quad Sm(C_5Me_5)_2 \end{array} + 4\ THF \qquad (10)$$

The Sm-N distances are consistent with single σ bonds suggesting the
presence of a $(C_6H_5NNC_6H_5)^{2-}$ dianion. However, the 1.25(1) Å NN
distance is the same as the N=N double bond distances in a variety of
azobenzene structures. The molecule also contains distorted N-C
(phenyl) distances, which are stretched from a normal 1.42 Å distance
to 1.56-1.61 Å. In addition, the samarium atoms are displaced asym-
metrically such that both come within 2.29-2.34 Å of the ortho hydrogens of a single phenyl ring in a bonafide agostic (69) Sm-H interaction. The ability of Sm(II) to structurally distort azobenzene in
this way is remarkable.

Like $[(C_5Me_5)_2Sm]_2C_6(C_6H_5)_2$, $[(C_5Me_5)_2Sm]_2N_2(C_6H_5)_2$ reacts with
two molecules of CO. In this case, however, a different result is
obtained (Equation 11; Evans, W.J. Drummond, D.K. J. Am. Chem. Soc. in
press).

$$\begin{array}{c} (C_5Me_5)_2Sm \diagdown \\ N=N \\ \diagdown \\ Sm(C_5Me_5)_3 \end{array} + 2CO \rightarrow (C_5Me_5)_2Sm \begin{array}{c} C_6H_5 \\ | \\ N \\ O \diagdown \diagup \diagdown C \\ \diagup \quad \diagdown \\ C \quad Sm(C_5Me_5)_2 \\ N \diagdown O \\ | \\ C_6H_5 \end{array} \qquad (11)$$

In equation 11, two CO molecules have been inserted into the NN bond,
which was the shortest bond in the precursor complex! The overall
transformation accomplished by treating azobenzene with $(C_5Me_5)_2$-
$Sm(THF)_2$ and then CO is shown in equation 12.

$$
\begin{array}{c}
R \\
\diagdown \\
N = N \\
\diagdown R \\
C \quad C \\
||| \quad ||| \\
O \quad O
\end{array}
\quad\longrightarrow\quad
\begin{array}{c}
R \\
\diagdown \\
N \quad O \\
| \quad | \\
C - C \\
| \quad | \\
O \quad N \\
\diagdown R
\end{array}
\qquad (12)
$$

NN and CO multiple bonds are broken and 2 C–N and one C–C bond are formed in an unusual four center bond-breaking and bond-making sequence. Once again, Sm(II) is providing access to unusual chemistry.

Summary
    The high energy process of metal vapor chemistry has affected organolanthanide chemistry in several ways. It has demonstrated the viability of unsaturated hydrocarbons as reagents and substrates in lanthanide systems and has shown that these metals can effect homogeneous catalytic transformations of small molecules. It has identified a wide variety of hydrocarbon activation reactions that are possible with lanthanides. It has provided soluble organosamarium(II) complexes which have their own unique high energy chemistry. As is often the case with other high energy processes, these results have revealed new possibilities in structure and reactivity. Once the viability of a research goal is demonstrated, lower energy (in this case, solution) methods often can be found to get the same or related results. Metal vapor chemistry clearly has had a major impact on organolanthanide chemistry in this way.

Acknowledgments
    I thank the National Science Foundation for support of this research and the Alfred P. Sloan Foundation for a Research Fellowship.

Literature Cited
1. Skell, P. S.; McGlinchey, M. J. Angew. Chem., Int. Ed. 1975, 14, 195-199.
2. Timms, P. L.; Turney, T. W. Adv. Organomet. Chem. 1977, 15, 53-112.
3. Klabunde, K. J. Chemistry of Free Atoms and Particles Academic Press: New York, 1980.
4. Blackborow, J. R.; Young, D. Metal Vapor Synthesis in Organometallic Chemistry Springer Verlag: Berlin, Germany, 1979.
5. Moskovits, M.; Ozin, G. A. Cryochemistry Wiley: New York, 1976.
6. Marks, T. J.; Ernst, R. D. In Comprehensive Organometallic Chemistry Wilkinson, G.; Stone, F. G. A.; Abel, E. W., Eds., Pergamon Press: 1982; Chapter 21.
7. Schumann, H. Angew. Chem. Int. Ed. Engl. 1984, 23, 474-493.
8. Forsberg, J. H.; Moeller, T. In Gmelin Handbook of Inorganic Chemistry 8th Ed. Sc, Y, La-Lu Part D6, Moeller, T.; Kruerke, U.; Schleitzer-Rust, E., eds.; Springer-Verlag: Berlin, 1983, p. 137-282.
9. Evans, W. J. J. Organomet. Chem. 1983, 250, 217-226.
10. Evans, W. J. Adv. Organomet. Chem. 1985, 24, 131-177.
11. Moeller, T. In Comprehensive Inorganic Chemistry Bailar, J. C., Jr. et. al. Ed.; Pergamon Press, Oxford, England, 1973; Vol. 4; Chapter 44.

12. Freeman, A. J.; Watson, R. E. Phys. Rev. 1976, 9, 217.
13. Collman, J. P.; Hegedus, L. S. Principles and Applications of
    Organotransition Metal Chemistry University Science Books: Mill
    Valley, CA, 1980.
14. Skell, P. S. Proc. Int. Cong. Pure Appl. Chem. 1971, 23, 215.
15. McGlinchey, M. J.; Skell, P. S. in ref. 5, chapter 5.
16. Lagowski, J. J. Chem. Eng. News 1973, 51, December, 26.
17. Lanthanide metal vapor studies with CO have also been done:
    Slater, J. L.; DeVore, T. C.; Calder, V. Inorg. Chem. 1973, 12,
    1918; 1974, 13, 1808.
18. Lanthanide metal vapor studies with acetylacetone have also been
    done: Blackborow, J. R.; Eady, C. R.; Koerner Von Gustorf, E.
    A.; Scrivanti, A.; Wolfbeis, O. J. Organomet. Chem. 1976, 108,
    C32-C34.
19. Evans, W. J.; Engerer, S. C.; Neville, A. C. J. Am. Chem. Soc.
    1978, 100, 331-333.
20. Evans, W. J.; Engerer, S. C.; Piliero, P. A.; Wayda, A. L. In
    Fundamental Research in Homogeneous Catalysis; Tsutsui, M., Ed.;
    Plenum Press: New York, 1979; Vol. 3, pp 941-952.
21. Evans, W. J.; Engerer, S. C.; Piliero, P. A.; Wayda, A. L. J.
    Chem. Soc., Chem. Commun. 1979, 1007-1008.
22. Evans, W. J. Coleson, K. M.; Engerer, S. C. Inorg. Chem. 1981,
    20, 4320-4325.
23. Evans, W. J.; Engerer, S. C.; Coleson, K. M. J. Am. Chem. Soc.
    1981, 103, 6672-6677.
24. Evans, W. J.; Bloom, I.; Hunter, W. E.; Atwood, J. L. J. Am.
    Chem. Soc. 1981, 103, 6507-6508.
25. Evans, W. J. In The Rare Earths in Modern Science and Technology;
    McCarthy, G. J., Rhyne, J. J., Silber, H. E., Eds., Plenum Press:
    New York, 1982; Vol. 3, pp 61-70.
26. Evans, W. J.; Bloom, I.; Engerer, S. C. J. Catal. 1983, 84,
    468-476.
27. Evans, W. J.; Bloom, I.; Hunter, W. E.; Atwood, J. L. Organo-
    metallics 1985, 4, 112-119.
28. DeKock, C. W.; Ely, S. R.; Hopkins, T. E.; Brault, M. A. Inorg.
    Chem. 1978, 17, 625-631.
29. Holton, J.; Lappert, M. F.; Ballard, D. G. H.; Pearce, R.;
    Atwood, J. L.; Hunter, W. E. J. Chem. Soc. Dalton Trans. 1979,
    54-61.
30. Evans, W. J.; Dominguez, R.; Hanusa, T. P. Organometallics, 1986,
    5, 263-270.
31. Watson, P. L.; Parshall, G. W. Acc. Chem. Res. 1985, 18, 51-56
    and references therein.
32. Watson, P. L. J. Am. Chem. Soc. 1982, 104, 337-339.
33. Watson, P. L.; Roe, D. C. J. Am. Chem. Soc. 1982, 104, 6471-6473.
34. Watson, P. L. J. Chem. Soc., Chem. Commun. 1983, 276-277.
35. Watson, P. L. J. Am. Chem. Soc. 1983, 105, 6491-6493.
36. Evans, W. J.; Meadows, J. H.; Wayda, A. L.; Hunter, W. E.;
    Atwood, J. L. J. Am. Chem. Soc. 1982, 104, 2008-2014.
37. Evans, W. J.; Meadows, J. H.; Hunter, W. E.; Atwood, J. L. J. Am.
    Chem. Soc. 1984, 106, 1291-1300.
38. Jeske, G.; Lauke, H.; Mauermann, H.; Schumann, H.; Marks, T. J.
    J. Am. Chem. Soc. 1985, 107, 8111-8118.
39. Morss, L. R. Chem. Rev. 1976, 76, 827-841.
40. Watt, G. W.; Gillow, E. W. J. Am. Chem. Soc. 1969, 91, 775-776.

41. Evans, W. J. In The Chemistry of The Metal-Carbon Bond Hartley, F. R., Patai, S., Eds., Wiley: New York, 1982, Chapter 12.
42. Evans, W. J.; Hughes, L. A.; Hanusa, T. P. J. Am. Chem. Soc. 1984, 106, 4270-4272.
43. Lauher, J. W.; Hoffmann, R. J. Am. Chem. Soc. 1976, 98, 1729-1742.
44. Evans, W. J.; Grate, J. W.; Choi, H. W.; Bloom, I.; Hunter, W. E.; Atwood, J. L. J. Am. Chem. Soc. 1985, 107, 941-946.
45. Ortiz, J. V.; Hoffmann, R. Inorg. Chem. 1985, 24, 2095-2104.
46. Freyberg, D.P.; Robbins, J.L.; Raymond, K.N.; Smart, J.C. J. Am. Chem. Soc. 1979, 101, 892-897.
47. Evans, W. J.; Hughes, L. A.; Hanusa, T. P. Organometallics 1986, 5, 1285-1291.48. Andersen, R. A.; Boncella, J. M.; Burns, C. J.; Green, J. C.; Hohl, D.; Rosch, N. J. J. Chem. Soc. Chem. Commun. 1986, 405-407.
49. Evans, W. J.; Grate, J. W.; Hughes, L. A.; Zhang, H.; Atwood, J. L. J. Am. Chem. Soc. 1985, 107, 3728-3730.
50. Storch, H. H.; Golumbic, N.; Anderson, R. B. The Fischer-Tropsch Reaction and Related Syntheses Wiley: New York, 1951.
51. Ponec, V. Catal. Rev.-Sci. Eng. 1978, 18, 1515-1571.
52. Masters, C. Adv. Organomet. Chem. 1979, 17, 61-103.
53. Kung, H. H. Catal. Rev.-Sci. Eng. 1980, 22, 235-259.
54. Klier, K. Adv. Catal. 1982, 31, 243-313.
55. Dombek, B. D. Adv. Catal. 1983, 322, 325-416.
56. Wolczanski, P. T.; Bercaw, J. E. Accts. Chem. Res. 1980, 13, 121-127 and references therein.
57. Erker, G. Acc. Chem. Res. 1984, 17, 103-109 and references therein.
58. Katahira, D. A.; Moloy, K. G.; Marks, T. J. Organometallics 1982, 1, 1723-1726 and references therein.
59. Evans, W. J.; Grate, J. W.; Doedens, R. J. J. Am. Chem. Soc. 1985, 107, 1671-1679.
60. Buchner, W. Chem. Ber. 1966, 99, 1485-1492. Buchner, W. Helv. Chim. Acta 1963, 46, 2111-2120.
61. Bercaw, J. E.; Marvich, R. H.; Bell, L. G.; Brintzinger, H. H. J. Am. Chem. Soc. 1972, 94, 1219-1238.
62. Bottrill, M.; Gavens, P. D.; McMeeking, J. In Comprehensive Organometallic Chemistry; Wilkinson, G. W., Stone, F. G. A.; Abel, E. W., Eds., Pergamon Press: Oxford, 1982; Chapter 22.2.
63. Watson, P. L. J. Chem. Soc., Chem. Commun. 1980, 652-653.
64. Boncella, J. M.; Tilley, T. D.; Andersen, R. A. J. Chem. Soc., Chem. Commun. 1984, 710-712.
65. Evans, W. J.; Bloom, I.; Hunter, W. E.; Atwood, J. L. J. Am. Chem. Soc. 1983, 105, 1401-1403.
66. House, H. O. Modern Synthetic Reactions; W. A. Benjamin: Menlo Park, CA, 1972; pp. 205-209 and references therein.
67. Evans, W. J.; Hughes, L. A.; Drummond, D. K.; Zhang, H.; Atwood, J. L. J. Am. Chem. Soc. 1986, 108, 1722-1723.
68. Evans, W. J.; Wayda, A. L.; Hunter, W. E.; Atwood, J. L. J. Chem. Soc., Chem. Commun. 1981, 706-708.
69. Brookhart, M.; Green, M. L. H. J. Organomet. Chem. 1983, 250, 395-408.

RECEIVED November 25, 1986

Chapter 18

Photon- and Ion-Beam-Induced Reactions in Metallo-organic Films: Microchemistry to Microelectronics

M. E. Gross[1], W. L. Brown[1], J. Linnros[1], and H. Funsten[1,2]

[1]AT&T Bell Laboratories, Murray Hill, NJ 07974
[2]School of Engineering and Applied Sciences, University of Virginia, Charlottesville, VA 22901

The photothermal chemistry of thin palladium acetate films when irradiated with continuous wave Ar^+ laser and the non-thermal chemistry induced by high energy ion irradiation (2 MeV He^+ and Ne^+ ions) have been studied. The photothermal decomposition induced by scanned laser irradiation leads to pure metal lines that, however, may exhibit pronounced periodic structure. This results from coupling of the rate of heating by absorption of the laser radiation with the rate of release of the heat of reaction, giving an "explosive" reaction front that propagates ahead of the laser beam. In contrast, He^+ ion beam irradiation produces smooth metallic-looking features that contain up to 20% of the original carbon and 5% of the original oxygen content of the film. Films irradiated with Ne^+ ions contained slightly lower amounts of carbon and oxygen residues, but the films' appearance varied with thickness.

The decomposition reactions of acetic acid and metal acetates in solution induced by various energetic sources have been a subject of considerable study. Electrolytic oxidation and reduction and alpha (He^+) and gamma radiolysis provide interesting mechanistic comparisons of ionic and radical pathways (1,2). Most of the work in the solid state has focused on the thermal decomposition of metal acetates (3-6). Questions relating to formation of ions and radicals and the mobility of these species through the solid *vis à vis* further reactions or escape from the solid are but some of the issues of interest. Metallo-organic compounds in general are of current interest as precursors for metallization in microelectronics fabrication (7). Thus, studies of the reaction chemistries of metallo-organic compounds will contribute to the rational design of new molecular systems for these applications.

Palladium acetate, $[Pd(\mu-O_2CCH_3)_2]_3$, possesses a unique quality that makes it attractive for solid state decomposition studies as well as technological applications. It can be spin-coated from solution to form a homogeneous, apparently amorphous solid film. This provides large uniform areas over which we can study the effects of various irradiation sources on the chemical nature of the film. The bulky structure of palladium acetate, shown in Figure 1 (8), may offer a partial explanation of the molecule's ability to achieve an amorphous metastable phase upon rapid evaporation of solvent.

0097-6156/87/0333-0290$06.00/0
© 1987 American Chemical Society

Palladium itself is of interest for its metallic conductivity as well as its ability to catalyze electroless plating of copper. A schematic of the solid state decomposition process that we use is shown in Figure 2. Note that the difference in density between metallo-organic precursor and resultant metal leads to considerable reduction in thickness of the irradiated area as organic matter is lost. This decrease in thickness occurs in both thermal and non-thermal decompositions. Previous work on laser-induced photothermal gold deposition from metallo-organic films revealed additional structure in the deposited metal that resulted from coupling of the energy absorbed from the laser with the exothermic heat of reaction to give various forms of periodic structure (9-11). In this paper we present results on the spatially localized thermal and non-thermal decomposition of palladium acetate films induced by continuous wave (cw) Ar$^+$ laser or ion beam irradiation, respectively.

Experimental

Palladium acetate, purchased from Aldrich Chemical Co., is spin-coated from chloroform solution onto quartz, silicon or polished beryllium substrates to thicknesses of 0.1-1.5 μm depending on solution concentration and spin speed. After exposure to the laser or ion beam, films are developed in chloroform to remove unirradiated material. Samples should be developed promptly as degradation of the palladium acetate films may occur upon standing, giving rise to particulate residue as seen in the background of Figure 3. All analyses of laser-exposed materials were carried out on developed samples. Analyses of ion beam-irradiated regions were performed on undeveloped samples for consistency with in situ measurements. In either case, adhesion of the irradiated area to the substrate was good. Localized photothermal decomposition of the films is effected by translating the sample relative to a 20W cw Ar$^+$ laser irradiation focused to 0.8 μm FWHM in single line mode (5145Å) or ~50 μm in multi-line mode (principally 5145 and 4880Å). All laser exposures were done in air.

Ion beam irradiations were carried out in vacuum with 2 MeV He$^+$ and Ne$^+$ ions from a 3.75 MeV Van de Graaff accelerator at current densities of ≤1 μA/cm^2 using currents of ≤1 μA. Elemental composition of the exposed film was determined by Rutherford backscattering spectroscopy (RBS) with 2 MeV He$^+$ ions. To minimize decomposition by the analyzing He$^+$ ion beam itself, particularly for the low dose data points, RBS spectra were accumulated as sums of many individual spectra on different areas of the films. In the RBS measurements, a beryllium substrate was used to avoid spectroscopic overlap of substrate backscattering signal with signals from the carbon and oxygen constituents of the palladium acetate film. The data were corrected for impurities present at the substrate surface (primarily BeO) before spin-coating. The ejection of neutral molecules from the films during MeV ion bombardment was measured with a quadrupole mass spectrometer in an ultrahigh vacuum system using electron impact ionization at an electron energy of 70 eV. Infrared spectra were recorded on a Perkin-Elmer 683 IR spectrometer. In situ IR spectra were smoothed using the Perkin-Elmer PE680 SMOOTH routine, calculating a quadratic polynomial fit of successive sets of 13 data points. Film thicknesses were measured using a Sloan Dektak stylus profilometer. Infrared, profilometer and mass spectrometric measurements were made on palladium acetate films on doubly polished silicon substrates.

Results and Discussion

Laser-Induced Chemistry

Photothermal decomposition of palladium acetate by scanned cw Ar$^+$ laser irradiation produces metal features that exhibit pronounced periodic structure as a function of laser power, scan speed, substrate and beam diameter, as shown in Figures 3 and 4. The periodic structure is a function of the rate at which the film is heated by absorption of the incident laser radiation coupled with the rate at which the heat of the decomposition reaction is liberated. This coupling generates a reaction front that outruns the scanning laser until quenched by thermal losses, the process to be repeated when the laser catches up and reaches unreacted material. Clearly, such a thermal process is also affected by the thermal conductivity of the substrate, the optical absorption of the substrate in those cases where the overlying film is not fully absorbing,

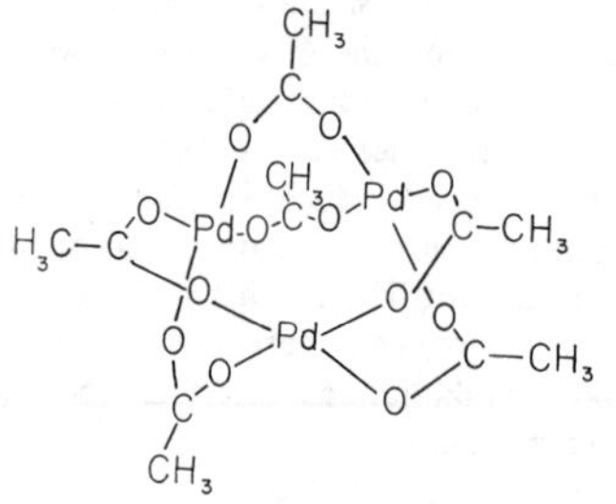

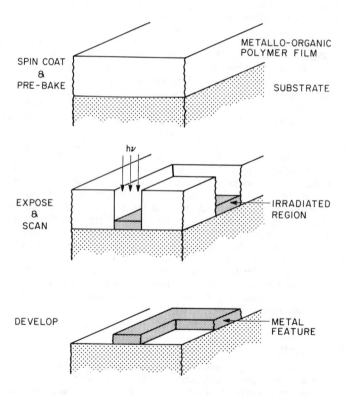

Figure 1. Structure of palladium acetate.

Figure 2. Schematic of direct-write process.

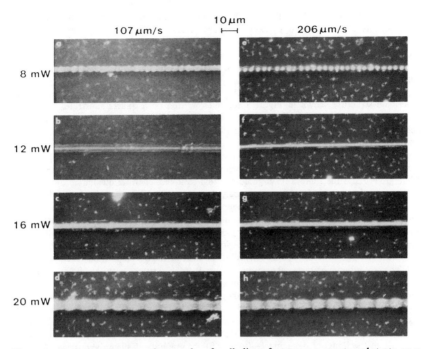

Figure 3. Scanning electron micrographs of palladium features on quartz substrate as a function of laser power (measured on target) and scan speed. Palladium acetate precursor film thickness is 1.5 μm (cw Ar$^+$ laser – 5145Å line, spot size ~0.8 μm FWHM).

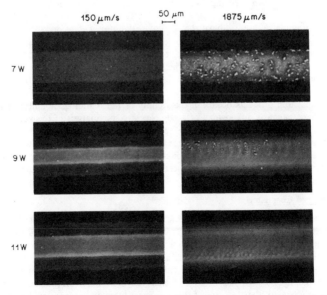

Figure 4. Scanning electron micrographs of palladium features on silicon substrate as a function of laser power (measured at laser) and scan speed. Palladium acetate precursor film thickness is 1.6 μm (cw Ar$^+$ laser – multiline mode, spot size ~50 μm).

the thermal conductivity of the film and the thermal conductivity of the metallic feature once the exposure is begun. Similar periodic structure in laser-crystallization of amorphous silicon has been labeled "explosive" crystallization (12,13). In that case, the additional heat comes from the amorphous-to-crystalline phase transition rather than chemical transformation. Our situation is further complicated by changes in the absorption of the laser radiation as the decomposition reaction proceeds.

It is interesting to consider the dynamics of the metallization reaction as manifested in the periodic structure shown in Figure 3 for lines written on a quartz substrate. The rate of heating of the palladium acetate film at the leading edge of the scanning laser beam is lower at lower incident powers. This rate of heating will in turn affect the rate of release of the heat of reaction. The structures of the top two features in Figure 3 written at a power of 8 mW are suggestive of palladium island formation within a matrix that prevents the islands from coalescing. In effect, there may be a competition between the rate of palladium nucleation and particle growth and the removal of organic reaction residue. Thermal quenching by the already metallized portion of the feature undoubtedly contributes to the structure as do the thermal conductivity of the substrate and surrounding film. The lines written at powers of 12 and 16 mW exhibit the least structure, suggesting the proper combination of laser heating and controlled release of the heat of the decomposition reaction. Finally, the lines written at the highest laser power of 20 mW exhibit the periodic structure associated with "explosive" reaction front propagation generated by rapid release of the heat of the decomposition reaction.

The palladium features shown in Figure 4 were written with a 50 μm laser beam diameter (multi-line mode) on silicon. While some periodic structure can be seen in the feature written at 9W at a speed of 1875 μm/s, the higher thermal conductivity of silicon compared with quartz contributes to greater spreading of the reaction zone and more rapid thermal quenching. The power densities of the irradiations on both substrates were comparable, on the order of 10^6 W/cm^2. Auger analysis of the metal features shown in Figure 4 revealed no carbon within the limits of detection ($\sim$0.1%). Slight amounts of silicon and oxygen are found in features written at powers up to 8W that are absent from the higher power features. These signals originate from the silicon substrate owing to the porosity of the lower power features. The increase in metal density for features written at powers greater than 8W is responsible for the electrical resistivities in the range 40-240 $\mu\Omega$-cm; the resistivity of bulk metallic palladium is 11 $\mu\Omega$-cm.

Thermogravimetric studies have shown that palladium acetate decomposes cleanly to the metal at 200-260°C in N$_2$ or air at a heating rate of 16°C/min (14). In air, further heating leads to oxidation of the metal above 400°C to form PdO. This is reduced back to the metal above 800°C. It is difficult to extrapolate the enthalpy of reaction from the thermogravimetric data to the laser exposure because of the large difference in heating rates, i.e. 16°C/min compared with roughly 10^{10} °C/min (15). At the latter rate, the extent to which oxidation of organic decomposition products occurs in the vicinity of the reation zone and contributes to the "explosive" phenomenon is unknown. Infrared spectroscopic analysis of the gaseous products of decomposition at the slow rates revealed the presence of acetic acid and the absence of acetone. Previous studies of the thermal decompositions of other metal acetates suggested that acetic acid is the dominant organic decomposition product in those cases where pure metal is formed. Those acetates whose decomposition produced metal oxides produced acetone as the major organic product (3,4).

Ion Beam-Induced Chemistry

Palladium acetate films irradiated with 2 MeV He$^+$ or Ne$^+$ ions were converted to smooth, reflecting, metallic-looking features without any of the structure produced in the photothermal decompositions. The range of the incident ions is much greater than the film thickness so that the energy deposition in the films is essentially uniform throughout their thickness. Direct momentum transfer to atoms of the film through elastic collisions with nuclei of the film constituents is negligible with these ions and energies. The main mechanism of energy loss is through interactions with the electrons of the film constituents in inelastic scattering processes.

This gives rise to electronic excitation or ionization of the palladium acetate which subsequently fragments to give a variety of gaseous and solid products. We have employed several spectroscopic techniques to monitor these ensuing reactions, as discussed below.

2 MeV He$^+$ Ion Irradiations

Figure 5 shows the dose dependence of the infrared spectrum for a 0.90 μm film exposed to 2 MeV He$^+$. The intense infrared COO$^-$ vibrations of the bridging acetate ligands at 1602 and 1433 cm^{-1} are useful indicators of the extent of decomposition of the palladium acetate trimer. Although these particular samples were exposed to air, in situ infrared spectra did not differ markedly. We were surprised at the appearance of isosbestic points for the spectra at lower fluence, suggesting a process having some specificity. The steadily decreasing background transmittance results from the increased metallic palladium content of the film with increasing dose. The separation of the symmetric and asymmetric COO$^-$ vibrations is usually a good indicator of the mode of bonding of the acetate group (16). We do not see any evidence for the formation of a palladium monomer with bidentate, non-bridging acetates, nor the formation of monodentate acetate ligands. The symmetric and asymmetric COO$^-$ vibrations decrease exponentially, as shown in the plot in Figure 6. Likewise, the film thickness decreases at a comparable rate as a function of dose as material is lost from the film. It does not however reach the limiting thickness of 0.11 μm expected for maximum densification to pure palladium.

Further examination of the infrared spectra reveals a decrease in the C-H vibrations at 3029 and 2946 cm^{-1}, appearing to occur at different rates. A new broad signal that appears at 1879 cm^{-1} may be due to CO adsorbed on metallic palladium. The position of this stretch is also suggestive of a carbonyl or anhydride group perhaps contained in a macromolecular residue, as discussed below.

RBS analysis of a palladium acetate film at intervals during irradiation begins to reveal a loss of carbon and oxygen at $\sim$3×10^{13} 2 MeV He$^+$ ions/cm^2, as shown in Figure 7. Although the films look metallic at a dose of 3×10^{14} ions/cm^2, there is still considerable oxygen and carbon present in the film. Beyond a dose of 3×10^{15} ions/cm^2, the films reach a limiting value of about 0.8 and 0.2 atoms of carbon and oxygen per palladium atom, respectively, compared to a ratio of 4:1 each initially. It is hard to imagine an organic moiety remaining intact at these irradiation doses, yet studies of ion beam decomposition of organic polymers show that both chain scission as well as cross-linking can occur upon irradiation (17,18). High doses often lead to graphitization of certain polymers (19). We suggest that a similar situation is found in the palladium acetate system. As material is lost from the film, ions and radicals in the film combine to form a material of sufficient molecular weight to preclude volatilization. This is supported by the fact that the intensities of the infrared acetate vibrations plotted in Figure 6 decrease at doses that are lower by a factor of 3 than the loss of carbon and oxygen from the film. Transmission electron micrographs indicate the presence of palladium as small metal particulates, presumably in a maxtrix of organic residue. The silvery metallic appearance of these films is clearly deceiving.

In situ measurements of electrical sheet resistance provide another clue to the composition of the films. A plot of the sheet resistance as a function of dose is shown in Figure 8. The decrease in sheet resistance trails the loss of carbon and oxygen by a factor of 10 in terms of ion dose and reaches a limiting value of 2×10^4 $\mu\Omega$-cm. This is two orders of magnitude greater than the lowest value measured in the laser-exposed material, at least qualitatively consistent with the relative purity of the metals in each case.

Gas evolution was monitored as a function of dose for individual masses up to m/e 63 using a quadrupole mass spectrometer. Three evolution patterns emerged and the species corresponding to each of these groups are listed in Table I. The corresponding spectra shown in Figure 9 are corrected for the background. Parent species were identified by comparing known cracking patterns with the observed yields of individual masses 2-63.

The species in Group I appear immediately at the start of irradiation and decrease slowly at different rates. The C$_4$ molecules of Group II also appear at the start of irradiation but have a slower rise, peaking at a dose of $\sim$1.75×10^{14} ions/cm^2. Finally, there is a threshold for appearance of the molecules in Group III at $\sim$1.75×10^{14} ions/cm^2, with maximum evolution at

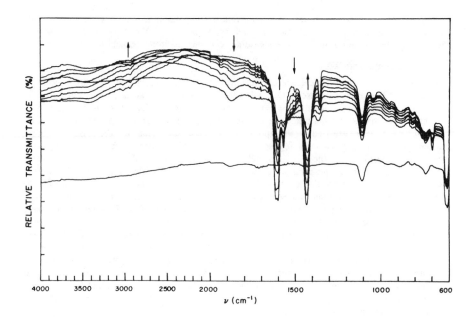

Figure 5. Infrared spectra of 0.90 μm palladium acetate film on silicon as a function of 2 MeV He$^+$ ion dose. Dose range: 0-6×10^{14} ions/cm^2.

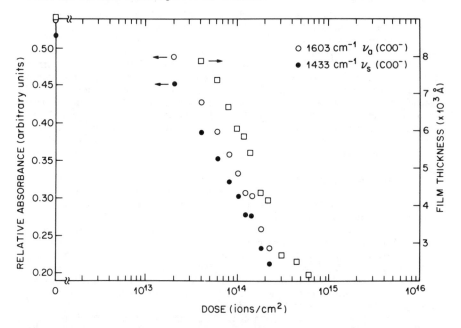

Figure 6. Relative absorbance of symmetric and asymmetric COO$^-$ vibrations of 0.90 μm palladium acetate film on silicon as a function of 2 MeV He$^+$ ion dose. Decrease in film thickness with dose also shown.

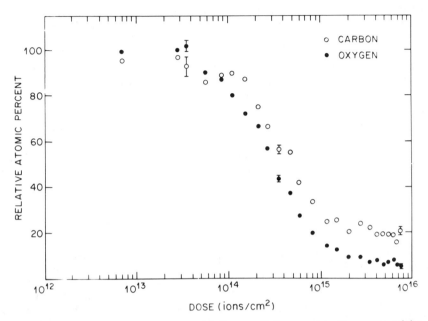

Figure 7. Carbon and oxygen content of 0.12 μm palladium acetate film, measured by Rutherford backscattering spectroscopy, as a function of 2 MeV He⁺ ion dose.

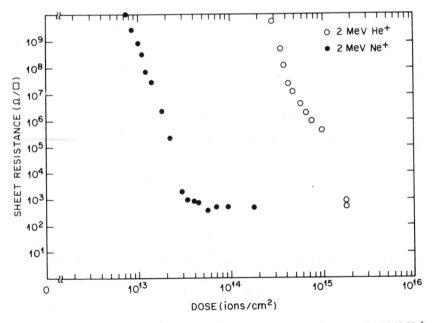

Figure 8. Sheet resistance of 0.90 μm palladium acetate film as a function of 2 MeV He⁺ and Ne⁺ ion dose.

a dose of $\sim3\times10^{14}$ ions/cm^2. It is important to remember that since we are dealing with a solid target, detection thresholds for species at the mass spectrometer reflect not only the sequence of chemical reactions in the film, but also the rate of diffusion of the resultant molecules and radicals through the film and into the vacuum.

Table I. Evolved Gases From 2 MeV He$^+$ Irradiation of 0.9 μm Palladium Acetate Film. Molecules arranged in order of decreasing abundance within each group

I		II		III	
CO_2, C_3H_8	(m/e 44)	C_4H_2	(m/e 50)	CH_3CHO	(m/e 44)
CO, C_2H_4	(m/e 28)	C_4H_4	(m/e 52)	CH_2CO	(m/e 42)
H_2O	(m/e 18)			CH_3CO_2H	(m/e 60)
$\cdot CH_3$, CH_4	(m/e 15,16)			C_2H_6	(m/e 30)
H_2	(m/e 2)				

The observed yields for some of the parent ion masses at m/e 44 and below contain significant contributions from fragments of higher mass species, making quantitative yield assignments difficult. Qualitatively, however, it is clear that CO_2 is the most abundant species evolved. Overall, the proposed parent species, in order of decreasing yield are: CO_2, C_3H_8 > CO, C_2H_4 > H_2O > $\cdot CH_3$, CH_4 > H_2 ~ CH_3CHO ~ CH_2CO > CH_3CO_2H ~ C_4H_2 > C_2H_6 > C_4H_4.

Many of the above products are also observed in α- and γ-radiolysis of glacial acetic acid (20,21). Proposed mechanisms involve ionization of the acid followed by cleavage of the C-C bond to produce methyl radicals and carbon dioxide as the main decomposition pathway. The methyl radicals can then combine with each other to form ethane, abstract hydrogen from a parent molecule to form methane or combine with oxygen-containing fragments to form acetone and methyl acetate. Water, hydrogen, carbon monoxide and acetaltehyde are also formed. Electrolytic oxidation and reduction of acetate ions in solution produce similar species (1).

The most significant difference between our studies of the He$^+$ ion-induced decomposition of palladium acetate and the above experiments is our observation of species in the mass spectra that we assign to be saturated and unsaturated C_3 and C_4 hydrocarbons. We propose that palladium, either in its oxidized form as palladium acetate or as fine metal particles, catalyzes the coupling of smaller hydrocarbon fragments (22). The relatively high amount of residual carbon and oxygen in the fully exposed palladium acetate films is consistent with this mechanism, being most probably present in the form of higher molecular weight molecules whose formation is catalyzed by the active palladium centers.

Thermal decomposition of palladium acetate, either by laser irradiation or conventional means, leads to complete volatilization of the organic components. The purity of the ion beam-irradiated samples is significantly improved by heating the samples in hydrogen at 300°C after removal of unirradiated palladium acetate.

2 MeV Ne$^+$ Ion Irradiations

The electronic stopping power of the 2 MeV Ne$^+$ ions in the palladium acetate films is much larger than that of 2 MeV He$^+$ ions. The most obvious difference between the effects of the two ions is in the appearance of the films at the high dose limit. A 0.90 μm thick palladium acetate film exposed to 2 MeV Ne$^+$ ion irradiation until no further spectroscopic changes occur looks black, compared with the metallic silvery films produced in the He$^+$ ion irradiation. However

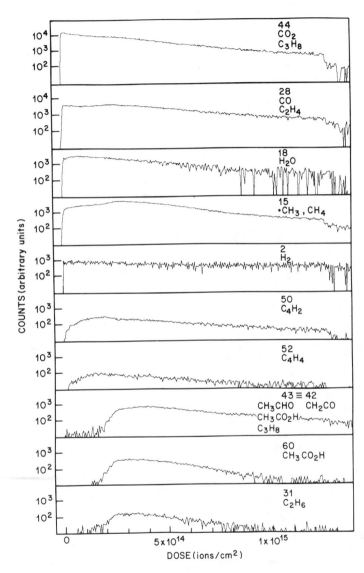

Figure 9.   The quadrupole mass spectrometer signal for volatile species released from 0.90 μm palladium acetate film as a function of 2 MeV He$^+$ ion dose. Mass 15 is shown for both ·CH$_3$ and CH$_4$ because of overlap at m/e 16 with oxygen. Mass 31 is shown for C$_2$H$_6$ (^{13}C isotope) because of overlap at m/e 30 with major fragments of other parent ions.

Ne$^+$ ion irradiation of a 0.13 μm thick film produces a metallic silvery film. A plot of the infrared COO$^-$ vibrations as a function of fluence in Figure 10 shows that the intensity decreases with approximately the same functional dependence as in the He$^+$ ion irradiation, but at a dose that is 17 times lower. In addition, a new band appears at 1616 cm^{-1}, peaking at a dose of $\sim$1.7×10^{12} ions/cm^2, then decreasing rapidly to the same level as the original acetate bands. This may represent the formation of some monodentate acetate species as the palladium acetate trimers are cleaved. In situ infrared spectra of the He$^+$ ion-irradiated films show a similar band of much smaller relative intensity.

RBS analysis of a palladium acetate film at intervals during irradiation reveals that loss of carbon and oxygen from the film, plotted in Figure 11, occurs at doses a factor of 50 lower than those required to achieve similar loss with He$^+$ ions. Significant loss of carbon and oxygen is already observed at the lowest doses that we could practically achieve. Note that the dose dependence of the loss of carbon and oxygen is more closely correlated with the decrease in intensity of the infrared acetate vibrations than was observed in the He$^+$ ion irradiations. A limiting residue content of 0.6 and 0.2 atoms of carbon and oxygen per palladium atom, respectively, is achieved beyond a dose of 4×10^{13} ions/cm^2. Thus, although the final film appears black, it has a slightly lower carbon content than the silver-looking film produced by He$^+$ ion irradiation. This suggests a difference in nucleation kinetics and particle growth relative to removal of organic residue.

In situ measurements of sheet resistance as a function of Ne$^+$ ion dose reveal a similar dose dependence relative to He$^+$ irradiation as the RBS data namely a factor of $\sim$50 lower dose for comparable resistivity values. The limiting resistivity value of 2×10^4 $\mu\Omega$-cm is the same as for the He$^+$ ion irradiation despite the difference in the film's final appearance. Assuming comparable decomposition chemistry in the He$^+$ and Ne$^+$ ion irradiations, the higher electronic stopping power of the Ne$^+$ ions gives rise to a higher density of events along individual ion paths. We propose that this higher density of events produces a fluffy film with deposited metal particles that are responsible for the film's black appearance. The above analyses offer little information regarding the chemical nature of the organic residue. Preliminary X-ray photoelectron spectroscopy analyses reveal that the palladium particles formed by both He$^+$ and Ne$^+$ ion irradiations are metallic and contain no palladium oxide.

Conclusion

Comparison of the photothermal laser-induced and non-thermal ion beam-induced decompositions of palladium acetate films provides immediate contrasts. The thermal reaction produces pure palladium metal whose morphology negatively affects the feature definition. Periodic structure arising from the coupling of absorbed laser energy with the heat of reaction, and porosity of the deposited metal are responsible for the increased electrical resistivity of the deposits relative to bulk metal. In contrast, He$^+$ ion irradiation produces silvery, well defined, smooth deposits. The appearance of films irradiated with Ne$^+$ ions varies from black to silver, depending on starting film thickness. Unfortunately, all features contain significant carbon and oxygen residues that are responsible for the high electrical resistivities of these films. Mass spectrometry of the volatile products of the He$^+$ ion-induced reaction reveal the formation of saturated and unsaturated C$_3$ and C$_4$ hydrocarbon molecules not previously observed in the α- and γ-radiolysis of acetic acid. We propose that palladium is catalyzing the coupling of smaller hydrocarbon species to form not only the higher molecular weight volatile species observed, but larger non-volatile molecules that comprise the solid organic residue in which metallic palladium particles are dispersed.

Acknowledgments

The authors thank A. Appelbaum for Auger analyses, J. M. Gibson for transmission electron microscopy, S. B. Dicenzo for X-ray photoelectron spectroscopy and G. K. Celler and L. E. Trimble for use of their laser.

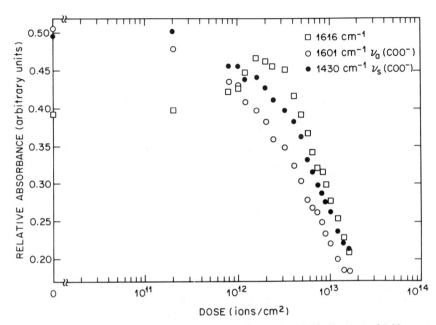

Figure 10. Relative absorbance of symmetric and asymmetric COO⁻ vibrations of 0.90 μm palladium acetate film, measured in situ, as a function of 2 MeV Ne⁺ ion dose.

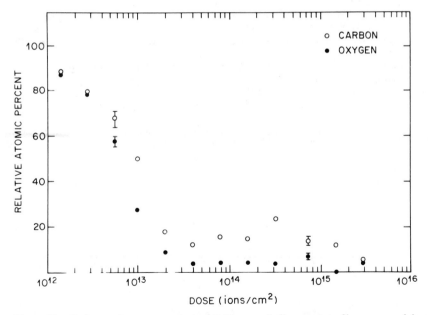

Figure 11. Carbon and oxygen content of 0.12 μm palladium acetate film, measured by Rutherford backscattering spectroscopy as a function of 2 MeV Ne⁺ ion dose.

Literature Cited

1. Eberson, L.; Utley, J. H. P. in Organic Electrochemistry; Baizer, M. M.; Lund, H., Eds.; Marcel Dekker: New York, 1983; Ch. 11, 14.
2. Swallow. A. J. Radiation Chemistry of Organic Compounds; Pergamon: New York, 1960; pp 111-113, and references therein.
3. Judd, M. D.; Plunkett, B. A.; Pope, M. I. J. Therm. Anal. 1974, 6, 555-63.
4. Cornejo, J. Afinidad 1981, 38, 525-9.
5. Eskildsen, S. S.; Sorensen, G. Nucl. Inst. Meth. Phys. Res. 1985, B7/8, 481-6.
6. Eskildsen, S. S.; Sorensen, G. Appl. Phys. Lett. 1985, 46, 1101-2.
7. In a strict sense, organometallic compounds are those in which the carbon atoms of organic ligands are bonded directly to the metal. Therefore we shall adopt the term metallo-organic to denote those compounds in which the hydrocarbon moieties in the ligand are coordinated to the metal through a heteroatom (O, S, N, P, etc.), a subset of classical inorganic coordination complexes.
8. Skapski, A. C.; Smart, M. L. Chem. Commun. 1970, 658-9.
9. Gross, M. E.; Appelbaum, A.; Schnoes, K. J. J. Appl. Phys. 1986, 60, 529-33.
10. Gross, M. E.; Fisanick, G. J.; Gallagher, P. K.; Schnoes, K. J.; Fennell, M. D. Appl. Phys. Lett. 1985, 47 923-5.
11. Fisanick, G. J.; Gross, M. E.; Hopkins, J. B.; Fennell, M. D.; Schnoes, K. D.; Katzir, A. J. Appl. Phys. 1985, 57, 1139-42.
12. Auvert, G.; Bensahel, D.; Perio, A.; Nguyen, V. T.; Rozgonyi, G. A. Appl. Phys. Lett. 1981, 39, 724-6.
13. Lemons, R. A.; Bosch, M. A. Appl. Phys. Lett. 1981, 39, 343-5.
14. Gallagher, P. K.; Gross, M. E. J. Therm. Anal., in press.
15. The instantaneous heating rate induced by the laser at the onset of the decomposition reaction is estimated to be 10^{10} °C/min on the basis of a minimum temperature for complete metallization in air of 1000°C and a reaction time on the order of a few microseconds.
16. Nakamoto, K. Infrared and Raman Spectra of Inorganic and Coordination; Compounds; Wiley: New York, 1978, pp. 232-3.
17. Brown, W. L.; Venkatesan, T.; Wagner, A. Nucl. Instrum. Methods Phys. Res., 1981, 191, 157-68.
18. Calcagno, L.; Sheng, K. L.; Torrisi, L.; Foti, G. Appl. Phys. Lett., 1984, 44, 761-3.
19. Orvek, K. J.; Huffman, C. Nucl. Instrum. Methods Phys. Res., 1985, B7/8, 501-6.
20. Newton, A. S., J. Chem. Phys., 1957, 26 1764-5.
21. Ayscough, P. B.; Mach, K.; Oversby, J. P.; Roy, A. K., Trans. Faraday Soc., 1971, 67, 360-74.
22. Maitlis, P. M., The Organic Chemistry of Palladium. Vol. II: Catalytic Reactions; Academic Press: New York, 1971, and references therein.

RECEIVED November 24, 1986

Chapter 19

Plasma and Ion Beam Synthesis and Modification of Inorganic Systems

W. Brennan, D. T. Clark, and J. Howard

ICI New Science Group, ICI Plc, P.O. Box 90, Wilton, Middlesbrough, Cleveland, TS6 8JE, England

The need for the production of high lateral and vertical definition patterns (viz spatially resolved) inorganic systems, has led to the development of a number of novel techniques for synthesis and modification involving plasmas, ion and electron beams, which may well have more general applicability. Whilst these needs have been in response to the demands of the Electronics and Information Technology Industries, the need for more cost effective processing in more traditional areas such as metallurgy and ceramics have seen interesting developments (eg in the plasma sphere) which again may hold interesting lessons for inorganic chemists. The talk will briefly review some of these developments ranging from high temperature equilibrium plasmas to cool plasmas, PECVD, ion implantation, ion beam mixing and ion assisted etching and deposition. Brief consideration will also be given to sputtering and ionised cluster beam deposition techniques in inorganic synthesis.

Modern industry has requirements for materials with extremely high performance in one or more properties (hardness, abrasion resistance, high melting points etc). This, together with ever increasing awareness of the potential of thin film technology and of the needs for increasing flexibility through the production of modified versions of established materials, has led to an upsurge in the use of plasma and ion beam techniques in inorganic synthesis.

Combination techniques are also of considerable interest and relevance. These methods enable an unusually high degree of control over spatial, physical and chemical characteristics to be achieved.

In this paper we will survey examples of some of the types of work done and their relevance, as well as give an outline of the major techniques used.

0097-6156/87/0333-0303$07.00/0

Background

Plasmas and ion beams can be used for (a) bulk synthesis, (b) the deposition of new thick or thin surface layers and (c) surface modification. One growing area of importance is in vacuum deposited thin films. The major growth area in the use of plasma and ion beam techniques has been micro circuit fabrication. As is shown in Figure 1 this is a complex multi-stage process. An improved understanding of and control over the relevant surface and gas-phase chemistry should enable some of these steps to be eliminated (eg maskless etching). There are continuing demands from the electronics and communications as well as other industries for new or modified inorganic materials for displays, integrated circuits, optical communications, data storage media, decorative coatings, solar controls, the cutting edge of machine tools etc.

Plasmas

Plasmas occur widely in nature, (Figure 2), however those of relevance to this paper are obtained by coupling D.C, A.C, r.f or microwave electrical energy into a gas to produce a complex environment containing electrons and ground and excited states of ions, neutrals, radicals etc (1). Electrical energy is transferred to the plasma by the acceleration of electrons in the applied electric field and subsequent collisions with other species in the plasma. At high pressures collision frequencies are such that the electron energy is the same as that of the other species in the plasma. However, at lower pressures the electrons acquire a greater energy than the other species giving rise to 'non-equilibrium' conditions where the electron temperature is 10^4-10^5K but the neutrals and ions are at ambient temperature (Figure 3). Thermal plasmas have been used extensively in a wide range of applications, including welding, plasma spraying and arc furnace technology (2). The applications of 'non-equilibrium' or 'cool' plasmas are more recent (3). In both areas the lack of fundamental understanding of the physical and chemical processes involved results in a substantially empirical approach to new applications and the optimisation of the desired material.

A recent report by the US Department of Energy states (4) "The creation of new (surface) materials is the result of extensive, persistent and enlightened empiricism". In the report and elsewhere there is increased awareness of the need for more fundamental research.

High Temperature Plasmas: High temperature plasmas are essentially used as heat sources. They are more efficient than fossil fuels and their high temperature limit is much greater.

Some advantages of plasma over conventional chemical routes include:
(a) a reduction in the number of steps;
(b) higher purity;
(c) continuous instead of batch operation;
(d) yields greater than predicted from the thermodynamic equilibrium conditions;
(e) the isolation of high temperature phases by fast quenching ($>10^6$Ks^{-1}).

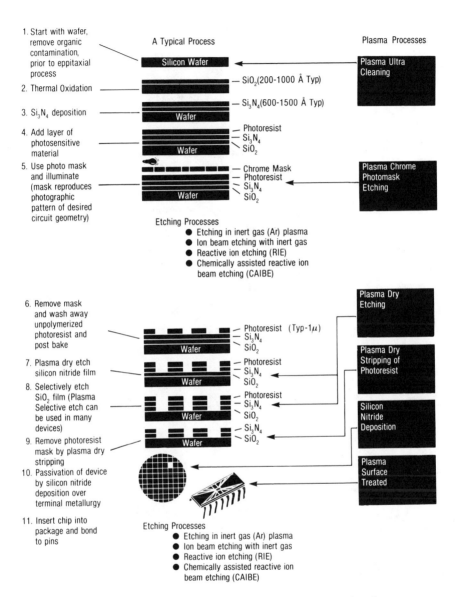

Figure 1. Microcircuit fabrication using plasma and ion beam techniques.

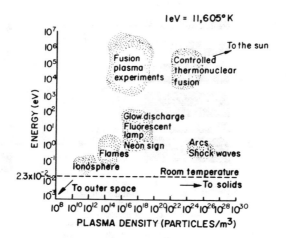

Figure 2. Plasma density (particles/m³) against energy for various plasma systems.

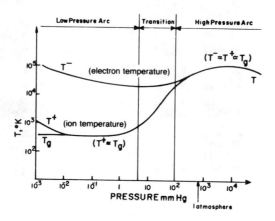

Figure 3. Variation of ion and electron temperatures with gas pressure in a plasma.

Some examples are:
Extractive Metallurgy:

Iron	Fe oxide in H_2/CH_4 plasma
Vanadium	V oxide in H_2/CH_4 Plasma
Ferroalloys	Mixed oxide in H_2/CH_4 plasma
Zinc	From waste ash in CO/H_2 plasma
Zr, Si, Ti-	Produce oxide then chloride then reduce to metal (see later)

Plasma Synthesis:

$$TiCl_4 + O_2 \longrightarrow TiO_2 + 2Cl_2$$
$$2NbCl_5 + 5H_2 + N_2 \longrightarrow NbN + 10HCl$$

The use of a plasma synthetic route can lead to materials with advantageous new physical characteristics such as high surface area (eg TiO_2), spherodised powders and sprayed-on coatings. A particularly interesting example is the use of plasmas to produce large monocrystals of refractory metals, directionally solidified eutectic alloys and high melting point carbides (Figure 4). The molten material is supplied by plasma melting of the corresponding alloy or pure metal. Continuous withdrawal of the ingot is conducted simultaneously with the solidification of the melted material. Metal carbides can be produced by using CH_4 or CO_2 as sources of C in the plasma torch. The design is compact, energy efficient and easy to control (5). In general, commercial plasma torch systems can be operated using powers between 1kW and 10MW, pressures between 100T and 400atm to give gas temperatures between 3000 and 30,000K. Gas flow rates are lower than $500gs^{-1}$. An example of the reduction in the number of processing steps achieved by using plasma methods as opposed to more conventional technology is provided by zirconium production (6). Starting from zircon the conventional process involves over sixty steps. The last stages, sometimes called the Kroll process, are shown in Figure 5. These represent the most expensive operations. Replacement of the Kroll process by a plasma-based route (Figure 6) has been studied and the technical feasibility demonstrated (6). An additional advantage over the conventional process is that dense metal, and not sponge, is produced directly.

Low Temperature (Non Equilibrium) Plasmas: The rigorous classification of plasma treatments is difficult, however, from the viewpoint of the treated surface there are three broad categories:
(a)   Inert, eg surface etching. The chemical nature of the surface is unchanged.
(b)   Chemically reactive non-depositionable, eg nitriding, boriding, carbiding. The surface is one of the reactants.
(c)   Depositionable, eg plasma enhanced chemical vapour deposition (PECVD). The deposited layer may be chemically totally independent of the substrate.
Only (c) will be discussed here since it is most relevant to inorganic synthesis.
The main user of these techniques at present is the electronics industry, where plasma equipment sales in 1987 are expected to

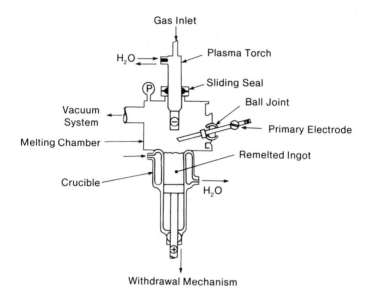

Figure 4. Plasma reactor for growing single crystals of refractory metals, alloys and high melting point carbides.

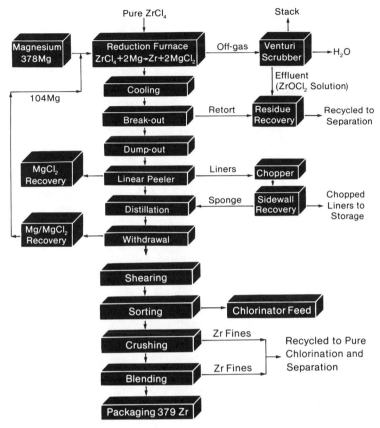

Figure 5.   The final stages in the production of zirconium sponge (the Kroll process).

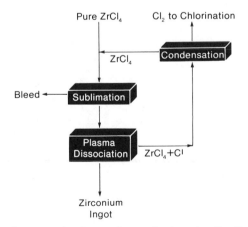

Figure 6.   Suggested scheme for replacing the Kroll process by plasma techniques.

exceed \$1 billion (7). Plasma reactors come in an almost infinite variety of configurations but all consist of:
(a) a vacuum system to operate from 10^{-3} - 10Torr;
(b) a gas inlet system;
(c) a method of coupling electrical power;
(d) a method of getting the substrate into the plasma.

A particularly attractive potential of plasma based methods is the ability to vary, continuously or discretely, the nature of the material deposited by varying the plasma parameters (eg flow rates, gas composition, power input, substrate temperature etc). This, of course, applies to organic as well as inorganic materials (8). This aspect of interface control is not yet well developed but is an exciting prospect.

Plasma Synthesis: The use of plasma methods has lead to a new range of materials having unique properties. An example is the family of amorphous elemental hydrides (eg α-C:H; α-Si:H;α-P:H) which contain a variable proportion of H from almost zero to 50 atomic %. The carbon films, known variously as "hard carbon", "diamond-like carbon", "α-carbon" etc (9). These layers are of considerable interest because of their optical and abrasion-resistant properties etc (Table I). The properties of these α-carbon films, can be tailored by modifying the plasma parameters.

Table I. Some Properties of α-Carbon

	α-C	Diamond	Graphite
Hardness (Kg/mm^2)	3000	7000	Soft
Band Gap (eV)	2.8-4.0	5.6	Metal
Conductivity (cm^{-1})	10^{-9}-10^{-5}	10^{-18}	Metal
Index of Refraction	1.8-2.0	2.42	Metal

Pressure and power are two such important parameters (10,11). Carbon prepared at high powers and low pressures tend to be hard (a Mohs hardness of 6), have refractive indices greater than 2.2 and are low in hydrogen content. Low power/high pressure films have as much as at 30% hydrogen, refractive indices of 1.8 or even lower and much less abrasion resistance since they can be scratched by steel. The latter films also have the greater tendency to peel since intrinsic stress is directly related to hydrogen content (12). These films can be deposited from a wide variety of hydrocarbon gases or gas mixtures (CH_4, CH_4/H_2, CH_4/Ar, C_2H_2 etc). The amorphous silicon and carbon films are finding a range of applications which exploit their physical or electronic properties. More recently a new form of amorphous black phosphorus (Figure 7) has been produced which has unusual chemical properties. For example this material, which contains a small proportion of H, is not only stable to atmospheric oxidation at room temperature but it is also stable for many hours at 150°C (13).

Given the relative paucity of work in this area there can be little doubt that many more interesting materials remain to be synthesised.

Plasma Enhanced Chemical Vapour Deposition (PECVD): Chemical vapour deposition (CVD) reactions commonly occur at high temperatures (Table II). The use of a plasma to generate chemically reactive species in conjunction with CVD overcomes one of the most common

Table II. A Comparison of CVD and PECVD Reactions

Material Deposited	CVD		PECVD		Application
	Reaction	Temp(°C)	Reaction	Temp	
BN	BCl_3+NH_3	1000-2000	$B_2H_6+NH_3$	400-700	High temp. ceramic Diffusion doping of Si
TiN	$TiCl_4+N_2+H_2$	650-1700	$TiCl_4+N_2+H_2$	400-600	Wear resistant layers
SiC	$CH_3SiCl_3+H_2$	1000-1600	$SiH_4+C_2H_4$	500	Oxidation resistant coatings. Machine tools, printer heads.
Si_3N_4	$SiCl_4+NH_3$	1100-1400	SiH_4+N_2/NH_3	350	High temp. ceramic. Encapsulation of semi-conductors
TiC	$TiCl_4+CH_4$	800-1400	$TiCl_4+CH_4$	500-700	Wear resistant layers
Al_2O_3			$AlCl_3+O_2$	250-350	Passive encapsulating layer
SiO_2			$SiCl_4+O_2$	1000	Optical films

objections to conventional CVD:- that the required substrate temperatures is incompatible with the substrate materials or device structure. In general the use of a plasma to enhance CVD results in:
(a)   reduced film deposition temperature;
(b)   improved crystallinity;
(c)   increased deposition rate;
(d)   increased stabilisation of solids with low configuration entropy, eg amorphous carbon, silicon etc.
Some examples are also included in Table II. When one compares PECVD with conventional chemical (thermal) routes it is clear that for some processes there are both thermodynamic and kinetic (Figure 8) advantages to the former (14). For example the effect of the plasma is to shift the equilibrium towards the solid, lowering the deposition temperature (Table II).

Such techniques could, in principle, be used to produce an entire memory device (Figure 9). The possible steps involved are explained in Table III, together with some of the general advantages

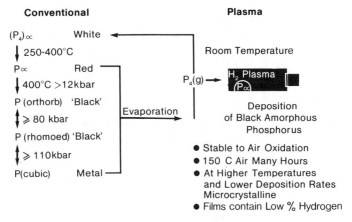

Figure 7. Simplified scheme for the allotropes of phosphorous formed by conventional and plasma methods.

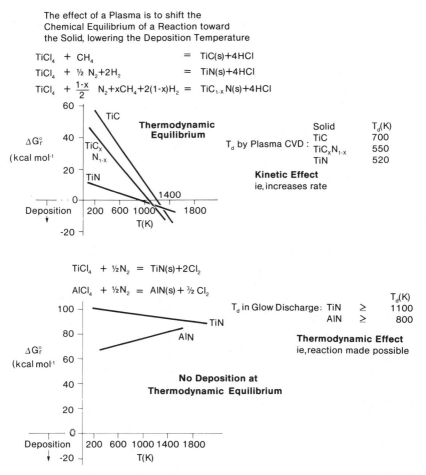

Figure 8. Kinetic and thermodynamic advantages of PECVD compared with conventional thermal processes.

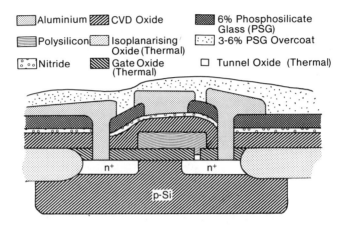

Figure 9. Schematic representation of an MOS non-volatile memory device and its components.

Table III. Components of an MOS Non-volatile Memory Device and Their Methods of Fabrication

Layer	Method
P-Si є pi	SiH_4/BH_3 PECVD
Gate Oxide ⎫ Tunnel Oxide ⎬	SiH_4 / N_2O PECVD Microwave Plasma Anodisation Laser CVD
Isoplanarising Oxide	SiH_4 / N_2O PECVD
CVD Oxide	SiH_4 / N_2O PECVD
Poly Si	SiH_4 PECVD
Nitride	SiH_4/NH_3 PECVD
PSG	$SiH_4/N_2O/PH_3$ PECVD

Advantages	
	1) Low Temperature
	2) Controllable Composition
	3) Step Coverage
	4) Pinhole Free
	5) Fast Deposition Rates
	6) Can Produce Graded Interfaces

of vacuum deposition. Plasma techniques are also being used in MOCVD processes, for example in the growth of epitaxial layers of GaAs at temperatures as low as 450°C (15).

A particularly elegant plasma deposition process has been developed by Philips for the production of optical fibre preform. Both high and low pressure plasmas are used for the fabrication of high quality telecommunication fibres (16). The efficiency of these plasma deposition processes (ca 100%) is superior to thermally induced deposition methods (40-70%). Advantageously high fluorine doping levels and highly sophisticated glass structures can be achieved uniquely with low pressure plasmas (Figure 10). The same techniques could be extended to produce a range of integrated optics (17).

A schematic of the Philips PCVD unit is shown in Figure 11. This is the only process used for the construction of optical waveguides which utilises a low pressure plasma to induce the chemical reactions for the deposition of both doped and undoped SiO_2. The deposition is performed inside a glass tube. The starting materials $SiCl_4$ and O_2, for bulk deposition, and $GeCl_4$ and C_2F_6 for refractive index modification by the introduction of impurities, are fed into the tube. The gas flows are carefully controlled. A microwave plasma is used to initiate the reactions. The tube is heated to 1150°C to avoid chlorine incorporation in the deposit and the cavity is oscillated so that on each pass a glassy layer is deposited. Any desired refractive index profile can be generated (Figure 10). On completion the tube is drawn to produce the fibre (ca 15-25Km is produced per tube). PCVD has the intrinsic advantage of high deposition efficiency for both SiO_2 (ca 100%) and dopants even at high deposition rates.

Ion Beams

Ion and plasma treatments of materials are related since plasmas contain a large proportion of ionised species. There is, however, a wide range of ion beam treatments ranging from monoenergetic monoatomic ions through to partially ionised supersonic beams of large clusters. Thus there are the possibilities of (a) radiation damage or modification of materials using inert gas ions, (b) implantation of inert or reactive elements, (c) ion beam deposition onto surfaces and (d) ion beam irradiation and deposition by some different technique.

Ion Implantation: A schematic of an ion implanter is shown in Figure 12. Some characteristics of implantation are summarised in Figure 13. The ion energy controls the depth of penetration and so by sequential implantation at different energies it is possible to build up a more uniform interface, if needed, than by single implantation. Note that the "damage" profile is different from the concentration profile of the implanted ion. Often this damage can be thermally annealed. Since implantation is a non-equilibrium process, (quench rates of $>10^{12} Ks^{-1}$) one can produce unusual phases, unusual lattice structures and compounds, notably alloys, which exceed the normal solubility of the materials (18). In addition one can produce graded interfaces and introduce a very high population of defects. The concentration of the implanted species can be varied from the parts per million range to virtually 100% in the implanted region.

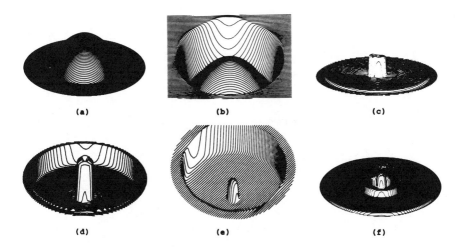

Figure 10. Tomographic representations of the refractive index profiles of telecommunications fibres prepared by plasma techniques. (Reproduced with permission from Ref. 16, Copyright 1985, IUPAC).

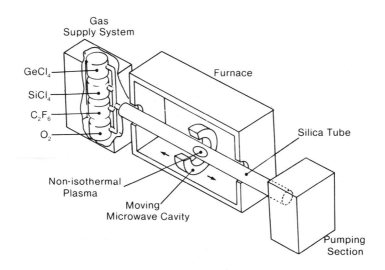

Figure 11. Schematic representation of a PECVD unit for the production of optical waveguides. (Reproduced with permission from Ref. 16, Copyright 1985, IUPAC).

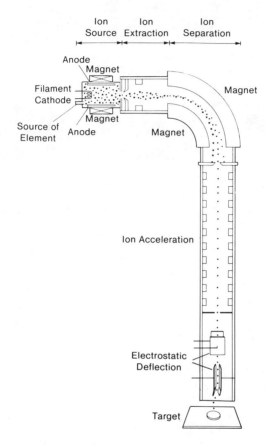

Figure 12. Schematic of an ion-implantation system.

The intrinsic flexibility involved in being able to select and vary the ion energy, ion type and ion dose enables complex interfaces to be "tailored" to specific requirements.

The Creation of Tailored Interfaces: Silicon oxynitride is superior to the oxide or nitride in both its mechanical stress characteristics and hardness to ionising radiation (19).   It also has excellent masking ability and high chemical stability.   Silicon oxynitride can be prepared by a variety of methods including CVD.   For VLSI applications one requires reliable very thin insulators with easily adjustable properties and low processing temperatures.   By simultaneous implantation of O and N into SiO_2 one can produce all phases from SiO_2 to Si_3N_4 and so this satisfies the stringent requirements of VLSI technology.   Depth profiling of the implanted surfaces (O, N and O+N) shows (Figure 14) that the silicon oxynitride/silicon interface is relatively sharp and that the surface layer is not simply the sum of what one would expect from the separate oxygen and nitrogen implantation curves.

The complexity of the phenomena involved is shown also by nitrogen implantation of α-Ti and Ti_2N (Figure 15).   As can be seen the lattice spacings of the α-Ti increase with implantation dose while those of Ti_2N decrease, after its initial formation (20).   The reasons for this behaviour are unclear and this lack of understanding is typical of the current state of the art.

Another example of the ability to tailor surface properties is the creation of high purity Al_2O_3 layers on Fe and steel.   The motivation is to reduce surface oxidation or hydrogen isotope permeation (21).   There are at least three approaches to this; (a) oxide an FeAl alloy: (b) coat iron with a thin layer of aluminium: (c) implant aluminium into iron, anneal then oxidise the surface.   The implantation method followed by selective oxidation has the following advantages:
(a)   Bulk alloy addition of greater than 5% Al in steel is not possible.   Using ion implantation the surface enrichment can be greater than this.
(b)   The surface layer is laterally uniform and reproducible.   The production of 10-30nm uniform layers by conventional coating techniques is difficult!
(c)   Ion implantation avoids any adhesion or interface problems. It is not currently economic to use high dose ($>10^{16}$ ions cm^{-2}) implantation for other than very high added value materials.   Doses used in the electronics industry are typically ca 10^{13} ions cm^{-2}. There may, however, be applications for small area high dose applications including eg patterned inorganic synthesis.   One example is the production of thin tracks of hard Si_xC_y from organosilanes (22).

Ion-Assisted Processes: An alternative use of ion beams generated from low cost sources is to assist particular chemical reactions, or vapour deposition.   An example here is in etching processes (Figure 16).   The simultaneous use of an argon beam with XeF_2 gas compared with the use of either separately, to etch silicon produces an etch rate of a factor of at least fourteen.   The use of ion beams can also increase the directionality (23) of the process (Figure 17).   Examples are given in Table IV of how ion bombardment during film formation modifies the final film.

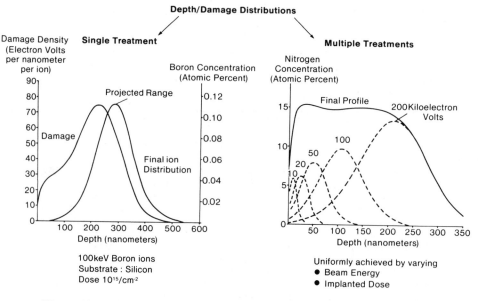

Figure 13. Damage and concentration profiles for single and multiple ion beam (boron) implantation into silicon.

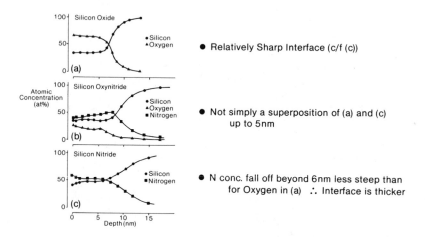

Figure 14. Depth analysis of silicon after implantation of (a) oxygen, (b) nitrogen and (c) oxygen plus nitrogen. (Reproduced with permission from Ref. 19, Copyright 1984, JAP).

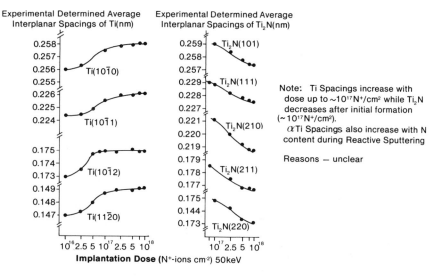

Experimental Determined Average
Interplanar Spacings of Ti(nm)

Experimental Determined Average
Interplanar Spacings of Ti$_2$N(nm)

Note: Ti Spacings increase with
dose up to ~10^{17}N$^+$/cm^2 while Ti$_2$N
decreases after initial formation
(~10^{17}N$^+$/cm^2).

αTi Spacings also increase with N
content during Reactive Sputtering

Reasons — unclear

Implantation Dose (N$^+$-ions cm^{-2}) 50keV

Figure 15. The variation of lattice spacings in α-Ti and Ti$_2$N with nitrogen implantation dose. (Reproduced with permission from Ref. 20, Copyright 1986, Chapman and Hall).

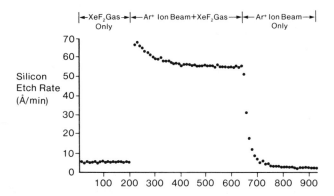

Figure 16. The influence of Ar$^+$ ion beams on the etch rate of silicon by XeF$_2$.

Table IV.   Examples of Thin Film Property Modification
By Ion Bombardment During Deposition

Film Material	Ion Species	Property Modified	Ion Energy (eV)	Ion/atom arrival rate ratio
GdCoMo	Ar^+	Magnetic anisotropy	1-150	0.1
Cu	Cu^+	Improved epitaxy	50-400	10^{-2}
BN	$(B-N-H)^+$	Cubic structure	200-1000	1.0
ZrO_2, SiO_2, TiO_2	$Ar^+O_2^+$	Refractive index amor Xtal	600	$2.5x10^{-2}$ to 10^{-1}
SiO_2, TiO_2	O_2^+	Refractive index	300	0.12
SiO_2, TiO_2	O_2^+	Optical transmission	30-500	0.05 to 0.25
Cu	N^+ Ar^+	Adhesion	50,000	10^{-2}
Ni on Fe	N^+ Ar^+	Hardness	10,000-20,000	0.25
Ge	Ar^+	Stress, Adhesion	65-3000	$2x10^{-4}$ to 10^{-1}
Nb	Ar^+	Stress	100-400	$3x10^{-2}$
Cr	Ar^+Xe^+	Stress	3,400-11,500	$8x10^{-3}$ $4x10^{-2}$
Cr	Ar^+	Stress	200-800	$7x10^{-3}$ to $2x10^{-2}$
SiO_2	Ar^+	Step coverage	500	0.3
SiO_2	Ar^+	Step coverage	1-80	4.0
AIN	N^+	Preferred orientation	300-500	0.96 to 1.5
Au	Ar^+	Coverage at 50A thickness	400	0.1

Ion Beam Deposition: The most commonly used vacuum method for the
rapid deposition of films (thin or thick) is sputtering (24). This
can be combined with ion beam techniques in a variety of ways (25)
including (Figure 18) ion beam sputter deposition (IBSD) eg of oxide
films or of hard carbon (26). In reactive systems the reactive gas
is added to the argon ion beam. The properties of the deposited
materials are modified substantially by varying the gas composition
(Figure 19).

The hard carbon produced by this method has a range of different
properties from those of plasma produced films (Table V). Note that
the maximum band gap achievable with ICBD is 1.2eV at maximum
hydrogenation (35 atomic %) while values up to 4eV can be obtained
by plasma deposition. These wide band-gap materials are soft and
easily scratched though they are more optically transparent.
Extreme hardness is associated with low hydrogen content. These
hard carbon films are often referred to as amorphous diamond;

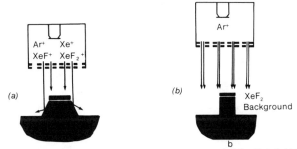

Silicon profile formed at 500eV.(a) An ion beam composed of Ar+, Xe+, XeF₂+;
(b) CAIBE with an Ar+ beam in conjunction with XeF₂ in the sample chamber.

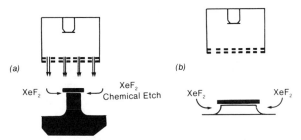

(a) Profile formed with high partial pressure of XeF₂ in the background (~4x10⁻⁴ torr)
under 900 eV Ar+ bombardment; (b) Profile produced by isotropic purely chemical
etching with XeF₂

Figure l7. The influence of gaseous XeF_2 on the ion beam etching profiles of silicon.

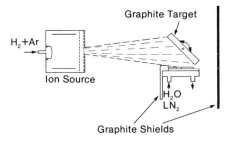

Figure l8. Ion beam sputter deposition apparatus for the production of hydrogenated amorphous carbon. (Reproduced with permission from Ref. 26, Copyright 1985, AIP).

Table V. Properties of α-Carbon and α-Silicon Produced
by Ion Beam Sputter Deposition

	Carbon Band Gap (eV)	Silicon Band Gap (eV)
Zero Hydrogenation	0.7	1.2
Maximum Hydrogenation	1.2	2.0
sp^3 Material	5.5	3.1

Amorphous silicon closer to sp^3 material than is Amorphous Carbon
Difference due to sp^3 states: from NMR sp^2/sp^3 = 3/2 for Maximum
Hydrogenation of C.

Plasma Deposited Films	• More Optically Transparent
	• Less Dense
	• Higher H ratio
	• Softer
	• Lower sp^2/sp^3 ratio

however, it is clear from Figure 19 that while the proportion of sp^3
carbon increases with hydrogenation both the film hardness and
density become less diamond-like. This indicates that the films
have only a remote relationship with diamond and that the physical
properties are a reflection of the unsaturated sp^2 bonding.

In the case of oxide films it appears that the oxide is formed
on the target (27,28) and is subsequently sputtered onto the
substrate. The oxide films formed (TiO_2, Ta_2O_5, SiO_2 etc) are
amorphous and have high refractive indices and low porosity which
are indicative of high density.

It can clearly be seen from Figure 20 that the physics and
chemistry of sputter deposition should be complex. Little
fundamental work has been done in this area.

Direct deposition from an ion beam (IBD) is normally a slow
process (28) which limits its range of commercial applications.
There has, however, been intense interest in using ionised cluster
beam deposition (ICBD) for inorganic systems (28) and recently
results on organic molecules have also been reported (29). The
essentials of the method are outlined in Figure 21. Clusters of the
material are produced by adiabatic expansion through a small nozzle
into a high vacuum. A proportion of the clusters (containing
typically 10^3 atoms each) are subsequently ionised by electron
bombardment and accelerated to the substrate. High deposition rates
(1um/min) are attainable over large areas. This technique has been
applied to produce films with applications in semi-conductors,
metals, dielectrics, optical films, magnetic films, solar cells etc.
When compared with ionised beam deposition (no adiabatic expansion
step) the mechanisms are comparable but with two important
differences. In ICBD the energy range (electron volts per atom) is
typically 10^{-2} smaller than in IBD (note that the clusters carry
only a single charge) and the growth process involves a simultaneous
flux of both thermal neutrals and ionised clusters.

If more than one chemical species is evaporated into the
chamber the nature of the final film is a function of which species
is ionised (Table VI). Reactive processing is possible if a

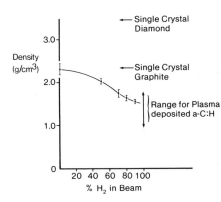

Film Characteristics

Very hard - 2kg on 50μm tip to scratch

Films are termed diamond-like but
the density is varying in the **wrong** direction,
therefore the films have only a remote
relationship to a diamond itself!

Figure 19. Film density as a function of hydrogen gas concentration
in the (H₂+Ar) beam during ion beam sputter deposition.
(Reproduced with permission from Ref. 26, Copyright 1985, AIP).

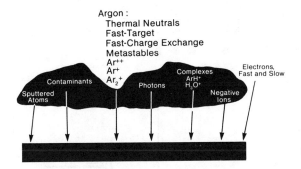

Figure 20. The range of particles bombarding the substrate
during sputter deposition.

Avoids

- Space Charge spreading of the Beam
- The need for Gases cf Plasma Processing

Features

- Ion Bombardment provides Continuous
  Substrate Cleaning
- Conventional Vacuum (10^{-7}Torr) requirements
- Epitaxial Growth of Semiconductors
- Reactive Processing via Enhanced Chemical
  Reactivity → Compound Materials
- Adatom Migration - Nucleation of Thin Films (100Å)
  - Excellent Step Coverage

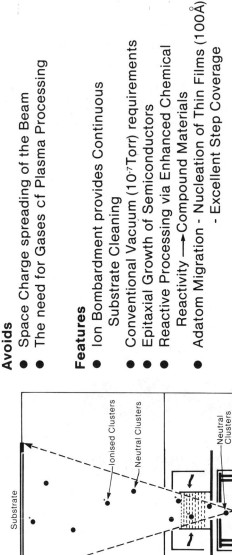

Figure 2l.   Schematic representation of ionised cluster beam deposition and some of its features.

suitable gas is introduced to the deposition chamber. Beryllum
oxide has been grown this way (30) onto sapphire and silicon
(eg c-oriented BeO on Si (100). Preliminary work on metal systems

Table VI. Some uses of Ionised Cluster Beam
Deposition

Uses	Aluminium on Silicon
. Metals	. Low Contact Resistance
. Elemental and Compound Semi-conductors	. Improved Corrosion Resistance c/f Evaporated Al
. Thermoelectric Materials	. Grain Size and Crystal Size
. Dielectrics - Insulating - Optical	Distribution is Function of Acceleration Voltage
. Magnetic Materials	. Crystal Orientation is strongly
. Organics	(111) under High Acceleration Voltage

Flexibility

Cadmium
Telluride
— Cadmium Ionised———Cubic Structure
Tellurium——Hexagonal Structure

Gallium
Arsenide
Ga and As Ionised ⎫
As Ionised   ⎬ Good Film Crystallinity
Ga only Ionised——— Polycrystalline

indicate substantial promise for advanced semiconductor
metallisation (31). Aluminium on Si has very low contact
resistance after annealing. The oxide penetration is reduced
below that in evaporation techniques and the grain size and crystal
size distribution can be controlled by varying the acceleration
voltage. A summary of the work done is available in the
literature (28).

Summary

Vacuum based techniques are capable of producing new materials or
new forms of existing materials either in bulk form or as surface
layers. There is a wide range of different techniques and
applications, only some of which have been covered here. It is
clear from the literature that much needs to be done in both the
fundamental and applied aspects of such work. For example,
relatively little is known about the effects caused by the arrival
of clusters or charged particles at a growing film.

Literature Cited
1.   Bell, A.T., In Techniques and Applications of Plasma Chemistry;
     Hollahan, J.R.; Bell, A.T., Eds.; Wiley: New York, 1974;
     Chapter 1.
2.   Roman, W.C., Mat. Res. Symp. Proc. 1984, 30, 61.
3.   Winters, H.F.; Chang, R.P.H.; Mojab, C.J.; Evans J.;
     Thornton, J.A.; Yasuda, H.K., Mat. Sci. Eng. 1985, 70, 53.
4.   Schwoebel, R.L., Mat. Sci. Eng. 1985, 70, 7.
5.   Rykalin, N.N., Pure Appl. Chem. 1976, 48, 179.
6.   Gauvin, W.H.; Choi, H.K., Mat. Res. Symp. Proc. 1984, 30, 77.

7. Industry News Update, Sol. St. Technol. 28(8), 40.
8. Yasuda, H.K., Plasma Polymerisation; Academic Press: New York,
    1985.
9. Woollam, J.A.; Chang, H.; Natarajan, V., Appl. Phys. Comm.
    1986, 5, 263.
10. Kobayashi, K.; Matsukura, N.; Machi, Y., J. Appl. Phys.,
    1986, 59, 910.
11. Moravee, T.J.; Lee, J.C., J. Vac. Sci. Technol, 1982, 20, 338.
12. Enke, K.; Th. Sol. Films, 1981, 80, 227.
13. Brunner, J.; Thuller, M.; Veprek, S.; Wild, R., J. Phys.
    Chem. Solids, 1979, 40, 967.
14. Benson, S.W., Pure and Appl. Chem., 1980, 52 1767.

15. Pande, K.P.; Seabaugh, A.C., J. Electrochem. Soc., 1984,
    131, 1357.
16. Bachmann, P., Pure App. Chem., 1985, 57, 1299.
17. Kuppers, D.; Schelas, K.H., Topical Meeting on Integrated and
    Guided Wave Optics, Th.C. 6-1: Kissimmee, USA, 1984.
18. Appleton, B.R.; Sartwell, B.; Peercy, P.S.; Schaefer, R.;
    Osgood, R., Mat. Sci. Eng., 1985, 70, 23.
19. Streb, W.; Hazel, R., J. Vac. Sci. Technol.B., 1984, 2, 626.
20. Rauschenbach, B., J. Mat. Sci., 1986, 21, 395.
21. Brown, D.W.; Musket, R.G.; Weirick, L.J., J. Vac. Sci.,
    Technol A, 1985, 3, 583.
22. Venkatesan, T., Nucl. Inst. Methods. Phys. Res. B., 1985,
    7/8, 461.
23. Coburn, J.W.; Winters, H.F., J. Appl. Phys., 1979, 50, 3189.
24. Vossen, J.L.; Cuomo, P.L., In Thin Film Processes; Vossen,
    J.L.; Kern, W.; Eds.; Academic: New York, 1978; p.11.
25. Martin, P.J.; J. Mat. Sci., 1986, 21, 1.
26. Jansen, F.; Machonkin, M.; Kaplan, S.; Hark, S., J. Vac.
    Sci. Technol. A, 1985, 3, 605.
27. Sites, J.R.; Demiryont, H.; Kerwin, D.B., J. Vac. Sci.
    Technol. A, 1985, 3, 656.
28. Gautherin, G.; Bouchier, D.; Schwebel, C.; In Thin Films
    from Free Atoms and Particles; Klabunde, K.L.; Ed.;
    Academic: Orlando, 1985; Chapter 5.
29. Usui, H.; Yamada, I.; Takagi, T., J. Vac. Sci. Technol A.,
    1986, 4, 52.
30. Takagi, T.; Matsubara, K.; Takaoka, M., J. Appl. Phys.
    1981, 51, 5419.
31. Inokawa, H.; Fukushima, K.; Yamada, I.; Takagi, T.,
    Proc. Symp. ISIAT, 6th, 1982, 1982, 355.

RECEIVED December 5, 1986

Author Index

Subject Index

Production by Paula M. Bérard
Indexing by Deborah H. Steiner
Jacket design by Pamela Lewis

Elements typeset by Hot Type Ltd., Washington, DC
Printed and bound by Maple Press Co., York, PA